Fortschritte der Chemie organischer Naturstoffe

Progress in the Chemistry of Organic Natural Products

63

Founded by L. Zechmeister
Edited by W. Herz, G. W. Kirby, R. E. Moore,
W. Steglich, and Ch. Tamm

Authors:
J. Cárdenas, B. Esquivel, M. Gupta,
A. B. Ray, L. Rodríguez-Hahn

Springer-Verlag
Wien New York 1994

Prof. W. Herz, Department of Chemistry,
The Florida State University, Tallahassee, Florida, U.S.A.

Prof. G. W. Kirby, Chemistry Department,
The University, Glasgow, Scotland

Prof. R. E. Moore, Department of Chemistry,
University of Hawaii at Manoa, Honolulu, Hawaii, U.S.A.

Prof. Dr. W. Steglich, Institut für Organische Chemie der Universität
München, München, Federal Republic of Germany

Prof. Dr. Ch. Tamm, Institut für Organische Chemie der Universität Basel,
Basel, Switzerland

© 1994 by Springer-Verlag/Wien
Softcover reprint of the hardcover 1st edition   1994

Library of Congress Catalog Card Number AC 39-1015

Typesetting: Macmillan India Ltd., Bangalore-25

Printed on acid free and chlorine free bleached paper

ISSN 0071-7886
ISBN-13:978-3-7091-9283-2   e-ISBN-13:978-3-7091-9281-8
DOI: 10.1007/978-3-7091-9281-8

Contents

List of Contributors

CÁRDENAS, J., Instituto de Química, Universidad Nacional Autónoma de México, Coyoacán, 04510 México, D.F., México.

ESQUIVEL, B., Instituto de Química, Universidad Nacional Autónoma de México, Coyoacán, 04510 México, D.F., México.

GUPTA, Dr. M., Department of Medicinal Chemistry, Institute of Medical Sciences, Banaras Hindu University, Varanasi 221 005, India.

RAY, Prof. A. B., Department of Medicinal Chemistry, Institute of Medical Sciences, Banaras Hindu University, Varanasi 221 005, India.

RODRÍGUEZ-HAHN, Prof. Dr. L., Instituto de Química, Universidad Nacional Autónoma de México, Coyoacán, 04510 México. D.F., México.

Withasteroids, a Growing Group of Naturally Occurring Steroidal Lactones

Anil B. Ray and Mohini Gupta, Varanasi, India

Contents

I. Introduction

Withasteroids are a group of naturally occurring C_{28}-steroidal lactones built on an intact or rearranged ergostane framework. Withaferin A (**64**), the first member of this group of compounds, was isolated (*1*) from the well-known Indian medicinal plant, *Withania somnifera* (Sanskrit: Aswagandha) and its structure was fully elucidated by LAVIE and coworkers in 1965 (*2*). The structural novelty and interesting biological activities (*3*) elicited by this compound led to a thorough chemical investigation of the plant and numerous compounds having similar structural features were isolated. These C_{28}-steroidal lactones characterised by a nine carbon side chain with a six membered ring lactone were designated as 'withanolides' after the name of the plant genus (*4*). The withanolide skeleton may therefore, be defined as 22-hydroxyergostan-26-oic acid 26,22-lactone (**I**) and using this trivial name, withaferin A (**64**) may be designated as 4β, 27-dihydroxy-5β, 6β-epoxy-1-oxowitha-2,24-dienolide, the nomenclature that has been most widely adopted. However, many structurally related C_{28}-steroidal lactones which do not conform to the above definition have also been referred to as withanolides (*5*), a practice giving rise to confusion. Also, there are many novel structural variants of withanolides with modifications either of the carbocyclic skeleton or of the side chain and these have often been

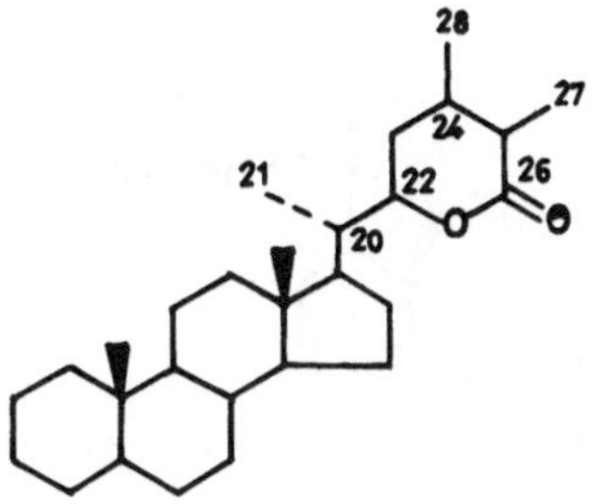

Withanolide skeleton (I) 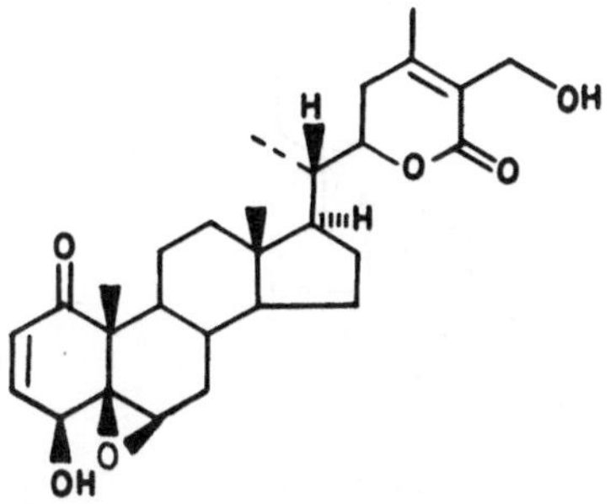   Withaferin A (64)

described as modified withanolides or ergostane-type steroids related to withanolides. Such expressions are far from satisfactory; the term 'withasteroid', originally proposed by Russian workers (6) appears to be more appropriate and can embrace all C_{28}-steroids built on an unaltered or a modified ergostane skeleton with a lactone or potential lactone ring in the C_9-side chain. Several review articles (3, 7–14) on withanolides and related ergostane-type steroids have appeared since the isolation of withaferin A. The present article aims to provide an overall picture of withasteroids highlighting some of the recent findings and covering aspects not properly dealt with in earlier review articles.

A. Features of Withasteroids and Their Classification

Withasteroids are generally polyoxygenated and it is believed that plants elaborating these compounds possess an enzyme system capable of oxidising all carbon atoms in a steroid nucleus. In fact, with the exception of C-8 and C-10, all the carbon atoms including C-13 (by fission of 13,14-bond) of the withasteroids have been found to bear oxygen functions. This wealth of oxygen functions has led to many modifications of the carbocyclic part as well as of the side chain. A C_9-side chain with a lactone or lactol ring is the characteristic feature of withasteroids but the lactone ring may be six-membered or five-membered and may be fused with the carbocyclic part of the molecule through a carbon–carbon bond or through an oxygen bridge. Appropriate oxygen substituents may lead to bond scission, formation of new bonds, aromatisation of rings and many other kinds of rearrangements resulting in compounds with novel features. Thus, in spite of the common framework on which all the withasteroids are built, the variety of structures exhibited by these steroidal lactones necessitates their classification into

Withanolide (I)

Withaphysalin (II)

Physalin (III)

Ring D aromatic Withanolide (IV)

Ring A aromatic Withanolide (V)

Acnistin (VI)

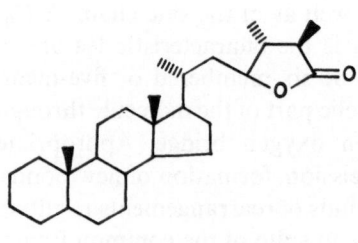

Ixocarpalactone (VII)

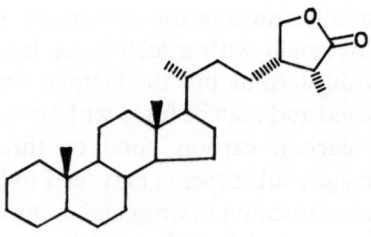

Perulactone (VIII)

Chart 1. Carbon-Oxygen Framework of Withasteroids

nine groups. These are: (i) withanolides, (ii) withaphysalins, (iii) physalins, (iv) nicandrenones or ring D aromatic withanolides, (v) jaborols or ring A aromatic withanolides, (vi) acnistins, (vii) ixocarpalactones, (viii) perulactones and (ix) miscellaneous withasteroids. Carbon-oxygen framework of each group of wistasteroid is shown in Chart 1.

Of the different groups of withasteroids, the withanolides (I) are most abundant and are regarded as possible precursors of withasteroids in groups (ii) to (vi). Subdivision of the withanolides on the basis of the orientation of the side chain is again possible as compounds with the normal 17β-orientated chain as well as the unusual 17α-orientated side chain are known, the former being the predominant group. At present, 132 withanolides with a 17β-side chain and 36 withanolides with a 17α-side chain are known. The physalins (III), the most complex molecules amongst the withasteroids, are 13,14-*seco*-16,24-cyclo-withanolides with an additional lactone ring, a γ-lactone, fused onto the D ring. The invariant presence in the physalins of a carbonyl function at C-15 suggests that the 16,24-bond is formed by an intramolecular Michael addition reaction (7) but it might be more appropriate to suggest that cyclisation is the result of intramolecular S_N2'-type reaction in withanolides having a good leaving group at C-27; the carbanion at C-16 attacks the electron deficient centre at C-24 with concomitant migration of the double bond, elimination of the leaving group at C-27 and formation of an α-methylene δ-lactone and a new 16,24 bond (Scheme 1). Scission of the 13,14-bond in the physalins is thought to be a result of oxidative cleavage by an appropriate biological oxidising agent with the hydroxyl group at C-14 and the carbonyl function at C-18 facilitating this cleavage (7, 10). Some Δ^{16}-withanolides with oxygen substituents at C-14 and C-15 have also been considered to have the desired functionality for cleavage of the 13,14-bond and occurrence of such withanolides (63, 93) in physalin-rich plants leads to the belief that these are either

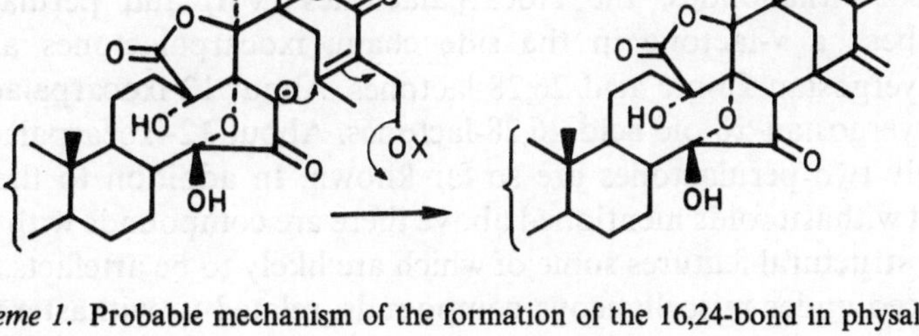

Scheme 1. Probable mechanism of the formation of the 16,24-bond in physalins

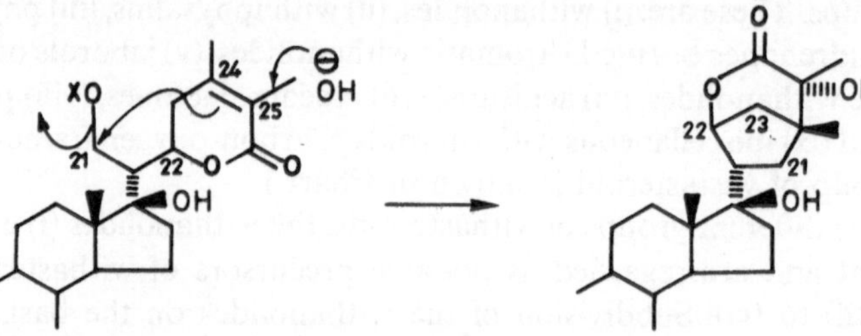

Scheme 2. Probable mechanism of the formation of bicyclic side chain in acnistins

precursors or shunt products in the biosynthesis of physalins (*13, 15*). About 18 physalins are known today; six of them are heptacyclic and the others are octacyclic. Physalin A (**173**) and physalin B (**175**) are representative members of these two types of physalins. Physalin P (**188a**) is a trislactone and corresponds to a benzilic acid type rearrangement product of a physalin (*16, 17*).

With the lone exception of withaphysalin C (**172**) which resembles the physalins in having the C-13, C-14 bond replaced by an oxide bridge, all withaphysalins retain the intact carbocyclic skeleton of ergostane and resemble the physalins only by the presence of an additional lactone or lactol ring fused with the D ring of the steroid nucleus. While the withaphysalins have been regarded as intermediates in the elaboration of the physalins from the withanolides (*7*), none of the five known withaphysalins bear the carbonyl function at C-15, essential for the formation of the 16,24-bond and common in the physalins.

The acnistins (**VI**) are withanolides with a bicyclic side chain. Formation of the new 21,24 bond is considered to take place by an S_N2'-type reaction in withanolides having a good leaving group at C-21 (Scheme 2).

Jaborol (**192**) and jaborosalactone Q (**193**) are the two known ring A aromatic withanolides while nicandrenone (**189**) and its relatives are ring D aromatic withanolides. The mechanism of aromatisation of these two groups has also been discussed (*7, 72, 73, 90*). Unlike the withanolides and modified withanolides, the ixocarpalactones (**VII**) and perulactones (**VIII**) bear a γ-lactone in the side chain; ixocarpalactones are 23-hydroxyergostan-26-oic acid 26,28-lactones. About 12 ixocarpalactones hydroxyergostan-26-oic acid 26,28-lactones. About 12-ixocarpalactones and only two perulactones are so far known. In addition to the eight types of withasteroids mentioned above there are compounds with rather diverse structural features some of which are likely to be artefacts. These are placed under miscellaneous compounds, related to withasteroids.

B. Distribution of Withasteroids

While biogenesis of withasteroids is almost a monopoly of solanaceous plants, withasteroids are not present in all members of the family Solanaceae. So far members of twelve solanaceous genera have yielded withasteroids. These are: *Acnistus* (Dunalia), *Datura, Deprea, Iochroma, Jaborosa* (Trechonaetes), *Lycium, Nicandra, Physalis, Salpichroa, Tubocapsicum, Withania* and *Witheringia*. The genera *Withania* and *Physalis* in particular have contributed in a major way to the list of this group of compounds. However, the occurrence of withasteroids is not completely restricted to Solanaceae, as was initially believed, and recent reports of

Table 1. *Sources of Withasteroids*

A. Solanaceous Plants

1a	*Acnistus arborescens* (*18–20*)	8a.	*Nicandra physaloides* (*80–98*)
1b.	*A. australis* (*21, 22*)	8b.	*N. physaloides* var. *albiflora* (*99*)
1c.	*A. breviflorus* (*23–29, 196*)	9a.	*Physalis alkekengi* (*100*)
1d.	*A. ramiflorum* (*30*)	9b.	*P. alkekengi* var. *francheti*
2a.	*Datura fastuosa* (*31*)		(*16, 101–109*)
2b.	*D. ferox* (*32*)	9c.	*P. angulata* (*110–117*)
2c.	*D. metel* (*33–46*)	9d.	*P. curasavica* (*118*)
2d.	*D. quercifolia* (*47a–50*)	9e.	*P. ixocarpa* (*119, 120*)
2e.	*D. stramonium* (*32, 51*)	9f.	*P. lancifolia* (*110–112*)
2f.	*D. stramonium* var. *violaceae* (*51*)	9g.	*P. minima* (*15, 121–126, 149*)
2g.	*D. tatura* (*52*)	9h.	*P. minima* var. *indica* (*127–130*)
2h.	*D. hybrids* (*32*)	9i.	*P. peruviana* (*83, 131–149*)
		9j.	*P. phyladelphica* (*118*)
3a.	*Deprea orinocensis* (*53*)	9k.	*P. pubescens* (*150–154*)
3b.	*D. procumbens* (*54*)	9l.	*P. viscosa* (*6, 100, 118, 155–158*)
4a.	*Dunalia australis* (*55–60*)	10a.	*Salpichroa origanifolia* (*159*)
4b.	*D. tubulosa* (*10*)	11a.	*Trechonaetes laciniata* (*160*)
5a.	*Iochroma coccinium* (*61*)	12a.	*Tubocapsicum anomalum* (*161, 162*)
5b.	*I. fuchsioides* (*62*)	13a.	*Withania aristata* (*163, 164*)
6a.	*Jaborosa bergii* (*63*)	13b.	*W. coagulans* (*165–169*)
6b.	*J. integrifolia* (*64–68*)	13c.	*W. frutescens* (*163, 170*)
6c.	*J. leucotricha* (*69–71*)	13d.	*W. obtusifolia* (*171*)
6d.	*J. magellanica* (*72–75*)	13e.	*W. somnifera*
6e.	*J. odonelliana* (*76*)		(*4, 135, 172–202, 214, 233*)
7a.	*Lycium chinense* (*77, 78*)	14a.	*Witheringia coccoloboides* (*203*)
7b.	*L. halimifolium* (*79*)		

B. Other Sources

15a.   *Tacca plantaginea* (Family: Taccaceae) (*204, 205*)
16a.   Soft coral (*Minabea* species) (*206*)
17a.   *Cassia siamea* (Caesalpiniaceae-Leguminosae) (*207*)

Table 2. *List of Withasteroids and Their Sources*

Name	Formula	Sources	Structure determination
A. Withanolides with 17β-Side Chain			
Daturalactone-1	**(1)**	2b, 2d, 2e, 2f, 2h	*(47a, 50)*
Daturalactone-2	**(2)**	2b, 2d, 8	*(47b, 50)*
Daturalactone-3	**(4)**	2b, 2d, 2h	*(48)*
Daturalactone-4	**(3)**	2d	*(49)*
Daturametelin A	**(17)**	2c	*(36)*
Daturametelin B	**(32)**	2c	*(36)*
Daturametelin D			
(= Datumetelin ?)	**(37)**	2c	*(41–44)*
Daturametelin E	**(33)**	2c	*(42)*
Daturametelin F	**(42)**	2c	*(42)*
Daturametelin G	**(38)**	2c	*(42)*
Daturataturin A	**(18)**	2g	*(52)*
Daturataturin B	**(47)**	2g	*(52)*
Daturilinol	**(39)**	2c	*(45)*
Dunawithagenin	**(48)**	13e	*(60, 200)*
Dunawithanin A	**(49)**	4a	*(55, 56)*
Dunawithanin B	**(50)**	4a	*(55, 56)*
Dunawithanin C	**(51)**	4a	*(59)*
Dunawithanin D	**(52)**	4a	*(59)*
Dunawithanin E	**(53)**	4a	*(59)*
Dunawithanin F	**(54)**	4a	*(59)*
Iochromolide	**(59)**	5a	*(61)*
Isowithametelin	**(43)**	2c	*(37)*
Isowithanolide E	**(78)**	13e	*(187)*
Ixocarpanolide	**(85)**	9e	*(120)*
14α-Hydroxy ixocarpanolide	**(86)**	9c	*(115)*
Jaborosalactone A	**(79)**	1c, 6b	*(64, 65)*
Dihydro-jaborosalactone A	**(74)**	1c	*(27b)*
Jaborosalactone B	**(88)**	6b	*(64, 65)*
Jaborosalactone C	**(94)**	6b	*(66)*
4β-Hydroxy-Jaborosalactone C	**(95)**	13c	*(170)*
Jaborosalactone D (= Acnistoferin)	**(89)**	1c, 6b	*(66)*
Jaborosalactone E	**(90)**	6b	*(66)*
Jaborosalactone F	**(91)**	6b	*(67)*
Jaborosalactone L	**(80)**	6c	*(69)*
Jaborosalactone O	**(97)**	6c	*(70)*
Lycium substance A (= Withanolide-A)	**(5)**	7a, 7b, 13b, 13c, 13e	*(77)*
Lycium substance B (= Withanolide B)	**(6)**	2b, 2c, 2d, 2h, 7a, 13b	*(77)*

Table 2 (*continued*)

Name	Formula	Sources	Structure determination
Minabeolide-1	(98)	16a	(206)
Minabeolide-2	(99)	16a	(206)
Minabeolide-3	(100)	16a	(206)
Nic-3	(101)	8a	(87)
Nic-7	(102)	8a	(87)
Nicalbin A	(103)	8b	(99)
Nicalbin B	(105)	8b	(99)
Nicandrin B (= Withaferoxolide)	(7)	2b, 2d, 2h, 8a	(32, 83)
Physalolactone B	(55)	9i	(143)
Physalolactone B glucoside	(56)	9i	(138)
Physangulide	(108)	9c	(117)
Physapubenolide	(60)	9k	(150)
Physapubescin	(104)	9k	(152)
Pubescenin	(106)	9k	(150)
Pubescenol	(109)	9k	(153)
Pubesenolide	(57)	9k	(154)
Secowithametelin (= Daturametelin C)	(34)	2c	(38, 42)
Sitoindoside IX	(61)	13e	(180)
Sitoindoside X	(62)	13e	(180)
Sominone	(58)	13e	(176)
Sominolide	(92)	13e	(176)
Vamonolide	(87)	9c	(116)
Viscosalactone A	(81)	9l	(155)
Viscosalactone B	(82)	9l	(155)
Withacnistin	(113)	1a, 5a	(19)
Withacoagin	(127)	13b	(168)
20-Deoxy-17α-hydroxy-withacoagin	(128)	13e	(184)
Withafastuosin A	(40)	2a	(31)
Withafastuousin B	(41)	2a	(31)
Withaferin A	(64)	1a, 1c, 5a, 9l, 13a, 13b, 13c, 13e	(2, 18, 19, 190)
5,6-Deoxywithaferin A	(114)	1c	(27a)
27-Deoxywithaferin A	(65)	1c, 13e	(183)
2,3-Dihydrowithaferin A	(75)	1c, 13a, 13c, 13e	(2)
2,3-Dihydro-5,6-deoxywithaferin A	(115)	1c	(27a)
2,3-Dihydro-27-deoxywithaferin A	(76)	1c	(27a)
24,25-Dihydro-27-deoxywithaferin A	(116)	1c, 13e	(183)
14α-Hydroxy-27-deoxywithaferin A	(66)	13e	(181)
17α-Hydroxy-27-deoxywithaferin A	(67)	13e	(184)
2,3,24,25-Tetrahydro-27-deoxywithaferin A	(117)	1c	(27a)

Table 2 (*continued*)

Name	Formula	Sources	Structure determination
Withaferin A chlorohydrin	**(118)**	1c, 13c	(*170*)
Withametelin (= Daturilin ?)	**(44)**	2a, 2c	(*33, 34, 37*)
Withametelin B	**(123)**	2c	(*39*)
Withametelin C	**(35)**	2c	(*40*)
Withametelin D	**(36)**	2c	(*40*)
Withametelin E	**(131)**	2c	(*40*)
Withametelin F	**(45)**	2c	(*46*)
Withametelin G	**(46)**	2a, 2c	(*46*)
Withaminimin	**(93)**	9g	(*15*)
Withangulatin A	**(63)**	9c	(*113*)
Withanicandrin	**(8)**	2b, 2d, 2h, 8a	(*97*)
Withanolide D	**(68)**	4a, 13e	(*4*)
7β-Acetoxywithanolide D	**(69)**	1a, 4a	(*20, 55*)
18-Acetoxywithanolide D	**(110)**	5b	(*62*)
18-Acetoxy-4,5,6-deoxy-5-withenolide D	**(111)**	5b	(*62*)
18-Acetoxy-5,6-deoxy-5-withenolide D	**(112)**	5b	(*62*)
4-Dehydro-withanolide D	**(119)**	13e	(*179*)
4-Dehydro-24,25-dihydro-withanolide D	**(120)**	13e	(*179*)
2,3-Dihydrowithanolide D	**(77)**	13e	(*179*)
24,25-Dihydrowithanolide D	**(121)**	121	(*183*)
24,25-Epoxywithanolide D	**(107)**	9c	(*115*)
7β-Hydroxywithanolide D	**(70)**	1b, 4a	(*22*)
14α-Hydroxywithanolide D	**(71)**	13e	(*173*)
17α-Hydroxywithanolide D	**(72)**	13e	(*173*)
27-Hydroxywithanolide D	**(73)**	13e	(*173*)
Withanolide D chlorohydrin	**(96)**	13e	(*196, 202*)
Withanolide G	**(23)**	13e	(*182, 187*)
Withanolide H	**(24)**	13e	(*182, 187*)
3β-Hydroxy-2,3-dihydro-withanolide H	**(122)**	13b	(*167*)
Withanolide I	**(124)**	13e	(*182, 187*)
27-Hydroxywithanolide I	**(125)**	13e	(*200*)
Withanolide J	**(25)**	13e	(*182, 187*)
Withanolide K	**(126)**	13e	(*182, 187*)
Withanolide L	**(26)**	13e	(*182*)
Withanolide M	**(27)**	13e	(*182*)
Withanolide N	**(28)**	13e	(*173*)
Withanolide O	**(29)**	13e	(*173, 187*)
Withanolide Q	**(30)**	13e	(*185*)
Withanolide R	**(9)**	13e	(*185*)

Table 2 (*continued*)

Name	Formula	Sources	Structure determination
5-Deoxywithanolide R	(10)	13e	(*175*)
Withanolide T	(11)	13e	(*186*)
(= 20-Hydroxywithanone)			
Withanolide U	(31)	13e	(*186, 187*)
Withanolide Y	(129)	13e	(*177*)
Withanone	(12)	13e	(*184*)
14α-Hydroxywithanone	(13)	13e	(*197*)
14β-Hydroxywithanone	(14)	13e	(*197*)
Withastramonolide	(15)	2f, 2h	(*51*)
12-Deoxywithastramonolide			
(= 27-Hydroxywithanolide B)	(16)	2c, 13e	(*39, 184*)
4α-27-Dihydroxy-5β,6β-epoxy- 1-oxowitha-2,24-dienolide	(83)	13d	(*171*)
7α,17α-Dihydroxy-5β,6β-epoxy- 1-oxowitha-2,24-dienolide	(84)	13e	(*11*)
5α,14α,17α-Trihydroxy-6β-7β epoxy-1-oxowitha-2,24-dienolide	(130)	13e	(*197*)
4β,7β,20-Trihydroxy-1-oxo- witha-2,5,24-trienolide	(21)	1b, 4a	(*22*)
4β,17α,27-Trihydroxy-1-oxo- witha-2,5,24-trienolide	(22)	13c	(*170*)
7α,27-Dihydroxy-1-oxowitha- 2,5,24-trienolide	(19)	13e	(*184*)
17α,27-Dihydroxy-1-oxowitha- 2,5-24-trienolide	(20)	13e	(*184*)
B. Withanolides with 17α-Side Chain			
Jaboromagellone	(138)	6d	(*73*)
Jaborosalactol M	(159)	6a	(*63*)
2,3-Dehydrojaborosalactol M	(160)	6a	(*63*)
Jaborosalactol N	(163)	6a	(*63*)
Jaborosalactone M	(161)	6a	(*63*)
2,3-Dehydrojaborosalactone M	(162)	6a	(*63*)
Nic-2	(164)	8a	(*85*)
Nic-11	(165)	8a	(*87*)
Physagulin C	(167)	9c	(*114*)
Physalactone	(139)	91	(*100, 158*)
Physalolactone	(147)	9i	(*142*)
4-Deoxyphysalolactone	(148)	9i, 13e	(*134*)
23-Hydroxyphysalolactone	(149)	9i	(*133*)
Physalolactone C	(140)	9i	(*131*)
Physanolide	(141)	91	(*6*)
14α,17β,20-Trihydroxy-1-oxowitha- 3,5,24-trienolide	(142)	13e	(*200*)
Visconolide	(132)	91	(*156*)

Table 2 (*continued*)

Name	Formula	Sources	Structure determination
Withanolide C	(156a)	13e	(233)
Withanolide E	(150)	91, 13e	(135, 192)
2,3-Dihydrowithanolide E	(136)	9i	(135, 137)
4β-Hydroxywithanolide E	(151)	9i, 91	(137, 148)
Withanolide F	(152)	13e	(135)
3β-Hydroxy-2,3-dihydro-withanolide F	(137)	13b	(169)
Withanolide P	(143)	13e	(135)
Withanolide S	(153)	9i, 13e	(135)
Withaperuvin	(154)	9i	(134)
Withaperuvin B	(144)	9i	(144)
Withaperuvin C	(155)	9i, 91	(144)
28-Hydroxy-withaperuvin C	(135)	91	(156)
Withaperuvin D	(157)	9i	(146)
Withaperuvin E	(156)	9i	(83)
Withaperuvin F	(158)	9i	(139)
Withaperuvin G	(145)	9i	(139)
Withaperuvin H	(166)	9i	(141)
Withaphysanolide	(133)	91	(158)
28-Hydroxywithaphysanolide	(134)	91	(157)
Withasomniferin A	(146)	13e	(175)
C. Withaphysalins			
Withaphysalin A	(168)	9g, 9h	(121)
Withaphysalin B	(171)	9g	(121)
Withaphysalin C	(172)	9g	(122)
Withaphysalin D	(169)	9g	(127)
Withaphysalin E	(170)	9h	(129)
D. Physalins			
Physalin A	(173)	9a, 9g, 9i	(101, 108)
Physalin B	(175)	9a, 9b, 9c, 9f, 9g, 14a	(109)
Physalin C	(174)	9a, 9b, 9g	(102)
25-Epi-25, 27-dihydrophysalin C	(185)	14a	(203)
Physalin D	(176)	9c, 9g	(112, 123)
Physalin E	(177)	9a, 9c, 9f, 9g, 9k	(110)
Physalin E acetate	(178)	9b, 9k	(153)
Physalin F	(179)	9b, 9c, 9f, 9g	(111)
Physalin G	(180)	9c, 9f, 9g	(112)
Physalin H	(181)	9c, 9f	(110)
Physalin I	(182)	9c, 9f	(112)
Physalin J	(183)	9c, 9f	(111)
Physalin K	(184)	9a, 9b	(112)

Table 2 (*continued*)

Name	Formula	Sources	Structure determination
Physalin L	(186)	9a, 9b	(103)
Physalin M	(187)	9a, 9b	(106)
Physalin N	(184a)	9b	(105)
Physalin O	(188)	9b	(105)
Physalin P	(188a)	9b	(16)

E. Nicandrenones (Ring D Aromatic Withanolides)

Nicandrenone			
(= Nic-1)	(189)	8a, 17a	(80, 84, 88, 91)
Nicandrenolactone			
(= Nic-1 lactone)	(190)	8a	(92)
Salpichrolide A	(191)	10a	(159)

F. Jaborols (Ring A Aromatic Withanolides and Precursor)

Jaborol	(192)	6d	(72, 75)
Jaborosalactone Q	(193)	6c	(71)
Projaborol	(194)	6d	(73)

G. Acnistins

Acnistin A	(195)	1d	(30)
Acnistin E	(196)	1d	(30)
Tubocaposide A	(197)	12a	(161, 162)
Tubocaposide B	(198)	12a	(161, 162)
Withajardin A	(199)	3b	(54)
Withajardin B	(200)	3a	(53)

H. Ixocarpalactones

Ixocapalactone A	(201)	9d, 9e, 9j, 91	(119)
Ixocarpalactone B	(202)	9d, 9e, 9j, 91	(119)
Jaborochlorodiol	(203)	6d	(73)
Jaborochlorotriol	(204)	6d	(73)
Jaborolone	(205)	6d	(73)
Jaborosalactone P	(212)	6e	(76)
Jaborotetrol	(206)	6d	(73)
Taccalonolide A	(209)	15a	(204)
Taccalonolide B	(210)	15a	(204)
Taccalonolide D	(211)	15a	(205)
Trechonolide A			
(= Jaborosalactone M-magellanica)	(207)	6d, 11a	(73, 74, 160)
Trechonolide B	(208)	11a	(160)

I. Perulactones

Perulactone A	(213)	9i	(136)
Perulactone B	(214)	9i	(145)

Table 2 *(continued)*

Name	Formula	Sources	Structure determination
J. Miscellaneous			
Taccalonolide C	**(215)**	15a	*(205)*
TH-6	**(216)**	12a	*(162)*
TH-12	**(217)**	12a	*(162)*
Datumelin	**(218)**	2c	*(35)*

their isolation from marine organisms (soft coral) and from members of Taccaceae and Leguminosae suggest that they are much more widely distributed. The various sources so far known to yield withasteroids are listed in Table 1. Table 2 contains a list of withasteroids arranged alphabetically within each group together with the sources from which their isolation has been reported.

II. Structure Elucidation Methods

The structure determination of withaferin A **(64)** and of a few other withasteroids isolated in the early years of the discovery of this group of compounds was achieved mainly by degradative methods and/or extensive chemical reactions but a considerable shift in the methods of structure elucidation has occurred in recent years. Analysis of sophisticated spectral data has largely replaced classical methods of yesteryear but the structures deduced by comprehensive spectral analysis have often been confirmed either by conversion to compounds of established structure and stereochemistry or by X-ray crystallographic analysis. Different methods that have played important roles in the structure elucidation of withasteroids are summarized below.

A. Physical Methods

1. Ultraviolet Absorption Spectra

A vast majority of withasteroids contain two isolated chromophores – an enone (steroidal 2-en-1-one) and an α,β-unsaturated-δ-lactone. The summation of the UV absorption characteristics of these two chromophores is manifested by a single maximum near 220 nm with high molar

Table 3. *Ultraviolet Absorption Spectral Data of Withasteroids*

Compound	λ_{max} in nm	$\varepsilon/\log \varepsilon$	Reference
Withaperuvin (154)	214	18,000	(134)
Withangulatin A (63)	215	16,500	(113)
Physalolactone (147)	217	11,700	(142)
Physalin C (174)	218	8,000	(102)
4-Deoxyphysalolactone (148)	220	17,900	(134)
Physapubenolide (60)	223	14,000	(150)
Withacoagin (127)	225	12,000	(168)
Physalolactone B (55)	230	5,555	(143)
Dunawithanin D (52)	230	4,200	(58)
Isowithametelin (43)	233, 227 sh	13,000	(37)
Minabeolide-1 (98)	240	16,204	(206)
Withaphysalin E (170)	312, 228	4,800; 13,700	(129)
Withametelin B (123)	314, 206	4,430; 10, 241	(39)
Jaborosalactone Q (193)	220, 278	4.20, 3.51	(71)
Projaborol (194)	220, 285	3.09, 3.05	(73)
Jaborol (192)	224, 281, 292	4.29, 3.51, 3.34	(72)
Physapubescin (104)	214.5	9,860	(152)
Daturalactone-1 (1)	223	–	(50)

absorptivity. The absence of either of these two chromophores obviously modifies the spectral characteristics; saturation of the double bond of the lactone moiety does not appreciably affect the maximal position but reduces the molar absorptivity, while reduction of the enone double bond alters both the maximal position as well as the molar absorptivity. A homoannular conjugated dienone (2,4-dien-1-one) chromophore is also discerned in a number of withasteroids which exhibit an absorption maximum near 310 nm. UV spectral data of some withasteroids are detailed in Table 3.

2. Infrared Spectra

Virtually all kinds of oxygen functions are present in this group of oxygen-rich steroids and the characteristic spectral bands for these functional groups are discernible in the infrared spectra. Particular information about the nature of carbonyl functions besides the presence of hydroxyl groups can be secured from IR spectra. The enone and α,β-unsafurated-δ-lactone functions of typical withanolides exhibit bands near 1660 resp. 1710 cm^{-1}. Perulactone (213) which contains a saturated five-membered lactone and an ester function exhibits carbonyl absorption bands at 1762 and 1732 cm^{-1}, and jaborolone (205) with an enone, an α,β-unsaturated-γ-lactone and an α-hydroxy ketone has bands at 1740, 1720 and 1680 cm^{-1}. The physalins bear a variety of carbonyl functions; thus in the IR spectrum of physalin D (176) carbonyl absorption bands are visible at 1792, 1742 and 1665 cm^{-1}. Taccalonolide D (211) contains a five-membered enol lactone to which an IR band at 1810 cm^{-1} is due.

3. Nuclear Magnetic Resonance Spectra

(a) ^{1}H-NMR Spectra

Recognition of the withanolide skeleton, structure elucidation of novel withasteroids, and determination of the stereochemistry of the different chiral centres is frequently possible by detailed analysis of the ^{1}H-NMR spectral data. For example, the signal of the carbinyl hydrogen at C-22 of a withanolide exhibits a characteristic pattern in the range δ 4.00-4.90 and is often regarded as the withanolide 'fingerprint'; it is a diagnostic double triplet in typical withanolides which are unsubstituted at C-20 and C-23 and a double doublet when either of these positions is substituted. The signal acts as a sensitive probe, any change in the environment of H-22 and the side chain is reflected in its chemical shift and splitting pattern. In withametelin (44), in which both C-20 and C-23 are unsubstituted, the H-22 signal is a broad singlet because of the oxygen bridge between C-21 and C-24 (33, 37). The change in chemical shift and splitting pattern of H-22 with change of substituents and modification of the side chain is amply manifested by the data in Chart 2. Thus in jaborosalactol M (159) which bears an epoxylactol ring side chain, the H-22 signal appears as a double triplet at δ4.04 while in the corresponding lactone (161), this signal moves downfield to δ4.92.

The relatively common 2-en-1-one system of withasteroids and for that matter, the different AB ring substitution patterns exhibited by them can readily be identified by ^{1}H-NMR spectroscopy. A typical withanolide shows signals for five methyl groups; two angular C-methyls (18- and 19-Me), two vinylic C-methyls (27- and 28-Me) and one secondary C-methyl (21-Me), if C-20 is unsubstituted. Any change in the number of these methyl groups indicates substitution at these sites with or without modification of the typical withanolide skeleton. Analysis of ^{1}H-NMR spectral data also helps in distinguishing withanolides having a 17β-side chain from those with a 17α-side chain. The orientation of the side chain affects the chemical shift of the neighbouring hydrogens, an effect which

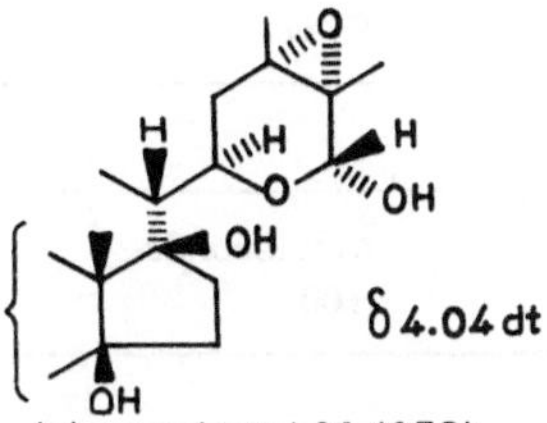

Jaborosalactol M (159) δ 4.04 dt

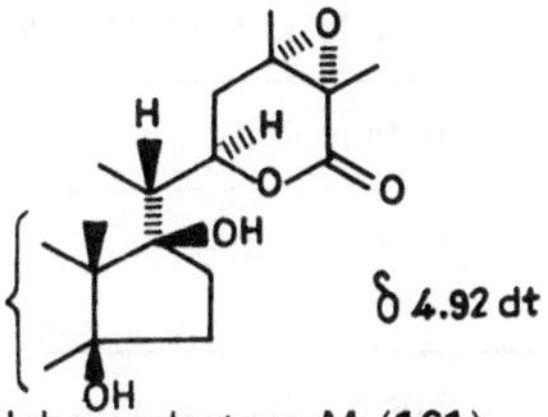

Jaborosalactone M (161) δ 4.92 dt

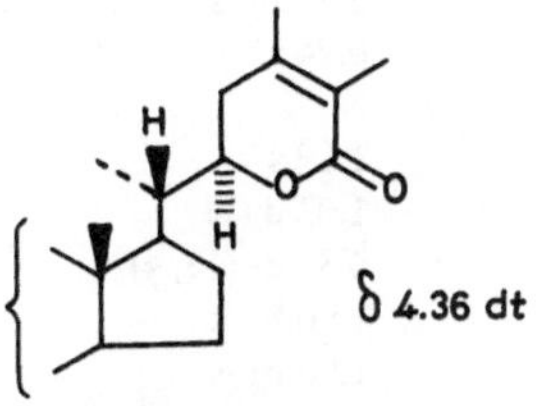

Minabeolide-1 (98) δ 4.36 dt

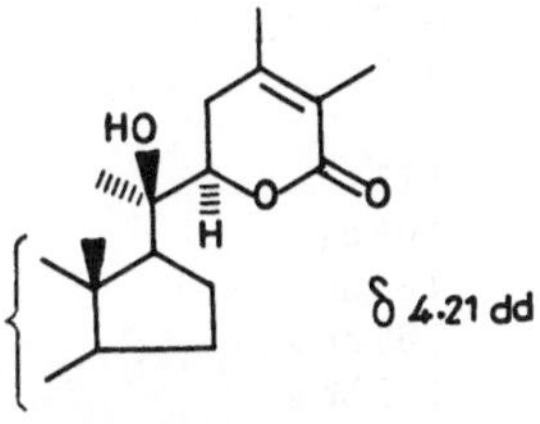

Withacoagin (127) δ 4.21 dd

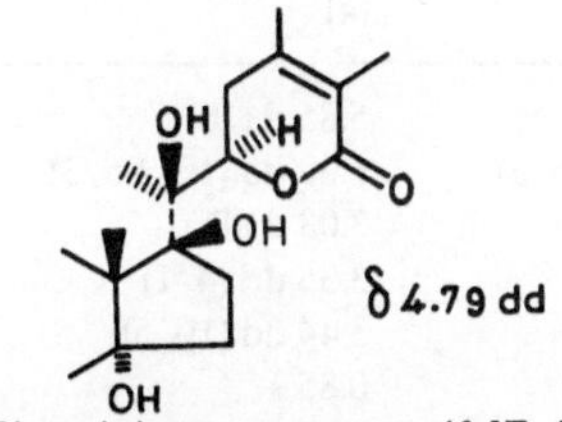

Physalolactone acetate (147-Ac) δ 4.79 dd

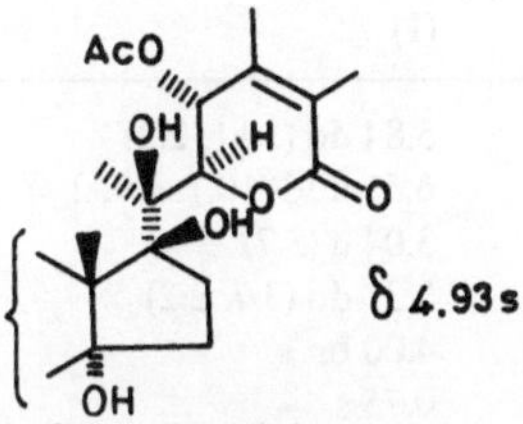

23-Hydroxyphysalolactone acetate (149 Ac) δ 4.93 s

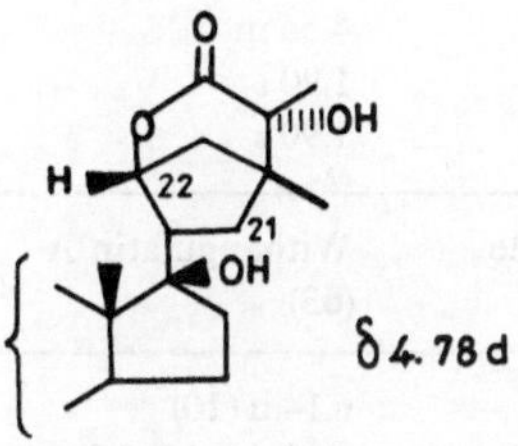

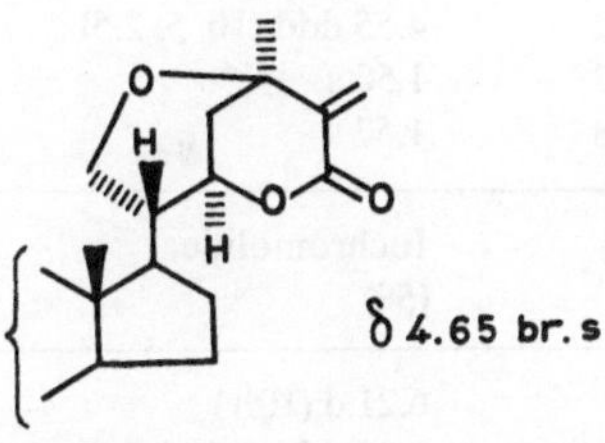

Acnistin E (196) δ 4.78 d

Withametelin (44) δ 4.65 br. s

Chart 2. Change of ¹H NMR Parameters of H-22 with Change of Side Chain

is demonstrated by comparing two C-17 epimers, withanolide E (150) and isowithanolide E (78). The H-18, H-21 and H-22 signals of isowithanolide E appear at δ0.98, 1.25 and 4.61, while the corresponding signals of withanolide E are seen at δ1.10, 1.42 and 4.88, respectively (187).

In the structure determination of withasteroids various 2D NMR techniques have been widely used in recent years. The chemical shifts and multiplicities of important hydrogen signals of withasteroids of different

Table 4. ^1H-NMR Data of Some Withasteroids (δ Values)

Hydrogens	Withanolides		
	Minabeolide-1 (98)	Minabeolide-2 (99)	Minabeolide-3 (100)
H-1	7.04 d (10)	7.01 d (10)	–
H-2	6.23 dd (10, 2)	6.24 dd (10, 2)	–
H-4	6.07 br. s	6.07 br. s	5.72 br. s
H-18	0.77 s	3.91 d (12) 4.35 d (12)	0.74 s
H-19	1.23 s	1.24 s	1.19 s
H-21	1.01 d (7)	1.12 d (7)	1.02 d (7)
H-22	4.36 dt (12.3, 4)	4.38 dt (12, 4)	4.37 dt (13, 4)
H-27	1.88 br. s	1.88 br. s	1.90 br. s
H-28	1.93 br. s	1.94 br. s	1.96 br. s
	Daturalactone-1 (1)	Daturalactone-2 (2)	Daturalactone-3 (4)
H-2	5.84 dd (10.1, 2.2)	5.85 d (10)	5.85 d (10)
H-3	6.61 ddd (10.1, 4.9, 2.4)	6.60 dq (10, 4.5, 3)	6.60 dq (10, 4.5, 3)
H-6	3.04 d (3.7)	3.08 d (4)	3.08 d (4)
H-7	3.31 dd (3.7, 2.2)	3.42 d (4.1)	3.35 dd (4, 1)
H-12	4.00 br. s	–	3.44 dd (10, 5)
H-18	0.75 s	1.1 s	0.85 s
H-19	1.17 s	1.26 s	1.18 s
H-21	1.05 d (5.5)	0.90 d (7)	1.02 d (6)
H-22	4.55 ddd (10, 5, 2.5)	4.55 m	4.55 m
H-27	1.50 s	1.52 s	1.90 s
H-28	1.57 s	1.58 s	1.90 s
	Iochromolide (59)	Physapubenolide (60)	Withangulatin A (63)
H-2	6.21 d (10.1)	6.18 d (10)	6.14 d (10)
H-3	6.93 dd (10.1, 5.8)	6.95 dd (10, 5.5)	6.97 dd (10, 5.8)
H-4	3.77 d (5.8)	3.79 d (5.5)	3.76 dd (5.8, 2)
H-6	3.24 br.s	3.36 m	3.34 br.s
H-15	–	4.97 d (4.4)	5.19 d (2.7)
H-16	4.90 t (9.0, 7.2)	–	5.63 d (2.7)
H-18	0.76 s	1.11 s	1.06 s
H-19	1.41 s	1.40 s	1.40 s
H-21	1.02 d (6.5)	1.02 d (6.8)	1.07 d (7)
H-22	4.17 dt (13.1, 4.3, 3.2)	4.32 dt	4.21 ddd (12, 7, 3)
H-27	1.88 s	1.87 s	1.81 s
H-28	1.93 s	1.94 s	1.90 s

Table 4 (*continued*)

Hydrogens	Withanolides		
	2,3-Dihydrowithanolide D (77)	Viscosalactone B (82)	Pubescenol (109)
H-3	–	3.25 br.	–
H-4	3.50 t (3.2)	3.38 d	4.28 m
H-6	3.13 br. s	3.03	–
H-7	–	–	3.64 m
H-18	0.81 s	0.66 s	0.68 s
H-19	1.30 s	1.13 s	0.91 s
H-21	1.25 s	0.96 d	0.84 d (6)
H-22	4.20 dd (12.7, 4.7)	–	4.38 m
H-27	1.87 s	4.33 br. s	1.45 s
H-28	1.97 s	2.03 s	1.45 s
	Visconolide* (132)	4-Deoxyphysalolactone (148)	Withaperuvin acetate (154-Ac)
H-2	6.45 d (10)	6.01 dd (10, 2.5)	6.04 dd (10.4, 2.1)
H-3	7.19 q (10, 6.4)	6.75 ddd (10, 5.5, 2.5)	6.36 dd (10.4, 2.3)
H-4	4.03 d (6.4)	3.00 dt (20, 2.5) 2.55 dd (20, 5.5)	6.44 t (2.2)
H-6	3.36 br.s	4.34 dd (11.5, 7)	5.21 dd (11.9, 4.9)
H-18	1.33 s	1.07 s	1.06 s
H-19	1.81 s	1.23 s	1.30 s
H-21	1.87 s	1.42 s	1.41 s
H-22	5.35 q (13.5, 3.1)	4.81 dd (10, 6.5)	4.80 dd (10, 6.5)
H-27	2.05 br. d (1.6)	1.86 s	1.86 s
H-28	4.43 d (14.4) 4.73 d (14.4)	1.93 s	1.93 s
	Withaperuvin D acetate (157-Ac)	Physalolactone C acetate (140-Ac)	Jaborosalactol N (163)
H-2	2.37 dd (18, 3.5) 2.82 d (18)	6.05 dd (8.5, 3.2)	6.01 dd (9.5, 1)
H-3	4.46 m	6.29 dd (8.5, 2.4)	6.92 dd (9.5, 6)
H-4	5.24 d (6.5)	6.27 br.	6.14 dd (6,1)
H-6	4.27 d (7.5)	4.40 dd (10.5, 6)	4.62 br. s
H-15	–	5.33 br. s	–
H-18	1.07 s	1.14 s	1.18 s
H-19	1.14 s	1.30 s	1.49 s
H-21	1.40 s	1.32 s	0.96 d (7)
H-22	4.48 dd (10, 4)	4.61 t (8.2)	4.07 m
H-26	–	–	5.00 d (11)
H-27	1.87 s	1.88 s	1.40 s
H-28	1.94 s	1.95 s	1.42 s

*In C_5D_5N

Table 4 (*continued*)

Hydrogens	Withanolides		
	Daturataturin A pentaacetate (18-Ac)	Daturataturin B heptaacetate (47-Ac)	Withanolide Q diacetate (30-Ac)
H-1	–	5.07 br.s	–
H-2	5.92 dd (9.9, 2.2)	–	5.90 dq (10, 3, 1)
H-3	6.81 ddd (9.9, 5, 2.6)	3.88 m	6.85 dq (10, 5, 2.5)
H-6	5.83 dd (5.9, 1.8)	5.77 d (4)	5.63 ($W_{1/2}$ 8)
H-7	4.90 m	4.92 m	–
H-18	0.73 s	0.70 s	0.79 s
H-19	1.25 s	1.26 s	1.24 s
H-21	1.04 d (6.6)	1.03 d (6.6)	0.95 d (7)
H-22	4.41 dt (13.2, 3.3)	4.41 br. d (9.9)	4.80 dd (2.5, 2)
H-27	4.47 d (11)	4.86 d (11.5)	4.95 s
	4.59 d (11)	4.91 d (11.5)	
H-28	2.06 s	2.05 s	2.06 s
Glc-1'	4.65 d (8.1)	4.58 d (8)	–
	Isowithametelin (43)	Withametelin (44)	Daturametelin F* (42)
H-2	2.72 dd (20, 5)	5.88 dd (9.8, 2.5)	3.15 dd (12.7, 9.5)
	3.28 d (20)		3.26 dd (12.7, 5.7)
H-3	5.58 dt (10, 5)	6.77 ddd (9.8, 5.1, 2.5)	5.07 m
H-4	6.02 dd (10, 2.5)	2.84 dd (20, 5.1)	2.89 br. t (12.9)
		3.30 dt (20, 2.5)	3.08 dd (12.9, 5.5)
H-6	5.62 dd (5, 2.5)	5.58 t (5.8)	5.47 d (5.1)
H-18/19	0.68 s/1.33 s	0.71 s/1.22 s	0.52 s/1.20 s
H-21	3.70 dd (13, 2)	3.73 dd (12.7, 3)	3.76 dd (13.2, 3.2)
	3.89 d (13)	3.95 d (12.7)	3.87 d (13.2)
H-22	4.63 br. s	4.65 br. s	4.66 br. s
H-27	5.99 s	6.02 br. s	6.02 s
	6.73 s	6.76 br. s	6.87 s
H-28	1.42 s	1.44 s	1.48 s
	Nic-7* (102)	Nicalbin A (103)	Nicalbin B (105)
H-2	5.90 d (10)	5.82 dd 10, 2.5)	5.83 dd (10, 2.5)
H-3	6.52 dd (10, 3.5)	6.62 ddd (10, 5, 2.5)	6.59 ddd (10, 5, 2.5)
H-6	3.12 d (4)	3.04 d (3.5)	3.03 d (3.5)
H-7	3.28 br.	3.26 dd (3.5, 1)	3.27 dd (3.5, 1)
H-11	3.85 br. d (11.5)	–	–

* In C_5D_5N

Table 4 (*continued*)

Hydrogens	Withanolides		
	Nic-7* (**102**)	Nicalbin A (**103**)	Nicalbin B (**105**)
H-16	–	4.18 t	3.98 dd (7.5, 4.5)
H-18	0.97 s	0.76 s	0.82 s
H-19	1.45 s	1.18 s	1.17 s
H-21	1.45 d (6)	0.93 d (7)	1.03 d (6.5)
H-22	4.22 m	3.97 dt	3.57 t (6.5)
H-26	5.43	5.08 s	5.02 s
H-27	1.26 s	1.40 s	1.36 s
H-28	1.36 s	1.42 s	1.43 s
	Withametelin B (**123**)	Withanolide K (**126**)	Withacoagin (**127**)
H-2	6.04 d (9.7)	–	5.89 dd (10, 2.4)
H-3	6.93 dd (9.7, 5.9)	5.64 dt (10)	6.58 ddd (10, 5.2, 2.2)
H-4	6.16 d (5.9)	6.08 dt (10)	2.36 dd (19, 5.2) 2.63 dt (19, 2.3)
H-6	4.59 br. s	5.70 t	5.73 dd (9.9, 1.7)
H-7	–	–	5.57 dd (9.9, 2.7)
H-18	0.77 s	1.07 s	0.95 s
H-19	1.42 s	1.39 s	1.20 s
H-21	3.72 dd (13.3, 2.4) 3.89 d (13.3)	1.32 s	1.32 s
H-22	4.64 br. s	4.66 dd (10, 5)	4.21 dd (13.3, 3.5)
H-27	6.02 br. s 6.76 br. s	1.88 s	1.89 s
H-28	1.48 s	1.95 s	1.95 s
	Withasomniferin A (**146**)	Physagulin C (**167**)	23-Hydroxyphysalolactone[†] (**149**)
H-2	–	6.40 d (9.8)	5.90 dd (10.2, 2)
H-3	–	7.26 dd (9.8, 6.2)	6.55 dd (10.2, 2.5)
H-4	5.53 t (1.6, 1.5)	3.96 d (6.2)	4.99 t (2.3)
H-6	3.15 d (1.9)	3.26 br.s	4.38 m
H-7	3.52 dd (6.1, 3.6)	–	–
H-15	–	5.21 s	–
H-16	–	3.72 s	–
H-18	0.76 s	1.26 s	1.04 s
H-19	1.33 s	1.78 s	1.22 s
H-21	1.10 d (7)	0.99 d (7)	1.35 s
H-22	4.35 m	4.43 m	4.67 d (8.5)
H-23	–	–	4.34 d (8.5)
H-27	1.87 s	1.68 s	1.85 s
H-28	1.92 s	1.87 s	1.97 s

* In C_5D_5N
† In CD_3OD

Table 4 (*continued*)

Hydrogens	Withaphysalin and Physalins		
	Withaphysalin E (170)	Physalin L* (186)	Physalin A* (173)
H-2	5.96 d (9.7)	2.68 dd (20,3) 3.47 br. d (20)	5.83 dd (10, 2)
H-3	6.93 dd (9.7, 6)	5.88 br. d (10)	6.93 ddd (10, 5, 2.5)
H-4	6.14 d (6)	6.14 d (10)	2.91 dd (22, 5) 3.26 br. d (22)
H-6	4.54 t (3)	5.76 br. d (5)	5.69 dd (6, 1.5)
H-7	–	4.61 m	4.47 br. t (5)
H-9	–	3.12 dd (12, 9)	3.00 dd (12, 9)
H-19	1.48 s	1.16 s	1.02 s
H-21	1.52 s	1.71 s	1.71 s
H-22	4.60 dd (13.4, 3.6)	4.57 m	4.59 dd (4,2)
H-25	–	2.62 q (8)	–
H-27	1.88 s	1.17 d (8)	5.59 s 6.43 s
H-28	1.96 s	1.32 s	1.55 s

Hydrogens	Ring D Aromatic Withanolides		
	Nicandrenone (189)	Salpichrolide A (191)	Nicandrenolactone (190)
H-2	5.88 dt (10)	6.00 ddd (10.1, 2.8, 1)	5.88 dt (10)
H-3	6.62 m	6.77 ddd (10.1, 5, 2.5)	6.65 m
H-6	3.22 d (4)	3.24 d (4.9)	3.25 d (4)
H-7	4.0 m (narrow)	–	4.04 m (narrow)
H-15	7.38 d (8)	6.99 d (8)	7.39 d (8)
H-16	7.05 d (8)	7.12 dd (8, 1)	7.05 d (8)
H-17a	7.0 s	6.90 d (1)	7.01 s
H-21	1.24 d (7)	1.24 d (7)	1.35 d (7)
H-22	3.87 m	3.84 ddd (11.3, 5.6, 2.6)	4.65 dt (12,5)
H-26	4.98 s	4.99 d (9.5)	–
H-19	1.22 s	1.38 s	1.24 s
H-27	1.33 s	1.37 s	1.50 s
H-28	1.35 s	1.35 s	1.40 s

*In DMSO d_6

Table 4 (*continued*)

Hydrogens	Ring A Aromatic Withanolides & Precursor		
	Jaborol* (192)	Jaborosalactone Q (193)	Projaborol* (194)
H-2	7.03 dd (7.6, 1.3)	6.66 dd (8, 2)	–
H-3	6.99 dd (7.6)	7.06 t (8)	–
H-4	6.68 dd (7.4, 1.3)	6.91 dd (8, 2)	–
H-6	4.94 dd (10.4, 4.5)	4.75 br. t (3)	5.30 m
H-9	4.59 m	–	4.52 m
H-11	2.73 dd (16.3, 5.9)	–	2.75 m
	2.89 dd (16.3, 9)	–	2.98 m
H-18	0.99 s	0.80 s	1.00 s
H-19	2.11 s	–	2.35 s
H-21	0.82 d (7)	1.04 d (6.5)	0.86 d (7)
H-22	4.65 dm	4.47 ddd (11, 6, 3)	4.65 m
H-27	1.79 s	4.38 br. s	1.78 s
H-28	1.89 s	2.04 s	1.87 s

Hydrogen	Acnistin and Ixocarpalactones		
	Acnistin E (196)	Jaborolone (205)[†]	Jaborochlorodiol (203)
H-2	6.26 d (10)	5.85 dd (10.2, 2.3)	5.95 d (10.3)
H-3	7.04 dd (10, 6)	6.57 m	6.94 dd (9.7, 5.7)
H-4	3.76 dd (6)	–	6.15 d (5.7)
H-6	3.20	–	4.55 br. s
H-18	0.85 s	1.00 s	1.07 s
H-19	1.40 s	1.32 s	1.40 s
H-21	–	1.26 d (6.6)	0.98 d (6.7)
H-22	4.78 d	4.69 dd (10.9, 2.0)	4.04 dd (11, 1.7)
H-23	–	5.00 br. s	4.78 br. s
H-27	1.50 s	1.66 s	1.70 s
H-28	1.21 s	1.85 s	1.88 s

	Perulactones		
	Perulactone (213)		Perulactone B (214)
H-1	5.05 t (2.5)		–
H-2	–		5.87 ddd (10, 2.5, 1)
H-3	3.87 m		6.83 ddd (10, 5, 2.5)
H-6	5.52 d (5.5)		5.55 d (5.5)

* In CD_3CN
† In C_5D_5N

Table 4 (*continued*)

	Perulactones		
	Perulactone (213)		Perulactone B (214)
H-18	0.88 s		1.12 s
H-19	1.08 s		1.24 s
H-22	1.22 s		1.27 s
H-21	3.46 dd (10.5, 1.5)		4.02 dd (10.5, 1.5)
H-27	1.20 d (7)		1.20 d (7)
H-28	4.12 dd (9, 7.5)		4.13 dd (9, 7.5)
	4.45 dd (9, 7)		4.43 dd (9, 7)

Hydrogens	Ixocapalactones		
	Trechonolide A (207)	Jaborosalactone P (212)	Taccalonolide A (209)
H-1	–	–	4.67 d (5.5)
H-2	5.98 dd (10.2, 2.8)	5.83 dd (10, 2.5)	3.42 dd (5, 4)
H-3	6.82 ddd (10.2, 6.7, 2.3)	6.74 ddd (10, 5, 2)	3.32 ddd (4, 2, 2)
H-4	1.87 dd (19, 6.7)	2.81 dd (21, 5)	2.08 ddd (14, 6, 2)
	2.91 dt (19, 2.6)	3.28 br. d (21)	2.15 ddd (14, 10, 2)
H-6	3.10 br. d (2.3)	5.59 br. d (6)	–
H-7	–	–	4.00 d (10.5)
H-11	–	–	5.25 dd (12, 3)
H-12	–	–	5.20 d (3)
H-15	–	–	5.47 dd (10, 10)
H-18	1.00 s	1.15 s	0.93 s
H-19	1.17 s	1.20 s	0.78 s
H-20	2.29 dq (11.2, 6.7)	–	2.11 m
H-21	1.00 d (6.7)	1.21 d (7)	0.85 d (7)
H-22	4.02 dd (11.2, 2.3)	4.22 dd (12.5, 3.5)	5.00 d (2)
H-23	4.80 m	–	–
H-27	1.82 dq (1.9, 1.0)	1.91 br. s	1.58 s
H-28	1.94 quintet (1)	2.27 br. s	1.28 s

groups with different substitution patterns are listed in Table 4, the compounds selected being mainly of recent origin.

(b) ^{13}C-NMR Spectra

^{13}C-NMR spectroscopy has become an indispensable tool in structural and conformational analysis of natural products and application of ^{13}C-NMR data in the structural investigation of withasteroids has

Table 5. ^{13}C-NMR Spectral Data of Withasteroids

Carbons	Withanolides with 17β-Side Chain				
	Physapuben-olide (60)	Viscosa-lactone A (81)	Viscosa-lactone B (82)	Witha-minimin (93)	Withangu-latin A (63)
C-1	202.6	206.9	210.5	204.1	202.4
C-2	131.5	54.9	42.6	128.7	131.4
C-3	143.5	55.8	68.5	141.3	142.7
C-4	69.6	74.6	77.1	36.0	69.5
C-5	63.4	64.0	64.6	77.2	63.2
C-6	62.7	59.6	60.0	74.3	63.2
C-7	26.1	29.6	29.8	26.5	24.7
C-8	40.3	31.0	31.1	35.4	34.7
C-9	36.1	42.7	42.9	35.4	39.3
C-10	47.9	48.6	50.5	52.2	47.6
C-11	21.8	20.0	21.7	23.2	21.8
C-12	41.4	27.3	27.4	38.8	37.5
C-13	46.1	42.7	42.8	52.2	52.1
C-14	84.1	56.1	56.1	82.3	81.5
C-15	80.8	24.3	24.4	83.4	83.6
C-16	33.8	40.7	39.2	120.4	120.8
C-17	52.3	51.9	52.0	161.3	162.4
C-18	15.6	11.5	11.6	16.8	17.7
C-19	17.2	14.5	15.8	15.1	15.9
C-20	37.5	38.9	38.9	36.1	35.4
C-21	17.2	13.3	13.4	17.2	17.3
C-22	78.5	78.7	78.9	78.5	79.4
C-23	31.2	29.8	29.3	32.4	33.1
C-24	149.5	153.8	153.4	150.3	148.4
C-25	121.9	125.8	125.7	121.4	122.2
C-26	166.8	167.4	167.2	167.5	166.4
C-27	12.4	57.2	57.4	12.4	12.4
C-28	20.5	20.1	20.1	20.6	20.3
–OAc	169.8, 21.5	–	–	170.6, 21.4	169.7, 21.3

Table 5 (*continued*)

Carbons	Withanolides with 17β-Side Chain				
	Withametelin (44)	Isowithame-telin (43)	Secowithame-telin (34)	Daturame-telin A (17)	Daturame-telin B (32)
C-1	204.5	210.7	204.4	203.9	203.9
C-2	128.0	39.6	127.9	127.7	127.9
C-3	145.0	127.0	145.5	145.8	145.9
C-4	33.4	129.5	33.5	33.4	33.6
C-5	135.9	141.2	135.4	136.2	136.3
C-6	124.5	121.7	124.6	124.6	124.8
C-7	30.7	31.1	30.8	30.8	31.1
C-8	33.2	31.8	33.3	33.1	33.4
C-9	42.8	41.0	42.9	43.3	43.5
C-10	50.4	52.2	50.5	50.6	50.8
C-11	23.6	22.6	23.8	23.8	24.1
C-12	39.6	39.9	39.2	39.8	39.3
C-13	42.5	43.1	42.3	42.5	42.6
C-14	56.0	56.0	56.1	56.2	56.3
C-15	24.0	24.2	24.3	24.2	24.5
C-16	26.5	26.7	27.2	26.9	27.1
C-17	47.6	47.7	46.2	51.9	47.2
C-18	12.7	12.9	12.3	11.7	12.4
C-19	18.9	20.5	19.0	18.8	19.0
C-20	39.9	40.0	45.3	38.9	46.0
C-21	60.4	60.6	59.5	13.3	58.9
C-22	75.8	75.7	77.9	78.2	77.9
C-23	33.2	33.4	32.7	29.9	33.3
C-24	69.2	69.5	157.6	157.0	157.9
C-25	138.9	139.1	123.2	122.8	123.1
C-26	165.2	165.4	165.9	166.0	166.2
C-27	129.7	130.0	65.7	63.2	63.4
C-28	25.6	25.7	20.6	20.4	20.5
−OMe	−	−	58.5	−	−
C-1′-6′	−	−	−	62.6–104.6	62.8–104.6

Carbon	Withanolides with 17β-Side Chain				
	Minabe-olide-2 (99)	Minabe-olide-3 (100)	Withacoagin (127)	Daturata-turin A (18)	Daturata-turin B (47)
C-1	155.4	35.7	203.8	203.3	72.3
C-2	127.7	34.0	129.6	127.9	37.6
C-3	186.2	199.4	140.4	145.6	73.8
C-4	124.0	123.9	36.5	33.6	38.2

Table 5 (*continued*)

Carbon	Withanolides with 17β-Side Chain				
	Minabe-olide-2 (99)	Minabe-olide-3 (100)	Withacoagin (127)	Daturata-turin A (18)	Daturata-turin B (47)
C-5	168.5	170.4	74.9	139.4	142.7
C-6	32.7	32.0	133.3	128.1	127.8
C-7	33.6	32.8	129.4	63.7	64.7
C-8	35.7	35.6	37.5	35.8	34.0
C-9	52.3	53.8	41.5	39.1	39.2
C-10	43.4	38.6	51.4	50.2	42.9
C-11	22.7	21.0	22.4	23.9	20.4
C-12	34.6	39.6	40.7	39.9	39.4
C-13	45.9	42.8	44.1	42.5	42.7
C-14	54.7	55.6	54.6	52.2	52.2
C-15	24.2	24.2	23.6	24.3	24.6
C-16	27.0	27.3	21.9	27.2	27.4
C-17	52.5	52.1	54.7	51.3	49.7
C-18	61.9	11.8	13.8	11.9	11.7
C-19	18.7	17.4	14.6	18.2	18.3
C-20	39.1	38.9	75.2	38.7	38.9
C-21	12.5	12.5	21.0	13.4	13.5
C-22	78.0	78.3	81.0	78.3	78.2
C-23	29.3	29.6	31.7	29.8	29.8
C-24	148.9	149.0	148.7	157.1	153.9
C-25	122.1	122.0	122.1	–	127.3
C-26	167.0	167.0	166.0	166.1	166.4
C-27	13.7	13.7	12.5	63.4	56.2
C-28	20.5	23.7	20.5	20.6	20.1
–OAc	171.2, 21.1	–	–	–	–
C-1'-C-6'	–	–	–	62.8–104.8	62.7–102.9

Carbons	Withanolides with 17α-Side Chain				
	Withaperu-vin (154)	Withaperu-vin C[a] (155)	Withaperu-vin D (157)	Withaperu-vin H[b] (166)	Physalo-lactone C acetate [a] (140-Ac)
C-1	202.4	207.9	214.0	202.0	198.6
C-2	127.2	118.2	42.3	127.7	128.2
C-3	145.9	142.7	77.6	145.3	140.9
C-4	67.0	126.5	74.5	64.9	66.9
C-5	79.4	160.2	79.8	79.7	78.6
C-6	74.5	74.7	77.3	47.9	63.8
C-7	32.0	37.4	26.4	29.4	35.3

[a] In CD_3OD [b] In C_5H_5N

Table 5 (*continued*)

Carbons	Withanolides with 17α-Side Chain				
	Withaperu-vin (**154**)	Withaperu-vin C[a] (**155**)	Withaperu-vin D (**157**)	Withaperu-vin H[b] (**166**)	Physalo-lactone C acetate [a] (**140**-Ac)
C-8	37.4	35.5	35.8	38.6	33.1
C-9	37.8	44.0	34.9	38.9	48.2
C-10	55.6	55.5	55.8	56.7	57.8
C-11	22.7	22.4	21.6	23.7	23.4
C-12	34.5	35.5	35.4	35.1	42.1
C-13	54.0	55.5	55.6	55.2	55.1
C-14	82.8	84.5	83.9	82.0	150.2
C-15	30.1	31.4	31.1	30.7	115.3
C-16	37.4	37.4	37.4	37.2	35.9
C-17	87.6	88.3	88.7	87.9	86.7
C-18	20.7	21.2	20.6	21.1	17.4
C-19	9.7	18.6	15.3	10.2	9.3
C-20	78.8	79.6	78.3	79.1	77.5
C-21	19.1	19.4	19.4	19.5	20.3
C-22	80.9	82.7	83.0	81.4	79.9
C-23	32.2	33.3	32.8	32.8	34.6
C-24	151.9	153.2	154.2	151.0	150.3
C-25	121.2	121.7	121.7	121.3	121.6
C-26	167.7	168.9	165.7	166.8	165.7
C-27	12.2	12.3	12.3	12.5	12.4
C-28	20.7	20.6	20.8	20.3	20.6
Extra	–	–	–	96.6, 31.2	–
–OAc	–	–	–	–	169.8, 21.5

Carbons	Withaphysalin, Physalins and Ring D Aromatic Withanolide			
	Withaphysalin E[a] (**170**)	Physalin M[b] (**187**)	Physalin N[b] (**184a**)	Salpichrolide A (**191**)
C-1	206.1	209.6	201.5	202.4
C-2	117.5	39.5	126.9	128.6
C-3	140.4	122.5	146.2	142.2
C-4	126.0	126.4	32.3	33.5
C-5	160.0	140.4	139.1	64.5
C-6	72.8	128.0	125.5	58.9
C-7	37.1	25.7	61.4	30.5
C-8	37.8	40.8	44.2	33.1

[a] In C_5D_5N
[b] In DMSO-d_6

Table 5 (*continued*)

Carbons	Withaphysalin, Physalins and Ring D Aromatic Withanolide			
	Withaphysalin E[a] (**170**)	Physalin M[b] (**187**)	Physalin N[b] (**184a**)	Salpichrolide A (**191**)
C-9	44.7	32.4	27.6	36.3
C-10	53.8	55.1	52.7	48.7
C-11	22.7	24.2	24.0	25.4
C-12	36.3	26.1	25.7	30.3
C-13	60.0	78.8	81.0	137.6
C-14	83.1	101.2	106.3	136.8
C-15	34.8	215.8	208.7	125.4
C-16	25.7	54.0	52.9	126.3
C-17	57.7	82.0	78.0	140.3
C-18(17a)	177.3	171.7	171.7	128.6
C-19	20.0	20.8	15.6	14.8
C-20	83.9	82.3	80.3	42.9
C-21	26.3	25.5	21.9	17.2
C-22	78.4	76.4	76.2	67.4
C-23	31.9	29.0	31.3	33.7
C-24	148.7	34.3	30.7	64.7
C-25	121.9	40.9	49.3	63.5
C-26	165.4	172.2	167.4	91.4
C-27	12.5	16.6	61.1	16.5
C-28	20.7	18.0	24.3	18.7
	–	–	–	–

Carbons	Ixocarpalactones			
	Ixocarpalactone (**201**)	Trechonolide A (**207**)	Jaborosalactone P (**212**)	Jaborolone (**205**)
C-1	202.0	202.2	202.8	201.5
C-2	132.6	129.5	127.3	127.7
C-3	143.2	143.9	144.8	141.6
C-4	73.0	32.7	33.3	31.2
C-5	64.1	61.8	136.0	82.1
C-6	61.2	63.2	124.2	209.1
C-7	31.2	30.1	34.0	30.2
C-8	29.0	29.3	32.6	37.4
C-9	44.2	41.9	40.0	37.9
C-10	48.1	47.3	50.0	55.4
C-11	21.3	36.4	30.0	39.9
C-12	40.0	98.9	82.0	101.7
C-13	43.3	47.8	51.2	47.8

Table 5 (*continued*)

Carbons	Ixocarpalactones			
	Ixocarpalactone (**201**)	Trechonolide A (**207**)	Jaborosalactone P (**212**)	Jaborolone (**205**)
C-14	54.5	45.8	44.1	46.3
C-15	37.2	22.8	23.2	22.5
C-16	70.0	34.1	32.1	33.6
C-17	57.6	80.3	82.9	79.5
C-18	14.1	9.9	12.7	10.5
C-19	16.8	14.7	18.2	13.8
C-20	79.5	35.3	40.5	35.2
C-21	22.1	11.9	11.3	12.2
C-22	73.2	68.7	71.0	69.0
C-23	78.9	82.4	97.6	82.5
C-24	41.9	157.0	157.5	156.0
C-25	40.3	123.7	130.7	125.2
C-26	181.6	175.0	173.3	175.4
C-27	14.6	8.2	9.5	8.4
C-28	13.0	11.9	18.7	12.2

Carbons	Perulactones and Ring A Aromatic Withanolides			
	Perulactone (**213**)	Perulactone B (**214**)	Jaborosalactone Q (**193**)	Jaborol (**192**)
C-1	75.3	205.2	155.1	149.7
C-2	35.4	127.8	115.5	114.0
C-3	66.4	146.1	126.8	126.5
C-4	41.3	33.6	122.5	116.5
C-5	137.3	135.4	140.0	140.6
C-6	124.1	125.2	68.2	78.1
C-7	31.5	25.5	34.2	39.6
C-8	31.1	35.8	32.9	43.5
C-9	42.1	37.2	44.4	76.2
C-10	40.3	50.9	126.7	121.0
C-11	20.3	23.2	25.8	58.9
C-12	40.1	32.3	40.7	212.8
C-13	43.3	54.4	43.8	45.6
C-14	56.7	83.4	54.6	46.9
C-15	24.0	31.7	23.9	24.4
C-16	22.1	37.2	27.4	34.4
C-17	54.8	88.5	52.3	83.3
C-18	13.5	20.5	12.5	16.1
C-19	19.5	19.0	–	11.0
C-20	77.2	78.8	38.8	40.8

Table 5 (*continued*)

Carbons	Perulactones and Ring A Aromatic Withanolides			
	Perulactone (**213**)	Perulactone B (**214**)	Jaborosalactone Q (**193**)	Jaborol (**192**)
C-21	20.2	18.8	13.3	13.0
C-22	74.9	73.1	78.8	78.3
C-23	29.0	30.3	29.8	33.7
C-24	38.0	38.4	153.5	153.9
C-25	37.9	38.0	125.2	121.7
C-26	180.9	181.8	167.0	166.1
C-27	10.6	10.5	57.2	12.4
C-28	72.6	72.6	20.1	20.5
-OAc	170.7, 21.2	–	–	–

become a regular feature in recent years. Two review articles dealing with this topic (*208, 209*) have appeared; the one authored by GOTTLIEB and KIRSON is particularly useful for investigators working in this area. The potential of ^{13}C-NMR spectrometry for structural investigation of withasteroids can be best appreciated by citing work which led to the revision of the structures of withanolide G (**23**) and a few related withanolides (*187*), the determination of the exact mode of fusion of A/B rings of withaperuvin (**154**) in the face of an anomalous Cotton effect (*134*) and the selection of the correct structure of withametelin out of the two possible alternatives (*33, 37*). ^{13}C-NMR chemical shifts of the carbons of withasteroids of different groups are recorded in Table 5.

4. Mass Spectra

While mass spectrometry is a very useful tool for the recognition of typical withanolides as well as of ixocapalactones, its application to the structure elucidation of different kinds of withasteroids is rather limited. A base peak at m/z 125 (ion *a*) formed by fission across C-20/C-22 bond is considered to be diagnostic for withanolides having an unsubstituted α, β-unsaturated-δ-lactone moiety. However, the presence of a hydroxyl group at C-20 facilitates cleavage of the 17,20 bond and gives rise to significant peaks at m/z 169 (*b*) and m/z 152 (*c*). In the mass spectrum of physalolactone (**147**), ion (*c*) appeared as the base peak and the relative intensity of the ions (*a*) and (*b*) was found to be 65% and 50%, respectively (*142*). The presence of a hydroxyl group in the lactone part of the molecule obviously shifts ion *a* from m/z 125 to m/z 141 (ion *a'*). A

similar observation has also been made in case of 24,25-epoxy withanolides (49). The ion peak at m/z 141 generated from 27-hydroxy-
withanolides like jaborosalactone D (89) readily loses a molecule of water
(66) and shows the ion peak at m/z 123 (d).

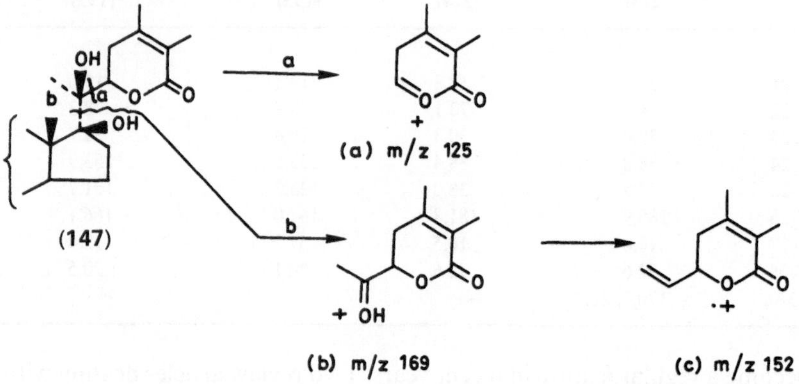

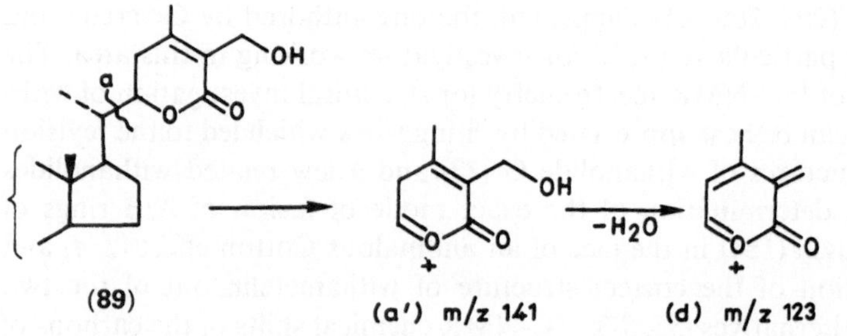

(a) m/z 125

(b) m/z 169

(c) m/z 152

(147)

(89)

(a') m/z 141

(d) m/z 123

It has, however, been observed that the base peak at m/z 125 may also
be derived from ring A of a withasteroid having a 5-hydroxy-2-en-1-one
system. The genesis of this ion (e) is probably initiated by a McLafferty-
type rearrangement involving the enone carbonyl followed by 5,6-bond
cleavage. The base peak at m/z 125 (e) appearing in the mass spectrum of
jaborosalactone F (91) is not derived by fragmentation of the lactone part
of the molecule but from its A ring (67) as shown below:

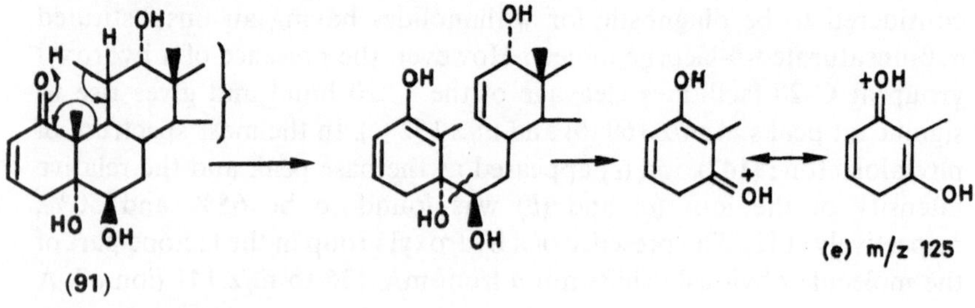

(91)

(e) m/z 125

The mere appearance of a base peak at m/z 125 should, therefore, not be construed as definite evidence of a withanolide skeleton with an unsaturated δ-lactone moiety. Though fission across 20,22-bond is the dominating feature of electron impact on withanolides, the mass peak corresponding to ion *a* is not discernible in the spectra of withanolides having a saturated δ-lactone ring. If, however, there is a hydroxyl group at C-20, cleavage across the 20,22-bond gives rise to a significant peak at (M$^+$-127) for the carbocyclic part of the molecule and fission across the 17,20 bond shows the ion peak for the side chain of the molecule. Thus 24,25-dihydrowithanolide D (121) shows the base peak at m/z 345 and a significant peak at m/z 171 (*183*).

Jaborol (192), a ring A aromatic withanolide, shows a significant ion peak at m/z 125 (ion *a*) from the lactone ring and a peak at m/z 121 from the aromatic A ring. The latter peak is presumably due to a tropylium cation bearing a hydroxyl and a methyl group formed by benzylic cleavage and not by fission of 5,6-bond as has been shown by the authors (*72*). The base peak of jaborol appears at m/z 109 which according to the authors originates from rings C and D of the molecule. Jaborosalactone Q (193), the other known withanolide with an aromatic A ring, shows a significant peak at m/z 157 which is considered to be due to ion **f**, formed by fission of the C ring. A similar peak has also been witnessed in the ring A aromatic withanolide (269) derived from withametelin (*210*). The

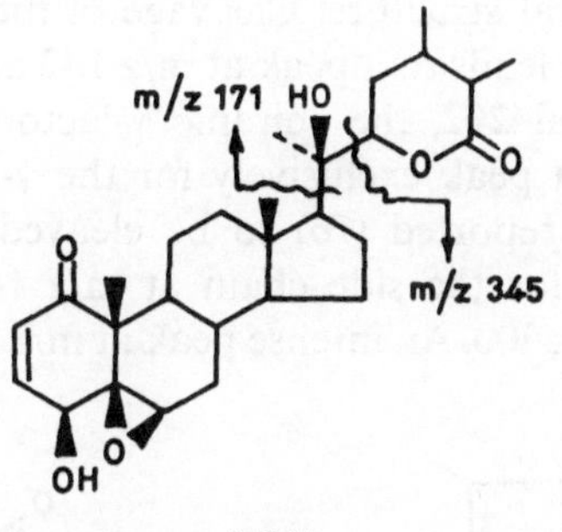

(121)

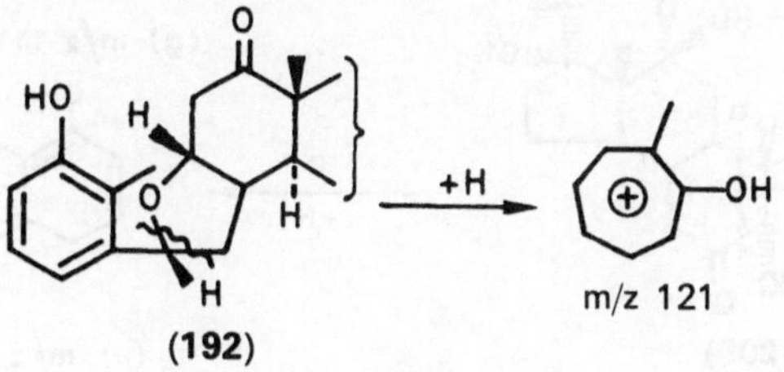

(192)

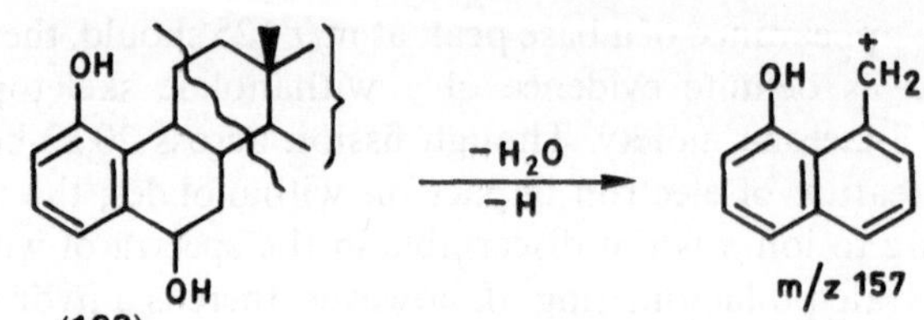

(193) $-H_2O$ $-H$ m/z 157
                                      (f)

genesis of different ion peaks appearing in the mass spectra of physapu-bescin (104), pubescenol (109) and some other withanolides has also been rationalised (*151, 152*).

Like withanolides, ixocarpalactones also exhibit a characteristic fragment ion due to the lactone part of the molecules; this peak appears at m/z 111 (ion **g**) if the γ-lactone is unsaturated and at m/z 113, if the lactone is saturated. All the ixocarpalactones isolated from *Jaborosa magellanica* (73) exhibit a peak at m/z 111 (**g**); this peak is quite intense in the spectra of jaborolone (205) and jaborochlorodiol (203). Jaborolone exhibits the base peak at m/z 125 obviously due to ion **e**. Ixocarpalactone A (201) in which the γ-lactone is saturated shows in its mass spectrum the characteristic peak at m/z 113 (ion **g′**) along with a number of other peaks the genesis of which has been rationalised by cleavage of the 17,20-, 20,22- and 22,23-bonds with subsequent loss of water (*119*).

The fragmentation mode of ixocarpalactone B (202) is slightly differ-ent due to its spiroketal structure. Cleavage of the 20,22-bond initiates the fragmentation and leads to a peak at m/z 142 and a peak at [M-142-OH]$^+$. Like spiroketal 202, the spiranic γ-lactone jaborosalactone P (212) fails to exhibit a peak exclusively for the γ-lactone moiety in its mass spectrum. It is reported (*76*) to be cleaved into two halves on electron impact, one for the side chain at m/z 168 and other for the carbocyclic part at m/z 300. An intense peak at m/z 135 is also discernible

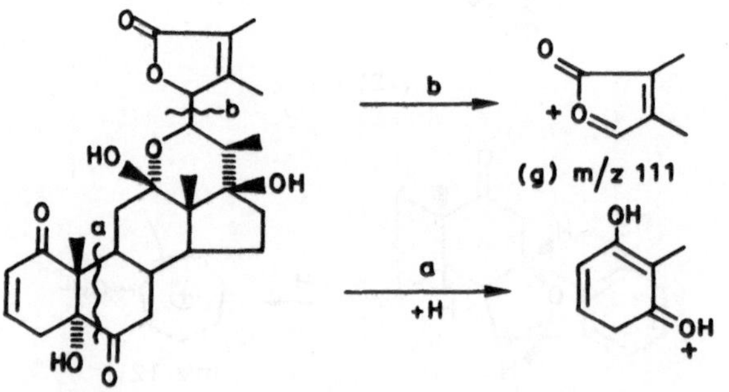

(205)   (g) m/z 111

               (e) m/z 125

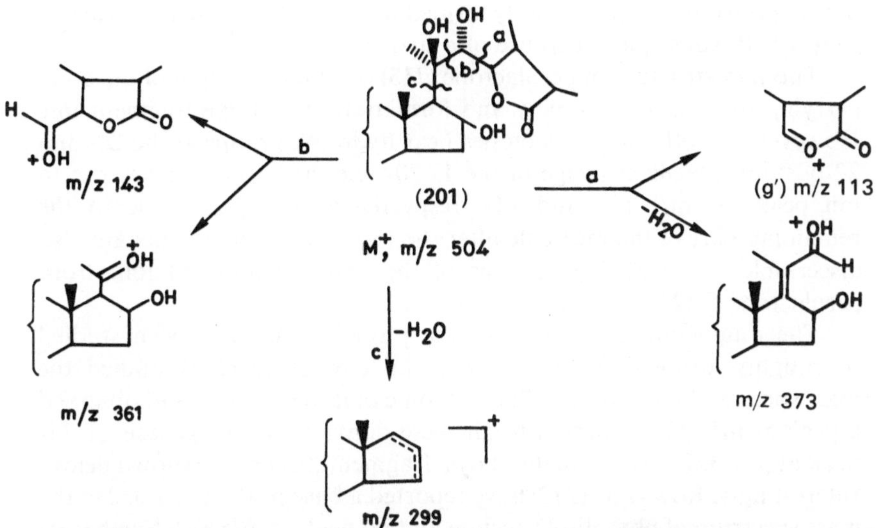
m/z 143
b
(201)
M⁺, m/z 504
a
−H2O
(g′) m/z 113
m/z 361
−H2O
c
m/z 299
m/z 373

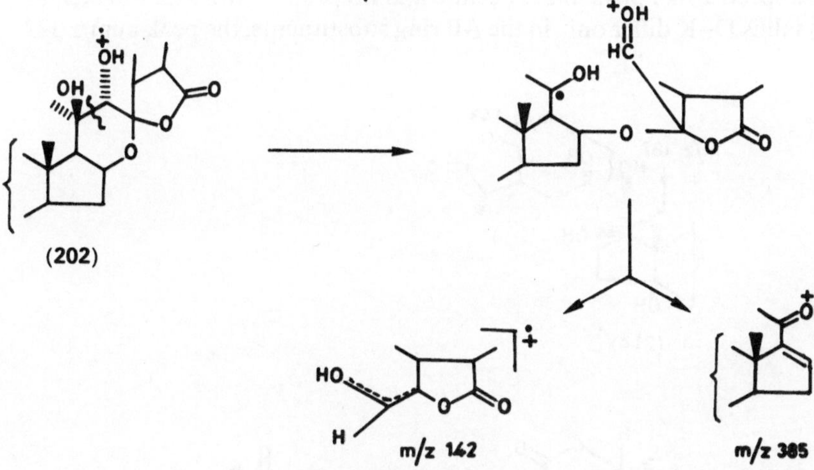
(202)
m/z 142
m/z 385

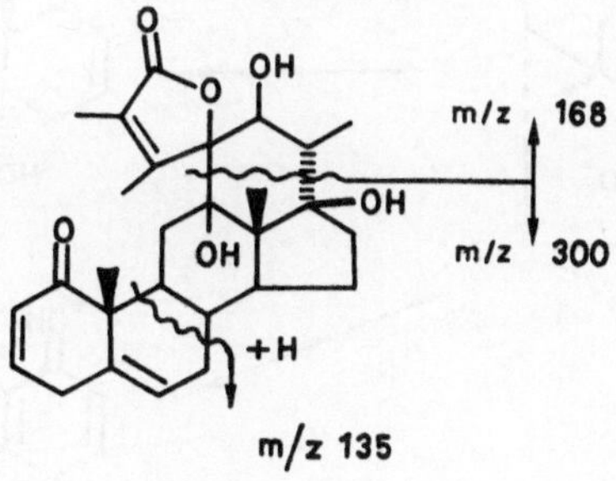
m/z 168
m/z 300
+H
m/z 135

(212)

in the spectrum; it is presumably formed by retro-Diels-Alder cleavage of ring B with subsequent capture of a hydrogen.

The very structure of perulactone (213) precludes scission of the bond carrying the γ-lactone moiety and formation of a stable fragment ion. However, as both the perulactones bear hydroxyl groups at the 20- and 22-positions, facile cleavage of the 17,20- and 20,22-bonds gives rise to ion peaks at m/z 187 and 143, respectively. The peaks due to the remaining part of the molecule after the loss of these fragments are also discernible (136,145). The genesis of the above mentioned ions from perulactone B (214) is depicted below.

The fragmentation pattern of the physalins has not been studied thoroughly although MULCHANDANI and coworkers (123) studied the mass spectra of physalin D (176) and some of its derivatives and observed a peak at m/z 322, common to all these derivatives. The genesis of this peak at m/z 322 was rationalised by a fragmentation mode shown below. Interestingly, Row et al. (112) have reported a base peak at m/z 332 in the mass spectrum of physalin D acetate and no peak at m/z 322. Further, as physalins D–K differ only in the AB ring substituents, the peak at m/z 322

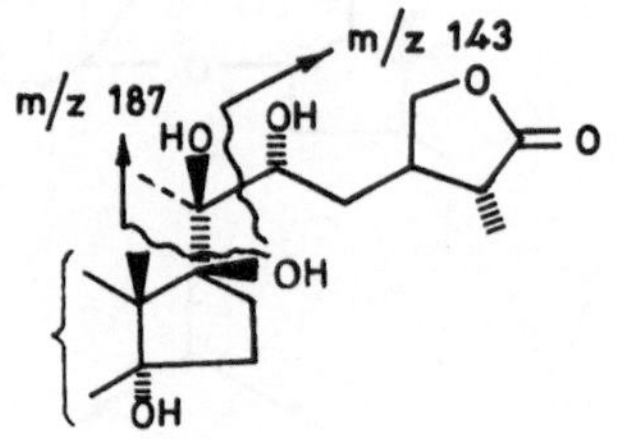

(214)

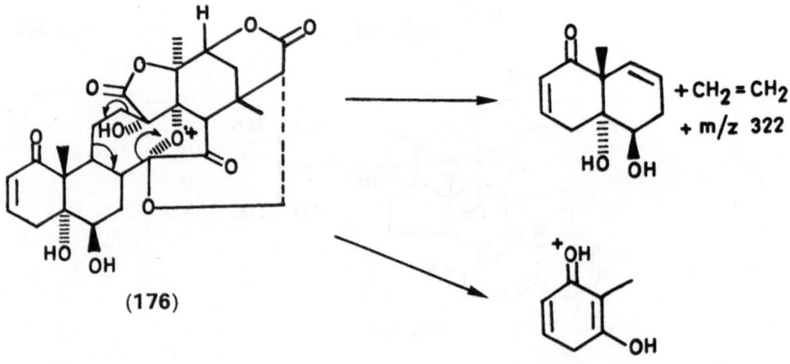

(176)

(e) m/z 125

should be present in the mass spectra of all these compounds, but the reported data fail to confirm it. Thus the proposed (123) mass peak at m/z 322 cannot yet be regarded as a diagnostic peak for octacyclic physalins.

5. Circular Dichroism Spectra

CD-data and the sign of Cotton effect have been frequently used for settling the configuration of withasteroids at appropriate centres, particularly in securing information about the mode of fusion of the AB rings and the configuration at C-22 of withanolides. Steroids with a 2-en-1-one system exhibit a Cotton effect around 340 nm and its sign helps in recognition of the mode of fusion of A and B rings; the sign is positive for cis-fused rings and negative for trans-fused ones. Thus withanolide E

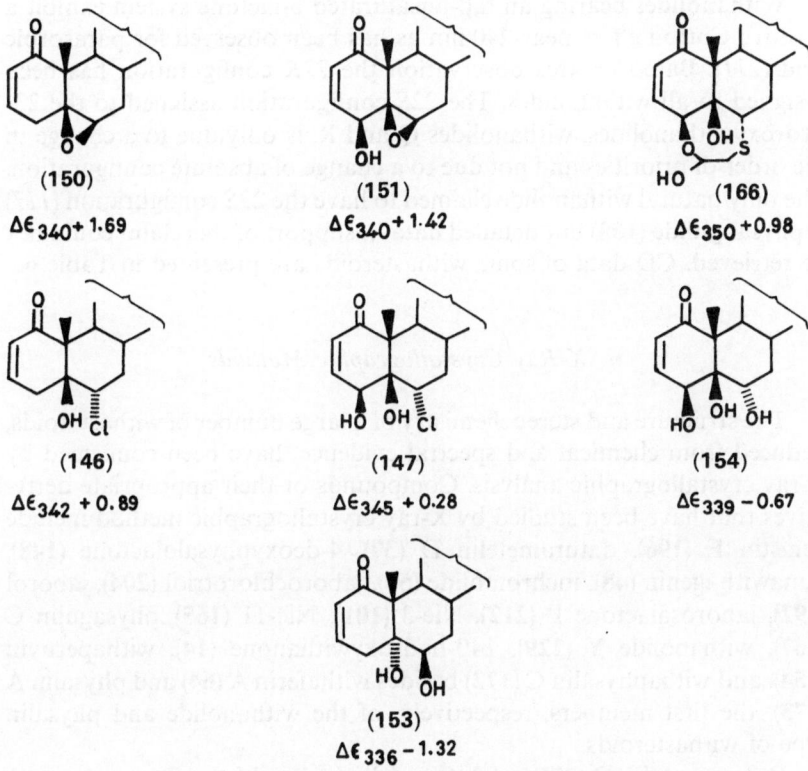

Chart 3. Change of Chiroptical Properties with Change of AB Ring Substituents

(150) shows a positive Cotton effect at 340 nm and withanolide S (153) shows a negative effect at 336 nm. An anomalous Cotton effect was, however, observed in the case of withaperuvin (154); this was proved to be due to a conformational change effected by steric compression of three contiguous hydroxyl groups at C-4, C-5 and C-6 of the molecule (134). The gradual change of chiroptical properties with change in AB ring substituents will be manifest from the data presented in Chart 3. Withanolide E with a cis-fused AB rings shows a positive Cotton effect; the value of Δε decreases progressively with change of AB ring substituents and becomes negative in withaperuvin, very much like withanolide S, the rings A and B of which are in a trans relationship. The relationship between the sign of Cotton effect and the mode of juncture of A and B rings is, however, reversed in withasteroids in which the enone double bond is absent. A negative Cotton effect around 285 nm is observed for 5β-withasteroids and a positive Cotton effect for their 5α-counterparts (182, 73).

Withanolides bearing an α,β-unsaturated δ-lactone system exhibit a positive Cotton effect near 240 nm as has been observed for parasorbic acid (211). Based on this observation the 22R configuration has been assigned to all withanolides. The 22S configuration assigned to the 23-hydroxywithanolides, withanolides Q and R, is only due to a change in the order of priorities and not due to a change of absolute configuration. The only natural withanolide claimed to have the 22S configuration (117) is physangulide (108) but detailed data in support of this claim could not be retrieved. CD data of some withasteroids are presented in Table 6.

6. X-Ray Crystallographic Methods

The structure and stereochemistry of a large number of withasteroids, deduced from chemical and spectral evidence, have been confirmed by X-ray crystallographic analysis. Compounds or their appropriate derivatives that have been studied by X-ray crystallographic method include acnistin E (196), daturametelin D (37), 4-deoxyphysalolactone (148), dunawithagenin (48), iochromolide (59), jaborochlorotriol (204), jaborol (192), jaborosalactone P (212), Nic-3 (101), Nic-11 (165), physagulin C (167), withanolide Y (129), 14β-hydroxywithanone (14), withaperuvin (154), and withaphysalin C (172) besides withaferin A (64) and physalin A (173), the first members, respectively, of the withanolide and physalin type of withasteroids.

Daturametelin D (37) and datumetelin isolated from Datura metel by two different groups of workers (41–44) were assigned the same gross

Table 6. *CD-Data of Withasteroids*

Compounds	Solvent	λ_{max} in nm ($\Delta\epsilon$)
Daturataturin A (18)	MeOH	325 (− 7.58), 255(+ 14.24)
Daturalactone-2 (2)	MeCN	346 (− 2.03), 338(− 2.09), 288(+ 1.00). 233 (− 3.14)
12-Deoxywithastramonolide (16)	Dioxan	340 (− 2.90), 258 (+ 2.81)
Jaborochlorotriol (204)	MeOH	325 (+ 0.04), 235 (+ 0.86)
Jaborolone (205)	MeOH	350 (− 0.16)
Jaboromagellone (138)	MeOH	280 (+ 0.86), 240 (+ 0.68)
Nicandrenone (189)	–	338 (− 1.7), 224 (+ 6.22)
Nicandrenolactone (190)	–	338 (− 2.25), 275.5 (+ 0.11), 268 (+ 0.10), 263 (+ 0.07), 244 (− 0.39), 224 (+ 5.29)
Physalin B (175)	Dioxan	342 (− 1.82), 335 (− 1.89), 275 (− 0.4), 238 (− 2.12), 208 (+ 6.95)
Physalin G (180)	Dioxan	415 (+ 0.78), 403 (+ 1.43), 380 (+ 2.14), 375 (+ 2.14), 355 (+ 1.43), 319 (− 4.01), 312 (− 4.80), 310 (− 4.87), 264 (− 0.64), 241 (+ 2.65), 213 (+ 0.64), 209 (+ 0.78),
Pubescenol (109)	Dioxan	295 (+ 0.5), 265 (+ 0.49), 238 (− 1.45), 235 (− 1.43), 233 (− 1.27)
Projaborol (194)	MeOH	300 (− 1.5), 255 (+ 5.7)

structure but different stereochemistry at C-22 and C-24, with the rather unusual 22*S* configuration being postulated for datumetelin. To settle its stereochemistry, daturametelin D was subjected to X-ray crystallographic analysis. This established that the chiral centres at C-17, C-20, C-22, C-24 and C-25 of the compound had the *R* configuration (*41*). The stereochemistry assigned to datumetelin remains to be confirmed (see p. 65). In the structure determination of withaperuvin (154), confusion arose regarding the nature of the juncture of A and B rings; while the chemical shift of the 19-methyl carbon suggested a *cis*-juncture, a *trans*-juncture was in agreement with the observed negative Cotton effect near 350 nm. In order to arrive at a definite conclusion, the androstan-1,17-dione derivative (221) obtained by Jones oxidation of withaperuvin diacetate was subjected to X-ray crystallographic analysis; this established that rings A and B were *cis*-fused (*134*).

B. Degradative Reactions

Classical degradative reactions for structure elucidation of complex organic molecules have lost much of their importance with the advent of modern spectroscopic methods on the one hand and the scarcity of many of the natural products, on the other. The only withasteroid that has been subjected to such degradative reactions is withaferin A (64). It was converted to bis-nor-cholanic acid (219) by sequential elimination of the oxygen functions and oxidative degradation of the side chain (190). Simpler degradative reactions have, however, been performed on other withasteroids to secure information about some sets of functional groups and the stereochemistry of certain centres.

Withanolides bearing hydroxyl groups at C-17 and C-20 positions undergo facile cleavage into two halves on Jones oxidation (148). The reaction was fully utilised in the structure determinations of physalolactone (147) and the withaperuvins isolated from Physalis peruviana (134, 142). Withaperuvin diacetate (154-Ac) yielded on Jones oxidation an androstan-1,17-dione (221) and a δ-lactone methyl ketone (220).

Similar oxidative cleavage of the vicinal glycol system has also been performed on perulactone (213) which yielded a mixture of pregnenones

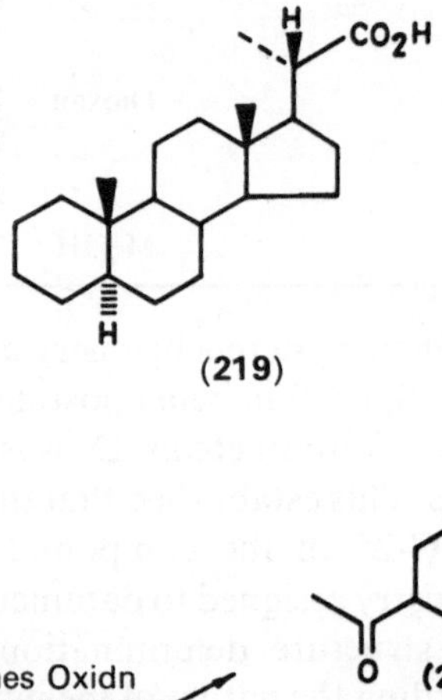

(219)

(154-Ac)   Jones Oxidn   (220)

(221)

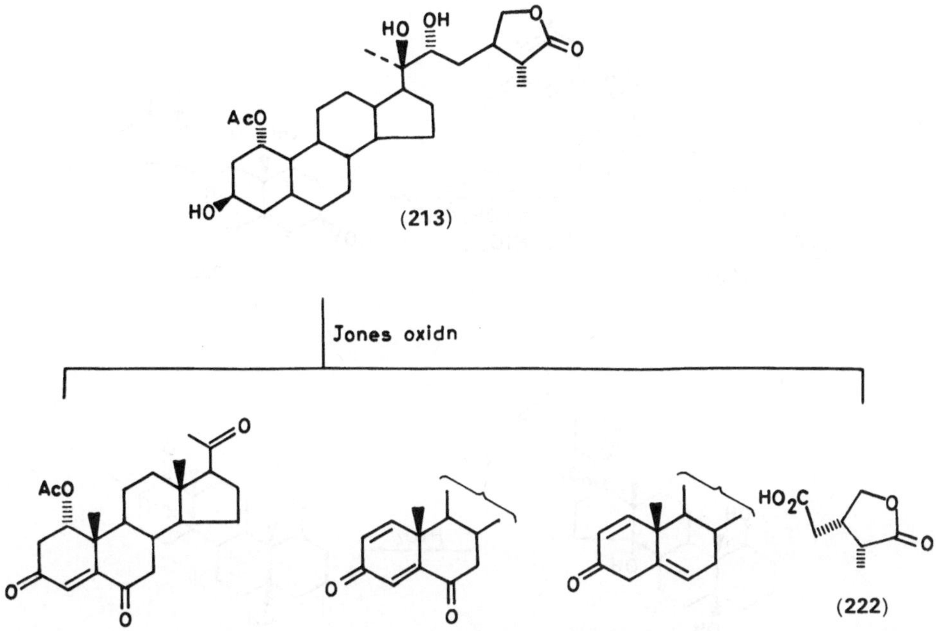

(213)

Jones oxidn

(222)

Chart 4. Oxidative Degradation of Perulactone

and a γ-lactone carboxylic acid (222) on chromium trioxide oxidation under Jones condition (Chart 4). The γ-lactone acid was fully characterised from detailed ^{1}H NMR spectral analysis of its methyl ester. Identification of the cleavage products facilitated structure determination of perulactone (136).

The major withanolides isolated from the roots and leaves of *Physalis peruviana* were proved to have a 17α-oriented side chain but spectral analysis of physalolactone B (55) isolated from the same source suggested that it might have the normal 17β-side chain. In order to adduce conclusive evidence in support of this deduction, it was converted to the pregnan-20-one derivative (223) by sequential catalytic reduction, methanolysis and periodic acid oxidation as shown below. The derived ketone exhibited a positive Cotton effect at 287 nm in agreement with a 17β-side chain (143). However, attempts to prepare a pregnan-20-one derivative from withanolide E (150) by a similar reaction sequence met with failure; the heptahydroxy compound (224) derived from withanolide E by catalytic hydrogenation followed by LAH reduction yielded a D-homo rearranged product (225) on periodic acid oxidation. The 14α-hydroxy

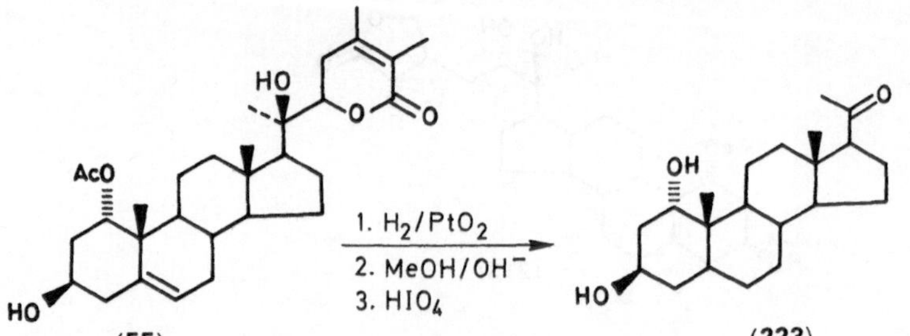

(55)   (223)

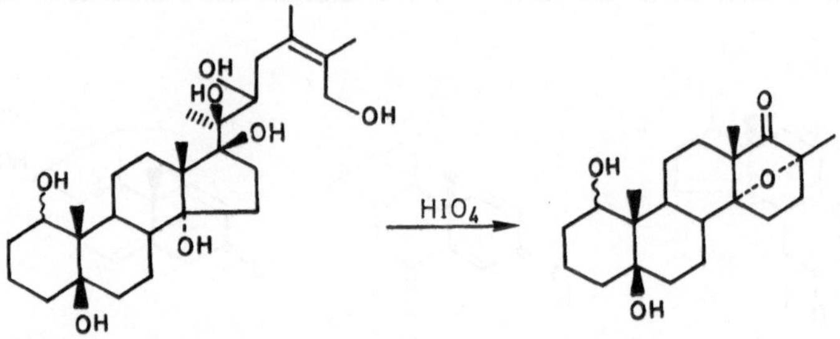

(224)   (225)

group was considered to trigger this reaction because the Δ^{14}-derivative of withanolide E gave the expected pregnan-20-one derivative (212).

Of the three tertiary hydroxyl groups present in withanolide E (150) and related withanolides the one at the 14α-position is easily eliminated under acidic conditions (Me_2CO-H_2SO_4). The conversion of physalolactone (147) and withaperuvin (154), to the corresponding Δ^{14} derivatives, physalolactone C (140) and withaperuvin B (144), by such acid treatment are clear examples of selective elimination of 14α-hydroxyl group (131, 144). It has, however, been reported that withanolide E gives, in addition to the Δ^{14}-derivative, a product (226) with a 14,20 oxide bridge concomitant with opening of the epoxide ring (213).

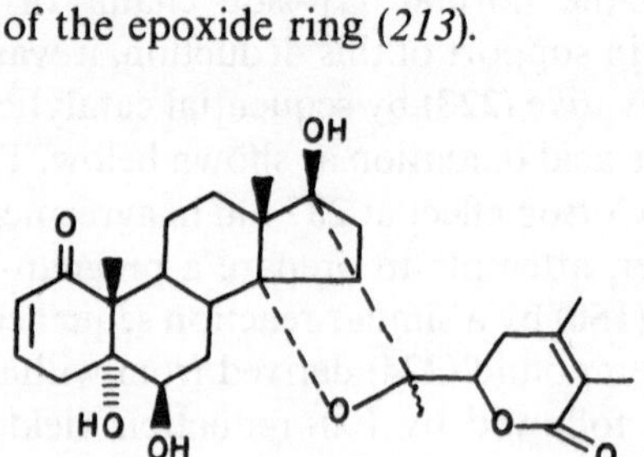

(226)

Acid-catalysed opening of the epoxides is often accompanied by rearrangements. Withaferin A (**64**) on treatment with H_2SO_4-AcOH gave a conjugated cyclopentenone derivative (**227**) by a pinacol-type re-arrangement and loss of C-4. Dihydrowithaferin A (**75**), however, gave the 5-aldehydo-derivative (**228**) without loss of carbon (*7*, *214*). NITTALA and LAVIE (*215*) in studying opening of the 5β,6β-epoxide established a

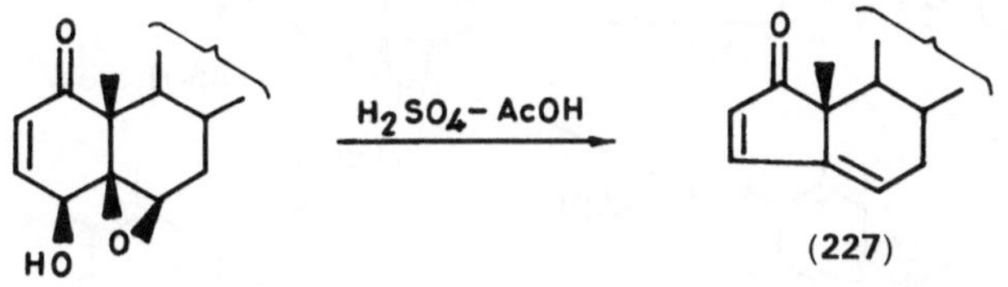

(**64**) H_2SO_4- AcOH   (**227**)

(**75**) H_2SO_4- AcOH   (**228**)

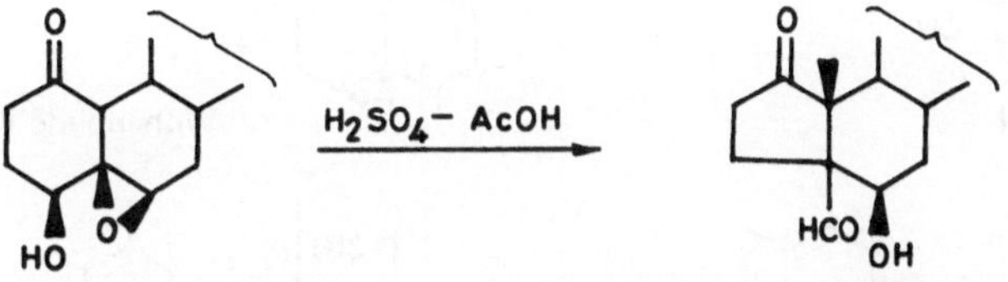

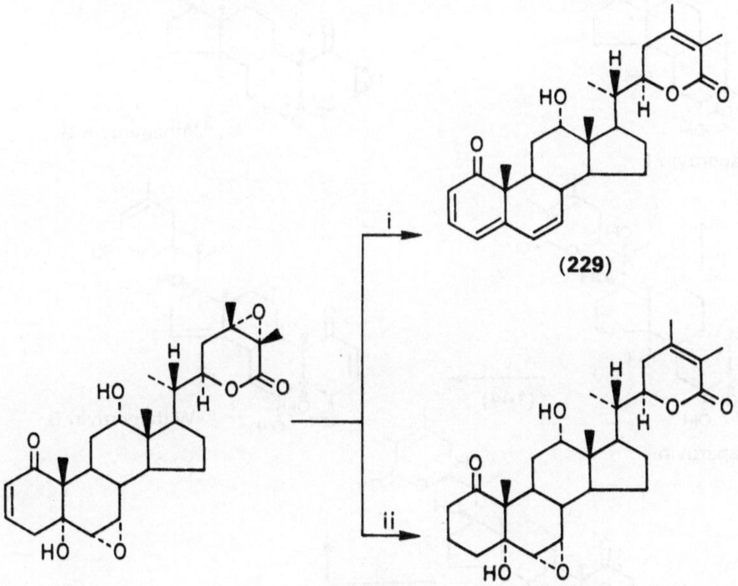

(**1**)   (i) HI   (ii) Zn-Cu couple   (**229**)   (**230**)

Chart 5. Interconversion of Withanolides

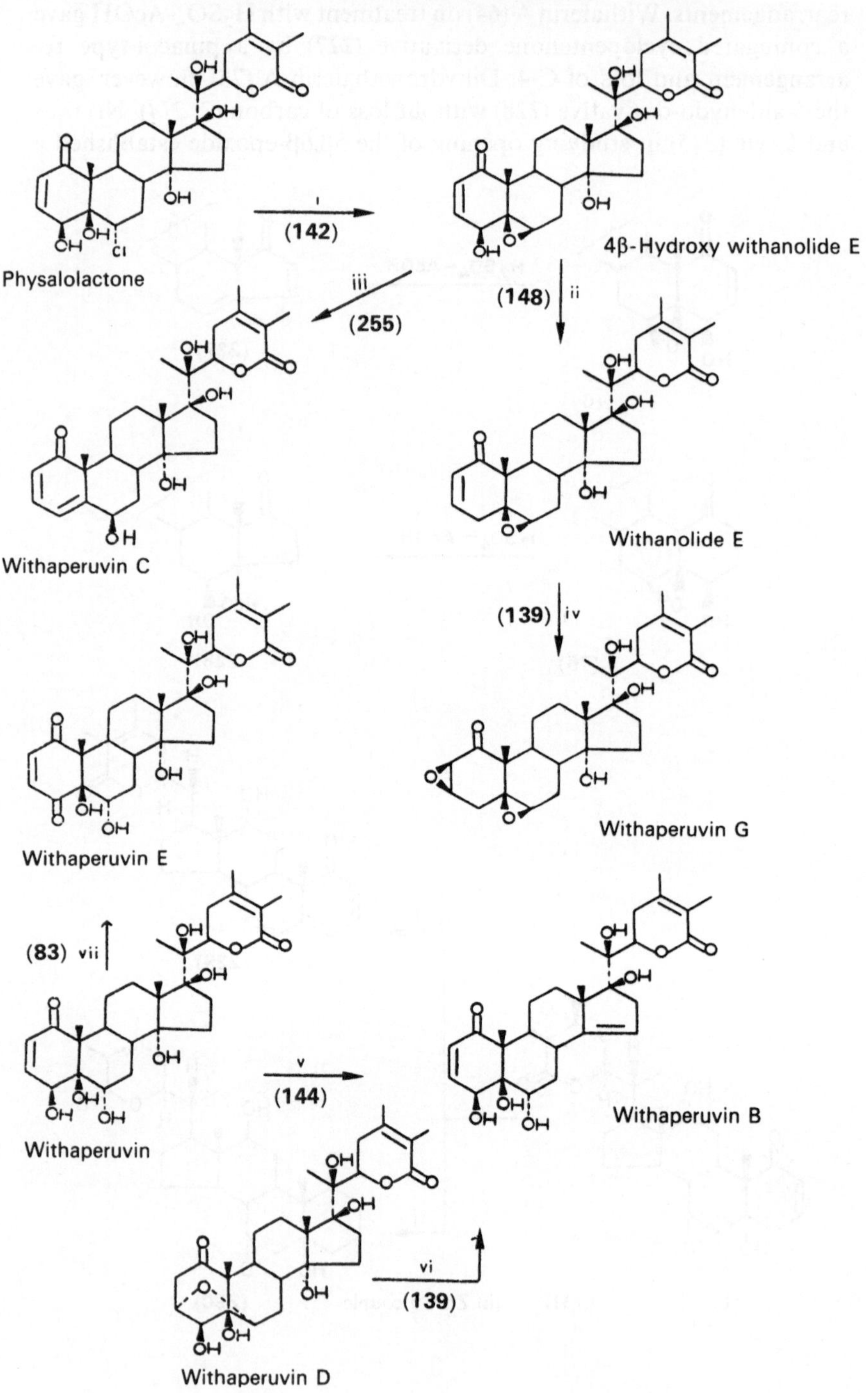

Physalolactone

(142)

4β-Hydroxy withanolide E

iii

(255)

(148) ii

Withaperuvin C

Withanolide E

(139) iv

Withaperuvin E

Withaperuvin G

(83) vii

(144)

v

Withaperuvin B

Withaperuvin

(139)

vi

Withaperuvin D

relationship between the reaction products and the reagents, reaction conditions and the presence or absence of the 2,3-double bond.

Daturalactone-1, the first withanolide isolated from *Datura* species, was initially (*47a*) formulated as (22*S*)-5α,12α,17α-trihydroxy-6α,7α-epoxy-1-oxowitha-2,24-dienolide which was later (*50*) revised to (22*R*)-5α,12α-dihydroxy-6α,7α: 24α,25α-diepoxy-1-oxo-witha-2-enolide (**1**) by correctly interpreting the spectral properties of the parent molecule as well as those of its degradation products. Reduction of daturalactone-1 (**1**) with hydroiodic acid gave a trienone (**229**) with removal of both the epoxide functions. On reduction with Zn-Cu couple, (**1**) gave an α,β-unsaturated δ-lactone (**230**) with reduction of the 2,3-double bond.

C. Chemical Transformations

Chemical evidence in support of spectroscopically deduced structures of a number of withasteroids has been provided by their transformation to compounds of known structure and stereochemistry as well as by establishing their interrelationship. The literature on withasteroids is replete with information on such transformations but this approach has been particularly suitable for compounds differing only in their AB ring substitution patterns. Precisely because of this reason, it has been possible to correlate the withaperuvins isolated from *Physalis peruviana* with withanolide E (**150**) of known structure and stereochemistry (Chart 5) and establish interrelationships between different physalins (Chart 6).

An interesting rearrangement of physalins has been reported rather recently by KAWAI and co-workers (*17*). When physalin A (**173**) was refluxed with acetic acid in the presence of ammonium acetate it underwent dehydration, an internal Michael addition and a benzilic acid type rearrangement to give a tris-lactone (**231**). Compounds having the skeletal pattern of this trislactone have been named by the authors as neophysalins. Physalin P (**188a**) isolated by these workers from *P. alkekengi* is also a neophysalin (*16*).

Chart 5: Reagents. (i) KOAc/EtOH/reflux (ii) H_2-Pd/C; TsCl/Py; Al_2O_3 (iii) Ac_2O/Py; Pd(PPh$_3$)$_4$/C_6H_6/30-min (iv) H_2O_2-KOH/DMF/5 min (v) Me_2CO/8N-H_2SO_4/8 h (vi) AcOH/90°/72 h (vii) MnO_2/Me_2CO/40 h

Chart 6. Interrelationship of Physalins

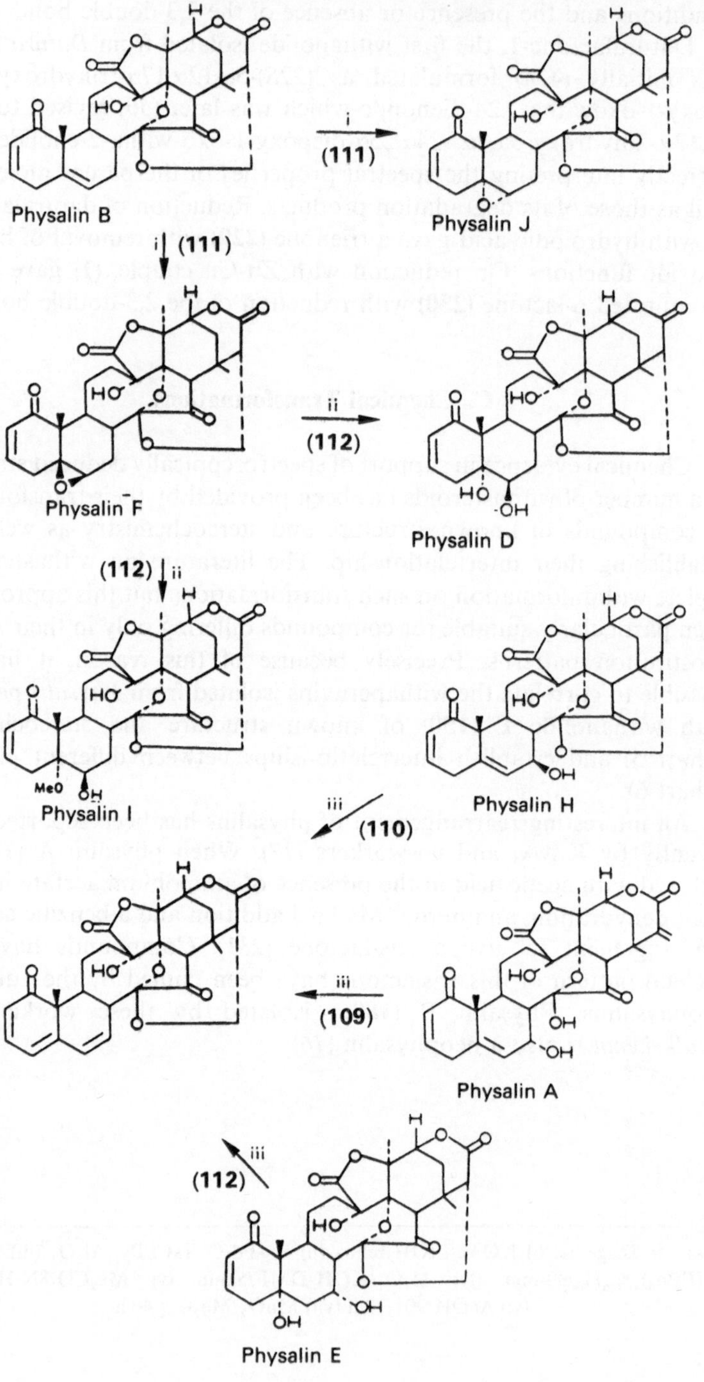

Physalin B

(111)

i

(111)

Physalin J

Physalin F

(112)

ii

(112)

ii

Physalin D

Physalin I

(110)

iii

Physalin H

(109)

iii

Physalin A

(112)

iii

Physalin E

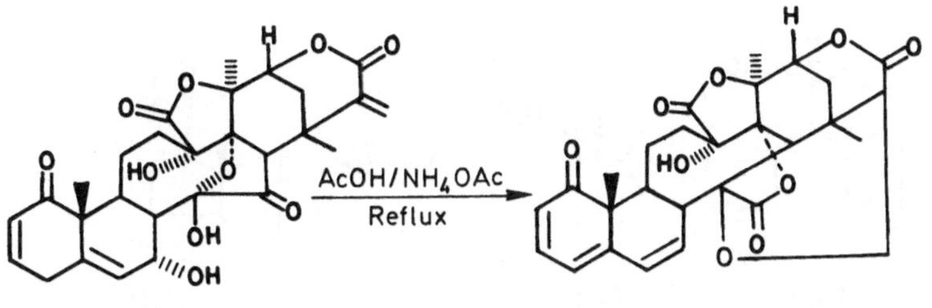

(173)   **(231)**

D. Synthesis of Withanolides

Among the withasteroids, only withanolides have so far been fully synthesised. As a prelude to the synthesis of this group of steroidal lactones, several model compounds having the commonly encountered AB ring substitution patterns of withanolides were prepared and attempts were made to build up the withanolide side chain on a steroid skeleton. The first synthesis of cholestane-type compounds with the AB ring portion of withaferin A (64) was achieved by IKEKAWA and coworkers (216, 217). This was followed by the work of GLOTTER and his group who devised a sequence of reactions to prepare cholestane derivatives having the AB ring substitution patterns of withaferin A, withanolide E (150) and several other naturally occurring withanolides (218, 219). An early attempt by Japanese workers (220) to synthesise the withanolide side chain by an aldol condensation of steroidal 22-aldehyde (232) with the lithium enolate of ethyl dimethylbutenoate (233) led to the synthesis of (234) which exhibited a negative Cotton effect at 258 nm indicating presence of the undesirable 22S configuration. Synthesis of a withanolide side chain with correct stereochemistry was, however, achieved by GLOTTER and coworkers (221) who adopted a strategy developed by WEIHE and McMORRIS (222) for the synthesis of 23-deoxyantheridiol (235). 3β-Acetoxyergosta-5,24-dien-22,26-olide prepared by the Israeli group showed a positive Cotton effect at 251 nm indicating the 22R configuration. The initial difficulty in securing the correct stereochemistry at C-22 was overcome almost concurrently by the IKEKAWA group who eventually succeeded in stereocontrolled synthesis of a number of 20-H and 20-OH withanolides having a normal 17β-side chain (223–225).

Chart 6: (i) m-CPBA   (ii) H^+/MeOH   (iii) H^+ (H_2SO_4)

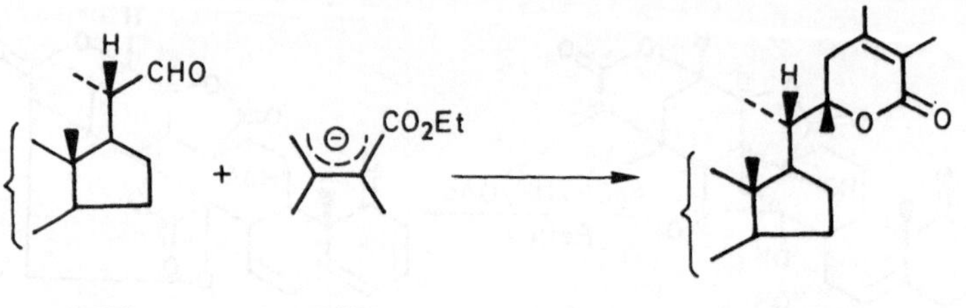

(232)   (233)   (234)

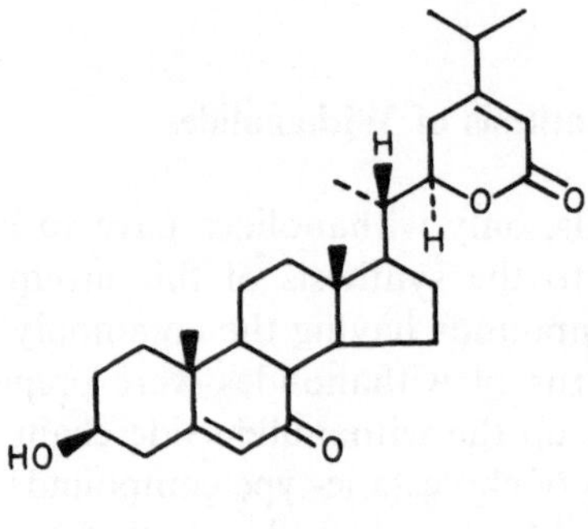

(235)

Withanolide E (150) is the only withanolide with a 17α-oriented side chain that has so far been synthesised as outlined in Chart 7. In addition to having a 17α-side chain the novelty of this compound lies in the presence of ten contiguous chiral centres six of which are oxygenated. The key steps in the recent synthesis of this substance by PEREZ-MEDRANO and GRIECO (226) were the insertion of a hydroxyl group at the 14α-position and introduction of the 17β-hydroxyl group. The former was achieved by a hetero Diels-Alder reaction between steroidal dienol acetate (252) and benzyl nitrosoformate and subsequent transformation of the desired adduct to 14α-hydroxy-17-oxo-androstane derivative (254). Attempts to introduce 17β-hydroxyl group by addition of appro-

---

Chart 7: Reagents. (i) TMSI, (TMS)$_2$NH, Et$_3$N, Cl–CH$_2$–CH$_2$–Cl, – 23 °C, 45 min (ii) Pd(OAc)$_2$, K$_2$CO$_3$, MeCN, 12 h (iii) Isopropenyl acetate, TsOH, reflux (iv) Benzyl nitrosoformate (v) Toluene, reflux, 20 min (vi) H$_2$, Pd-BaSO$_4$ (vii) CuCl$_2$·2H$_2$O, H$_2$O-THF 4 h (viii) 5% KOH, MeOH, reflux, 2 h (ix) TsCl-Py, 12 h (x) TBDMSOTf, Et$_3$N, CH$_2$Cl$_2$, 0 °C, 30 min (xi) MeOH, KOAc, reflux, 12 h (xii) Bu$_4$NF, THF, 4 days (xiii) CH$_3$–CH = P(Ph)$_3$, THF (xiv) MOMCl, iPr$_2$NEt, dioxane, 80 °C, sealed tube, 24 h (xv) OsO$_4$, Py, 24 h (xvi) TFAA, DMSO, Et$_3$N, CH$_2$Cl$_2$, – 78 °C, 1.5 h (xvii) CH$_2$ = CHLi, THF, – 78 °C, 1 h (xviii) O$_3$, MeOH, – 100 °C, Me$_2$S, 30 min (xix) α,β-Dimethylcrotonate, LDA, THF, HMPA (xx) 2M H$_2$SO$_4$ (xxi) Ac$_2$O, DMAP, Py, 15 h (xxii) Swern oxidn. (xxiii) DBN/CH$_2$Cl$_2$, m-CPBA, NaHCO$_3$, CH$_2$Cl$_2$

Chart 7. Outline of the Synthesis of Withanolide E

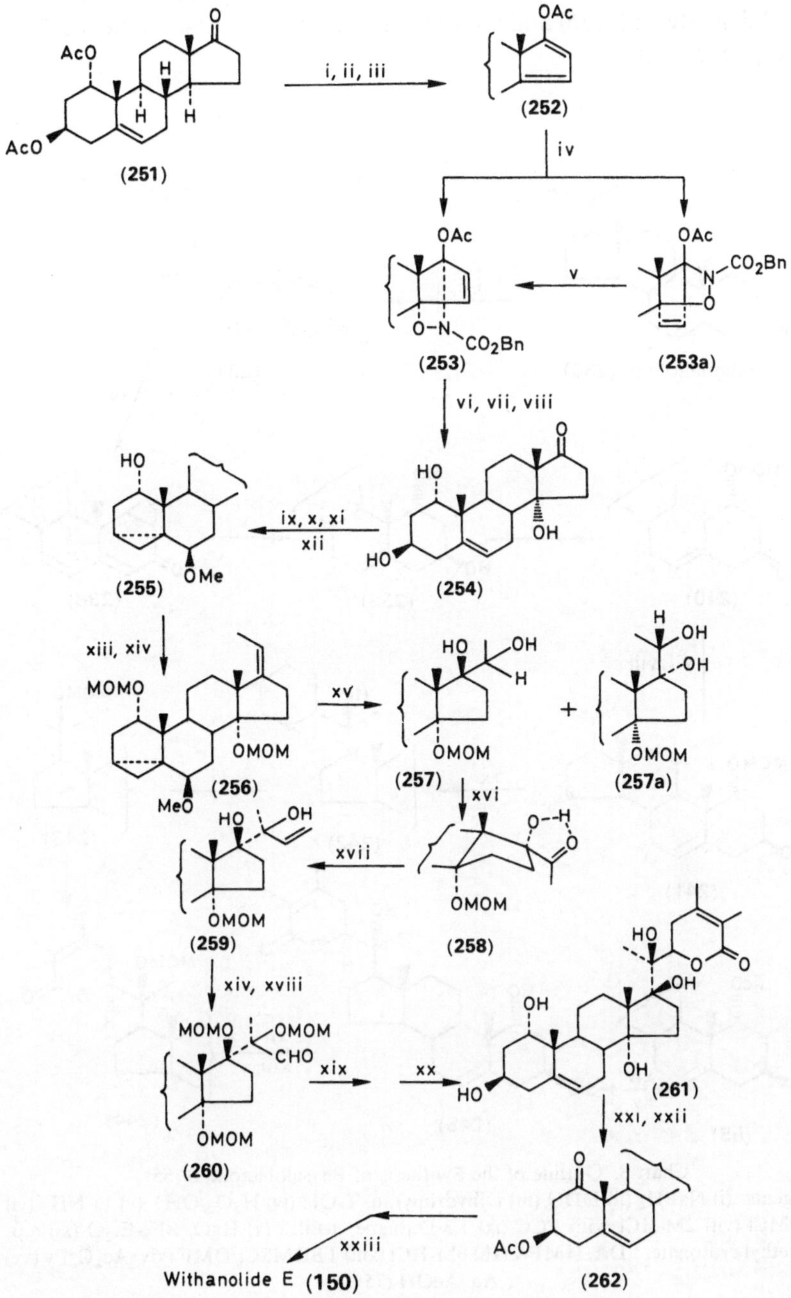

Withanolide E (150)

priate nucleophiles at C-17 gave rise to undesirable configuration at this centre but the difficulty was circumvented by preparation of (17Z)-ethylidene steroid (256) and its transformation to the desired diol (257) by OsO₄ oxidation.

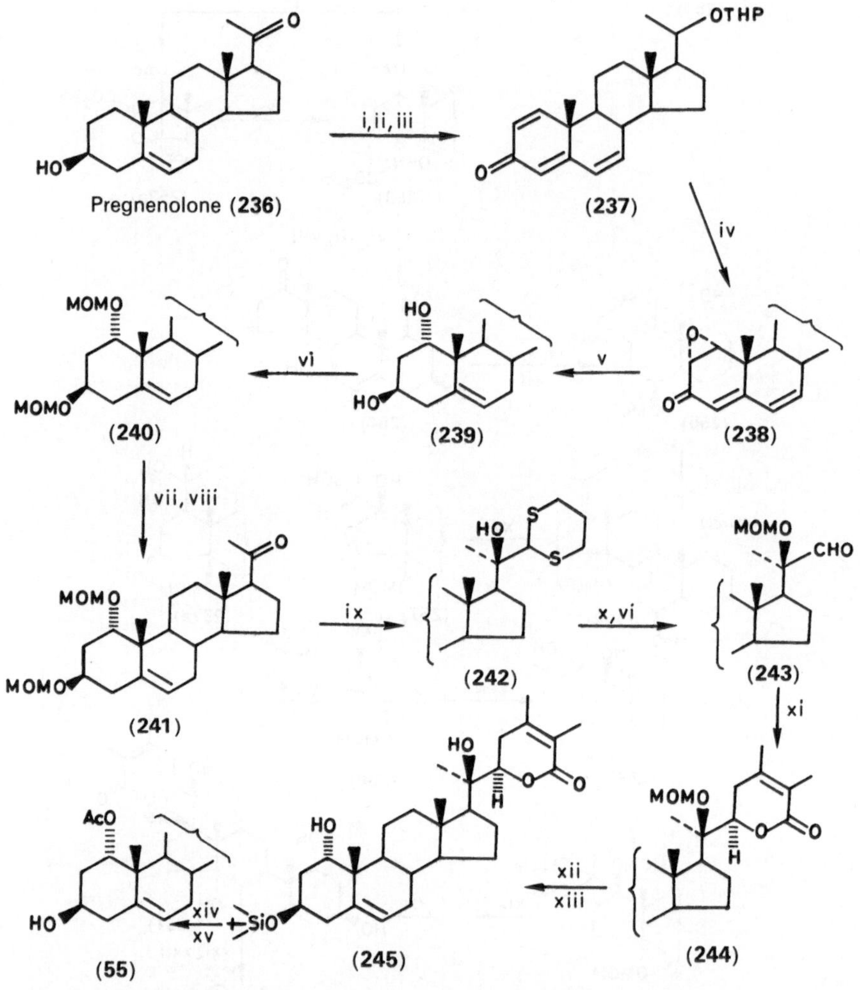

Chart 8. Outline of the Synthesis of Physalolactone B (55)

Reagents. (i) NaBH₄ (ii) DDQ (iii) Dihydropyran, TsOH (iv) H₂O₂/OH⁻ (v) Li/NH₃ (vi) MOMCl (vii) 2M-HCl (viii) PCC (ix) 1,3-Dithiane, n-BuLi (x) HgO, BF₃-Et₂O (xi) α,β-Dimethyl crotonate, LDA, HMPA (xii) 6M-HCl (xiii) TBDMSCl/DMF (xiv) Ac₂O/Py (xv) Aq. AcOH (75%)

Since syntheses of various other withanolides have been discussed by
GLOTTER in a recent review (*13*), the present article provides only an
outline of the synthesis of physalolactone B (**55**) and withanolide D (**68**) in
Charts 8 and 9, respectively.

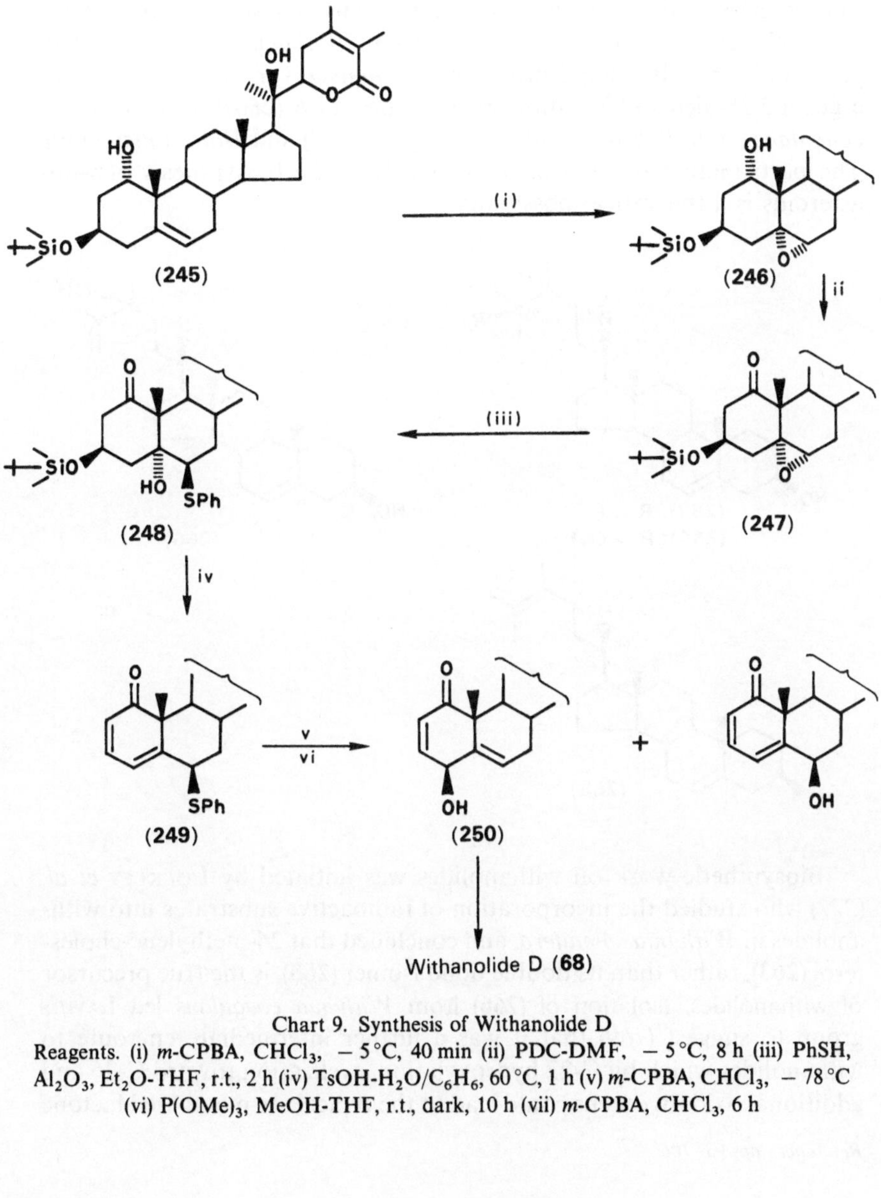

Chart 9. Synthesis of Withanolide D
Reagents. (i) *m*-CPBA, CHCl$_3$, − 5 °C, 40 min (ii) PDC-DMF, − 5 °C, 8 h (iii) PhSH,
Al$_2$O$_3$, Et$_2$O-THF, r.t., 2 h (iv) TsOH-H$_2$O/C$_6$H$_6$, 60 °C, 1 h (v) *m*-CPBA, CHCl$_3$, − 78 °C
(vi) P(OMe)$_3$, MeOH-THF, r.t., dark, 10 h (vii) *m*-CPBA, CHCl$_3$, 6 h

III. Biogenesis and Biosynthesis of Withasteroids

The occurrence of C_{28}-sterols in plants is not uncommon but the presence of C_{28}-sterols with an unsaturated C_9-side chain in with-asteroid-yielding plants becomes a subject of interest as these sterols are likely to play a part in the biogenesis of C_{28}-steroidal lactones. C_{28}-sterols with unsaturated C_9-side chains that have been reported from withasteroid-rich plants are 24-methylenecholesterol (**263**) and 24-methylcholesta-5,24-dien-3β-ol (**265**) from *Withania somnifera* (*193*), physalindicanol B (**264**) from *Physalis minima* var. *indica* (*237*) and ergosta-5,25-diene-3β,24ξ-diol [physalindicanol A (**266**)] from *Withania coagulans* (*169*), *Physalis minima* var. *indica* (*237*) and *Datura metel* (*39*). The participation of each of these sterols in the biosynthesis of with-asteroids is a theoretical possibility.

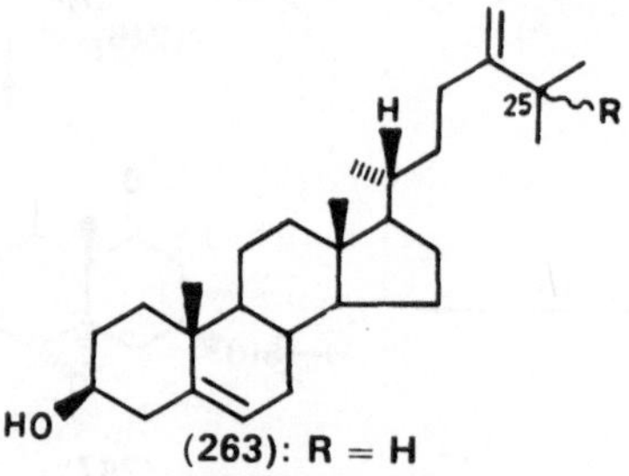

(**263**): R = H
(**264**): R = OH

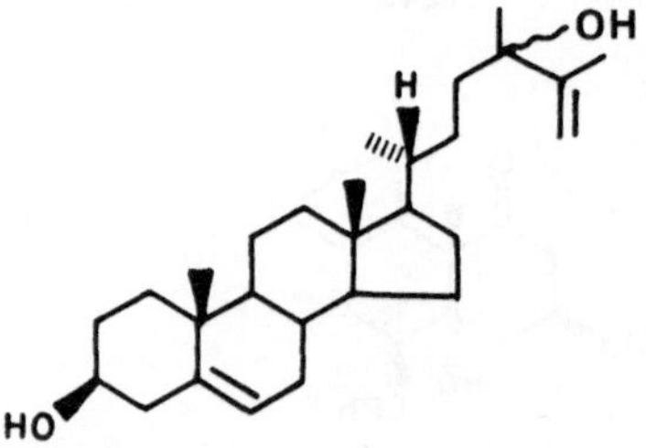

(**266**)

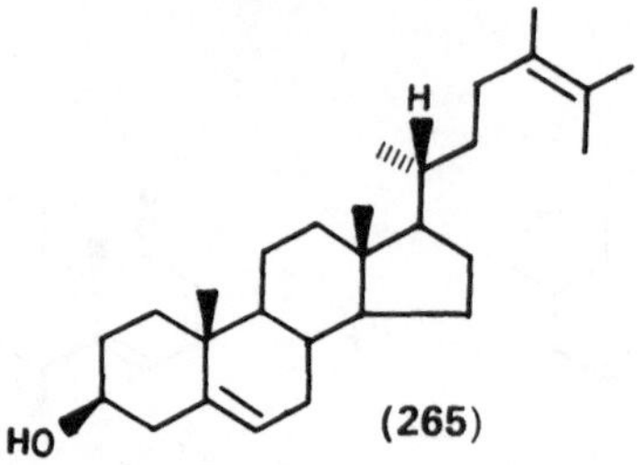

(**265**)

Biosynthetic work on withanolides was initiated by Lockley *et al.* (*227*) who studied the incorporation of radioactive substrates into with-anolides in *Withania somnifera*, and concluded that 24-methylene-choles-terol (**263**), rather than its double bond isomer (**265**), is the true precursor of withanolides. Isolation of (**266**) from *Withania coagulans* led Lavie's group to suggest (*169*) that it was a further intermediate en route to withanolides which by C-22 hydroxylation, cyclization toward C-26 and additional oxidative steps could lead to the typical six-membered lactone side chain of the withasteroids (however, see below). As for genesis of the functionalization in rings A and B, a reasonable pathway proposed by the same group (7) shown in Scheme 3 involves hydroxylation from the less hindered side of the 5-en-3-ol steroid precursor as the first step. Selective oxidation of the 1-hydroxyl of this intermediate and subsequent dehydration give rise to 2,5-dien-1-one. Isolation of withasteroids containing the 1α,3β-dihydroxy-5-ene system together with those having 2-en-1-one grouping from the same plants gives credence to this hypothesis (*136, 200*).

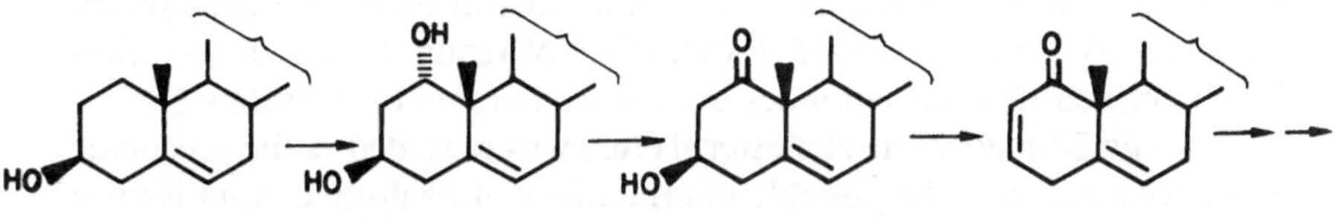

(263)

Scheme 3

While the radioactive withanolides isolated from *W. somnifera* by administration of 24-[28-^{3}H]-methylene cholesterol were not subjected to degradation by previous workers (*227*) to confirm that the label was still at C-28, GROS and coworkers (*28*) degraded the labelled withaferin A, obtained by administration of [2-^{14}C]-mevalonolactone to excised leaves of *Acnistus breviflorus*, to a C_{25}-hydroxy ketone and C_3-acid (glyceric acid) and observed that the maximum activity (about 98%) was retained by the C_{25} fragment. The glyceric acid moiety representing C-25, -26 and -27 of the withanolide molecule showed only one tenth of the expected 20% radioactivity. On the basis of this observation, the authors preferred to assume that in the biosynthesis of withanolides in *A. breviflorus*, the side chain of the sterol precursor is cleaved and rebuilt as is the case in the biosynthesis of cardiac glycosides (*228*). The exact site of cleavage, however, remains to be ascertained. Periodate oxidation of the C_3 fragment, $HOH_2C-CO-COOH$, yielded formaldehyde (C-27), formic acid (C-25) and CO_2 (C-26) and it was observed that the radioactivity of CO_2 obtained from C-26 of withanolide was ten times greater than that of CH_2O obtained from C-27. The results indicated that C-26 of withanolide was derived from C-2 of mevalonolactone and that the pro-*R*-methyl group of the precursor molecule was the source of C-26 of the withanolide (*229*). These authors have also demonstrated that jaborosalactone A (4-deoxywithaferin A) is the precursor of withaferin A (**64**) which indicates that 5,6-epoxidation precedes hydroxylation at C-4 (*230*).

The genesis of the 17α-side chain of withasteroids from the precursor sterol (263) with a normal 17β-side chain is a subject for investigation. Velde and Lavie (198) are of the opinion that Δ^{16}-withanolides are the precursors of withanolides with 17α-side chain. The occurrence of with-anolide E (150) and 14α,20-dihydroxy-1-oxowitha-2,5,16,24-tetraenolide in *W. somnifera* chemotype III, makes the authors believe that the latter is an intermediate in the biosynthesis of the former.

The biosynthesis of physalin D (176) was studied by Mulchandani and coworkers (231) who observed efficient incorporation of radioactive acetates [1-^{14}C and 2-^{14}C] and mevalonic acid [2-^{14}C] into physalin D on administration to *Physalis minima*. Administration of tritiated Lycium substance A (5) also yielded radioactive physalin D, an observation which verified that withanolides are the precursors of physalins (7).

Though 24-methylene cholesterol (263) was regarded as the precursor of the withanolides, the possible intermediacy of its double bond isomer (265) in the biogenetic process was not excluded outright by the early investigators (227). Recent work by Whiting and coworkers on the biosynthesis of ring D aromatic withanolides in *Nicandra physaloides* has, however, shown that both (263) and (265) take part in the elaboration of withasteroids and that they are probably adjacent in the biosynthetic sequence (81, 98).

The origin of the aromatic D ring in nicandrenone (189) has long been a subject of speculation. The co-occurrence of the epoxylactol with-anolide Nic-3 (101) and the ring D aromatic withanolide, nicandrenone, in *Nicandra physaloides* led Begley et al. (88) to speculate that ring D aromatic withanolides are derived from regular withanolides and that expansion of D ring and its aromatisation take place by inclusion of the 18-methyl group. Oxidative elimination of the 18-methyl and insertion of a C_1-unit from S-adenosyl-methionine for expansion of D ring and its subsequent aromatisation was entertained as second possibility. Isotope administration experiments in *N. physaloides* plant by Gill et al. demon-strated that [$3'$-C^2H_3]- and [$3'$-$^{14}CH_3$]-mevalonic acid are incorporated in nicandrenone and that trimellitic acid (267) obtained from radioactive nicandrenone by $KMnO_4$ oxidation retained 26% of the activity as can be expected from incorporation of C-18 in the aromatic D ring (90). It

(267)

was further observed that although nicandrenone obtained by administration of $[^{14}CH_3]$-S-adenosyl-methionine was radioactive, trimellitic acid derived from this withanolide was inactive. The presence of aromatic deuterium was also recorded in the ^{2}H-NMR spectrum of a withanolide sample obtained by administration of 3'-deuteriomethyl mevalonic acid. The findings pointed to a biosynthetic pathway in which the 18-methyl is oxidised and incorporated in ring D with retention of one hydrogen. The stages in the oxidative elaboration of the epoxylactol side chain in nicandrenone were also studied and it was shown that the lactol ring is not formed by reduction of the corresponding lactone but by the oxidation of C-26 to the aldehyde stage (*81, 98*).

Plausible biogenetic pathways for complex withasteroids like the jaborols, acnistins, withametelins, etc., have been postulated (*12, 13, 72, 73*) but verification of these postulates by further biosynthetic studies is required.

IV. Chemotypes and Chemogenetics

Plants which are morphologically indistinguishable but differ in their chemical constituents are termed chemotypes. Occurrence of three distinct chemotypes of *Withania somnifera* in Israel was discovered by ABRAHAM *et al.* (*172*) who examined populations collected in different regions. These were designated as chemotypes I, II and III of Israel which were different from Indian and South African chemotypes. There is evidence for the presence of more than one chemotype in India (*232*) but the exact number of such chemotypes has not yet been fully ascertained through systematic chemical analysis. Every chemotype has a characteristic feature in terms of its withanolide profile. The withanolides of chemotype I of Israel generally lack the hydroxyl group at C-20 (20-H withanolides); the major constituent of these plants is withaferin A (**64**) while chemotype II is rich in 20-hydroxywithanolides, with withanolide D (**68**) being its major constituent. 20-Hydroxywithanolides are also elaborated by chemotype III and most of these withanolides contain the 2,5-dien-1-one system. This chemotype also elaborates both 17α- and 17β-oriented side chains. The major constituent of the Indian chemotype is withanone (**12**); it also elaborates 20-H withanolides many of which contain a 5α-hydroxy-6α,7α-epoxy grouping (*184, 186*). The South African chemotype elaborates both 20-H and 20-OH withanolides some of which contain saturated side chain (*183*).

That the ability of different chemotypes to produce withanolides with different nuclear substitution patterns is genetically controlled was verified through results obtained by cross pollination of these chemotypes. The hybrids obtained in this manner yielded withanolides with new nuclear substitutions depending on the genetic inheritance from their parents (*7, 13, 177, 179, 185, 186, 197*). The cross of chemotypes I × III yielded withanolide D (**68**) as the major constituent although this substance was not detected in either of the parents. On the other hand withaferin A (**64**), the major constituent of the parent type I, was absent in the withanolides from this hybrid. The withanolide composition of the F_1 offspring of the cross between chemotype III and the Indian chemotype was significantly different from those of the parents. Most significantly, this hybrid yielded a withanolide with a 14β-hydroxyl group (**14**) which is rather unique within withanolides (*63, 195, 197*). The results suggest that it might be possible to develop hybrids with the ability to yield tailored withasteroids to meet specific needs. The occurrence of different chemotypes in other withasteroid-yielding plants is a possibility and there are indications that such chemotypes do exist at least in *Datura metel* (*39*), *Jaborosa leucotricha* (*71*) and *Acnistus breviflorus* (*28*).

V. Biological Activity of Withasteroids

Steroids are found throughout the evolutionary hierarchy, from microbes to man, and these are considered to act as architectural components of membrane apart from their other biological functions (*234*). However, while the specific role of complex steroid molecules with diverse structural features in the plants elaborating these compounds is often not very clear, many of these steroids have been proved to be very useful to man because of their beneficial biological activities. The withasteroids comprise one such group of natural products and some of the withasteroid-yielding plants are well known medicinal plants used in traditional systems of medicine. *Withania somnifera* is a valuable and very popular Ayurvedic drug which is rightly compared with 'ginseng' in view of the various therapeutic properties attributed to it (*238*). Because of the findings resulting from modern studies on its constituents, it may be labelled as a drug that possesses antistress, antiinflammatory, antitumor, antibiotic, anticonvulsant and CNS depressant activities (*235*). *Acnistus arborescens* has long been known for its use in the treatment of cancerous growths (*19*). Withaferin A (**64**), the first known withasteroid, was isolated from both of these plants and its structure was clarified almost concurrently by two different groups working in different parts of the globe.

References pp. 93–106

The biological activities of withasteroids have been reviewed (*3, 7, 13*) and antimicrobial, antitumor, antiinflammatory, hepatoprotective, immunomodulatory and insect-antifeedant properties of this group of natural products have been reported. The antibacterial activity of withaferin A (**64**) was established long before the structure of this compound was fully clarified (*1*). CHATTERJEE and CHAKRABORTI (*236*) who studied the antibacterial activity of withaferin A and relatives reported that they were active only against gram-positive bacteria with no activity against gram-negative bacteria and non-filamentous fungi. 20-Deoxywithanolide D (**65**) was reported to have the highest activity among the compounds tested. The variation of antibacterial activity with substitution pattern was also pointed out by these workers. The 16,24-cyclo-ixocarpalactone, taccalonolide A (**209**), from the Chinese medicinal plant *Tacca plantaginea* was effective against the malarial parasite *Plasmodium berghai* (*204*). Antifungal activity of withaferin A against *Aspergillus flavus, Epidermophyton floccosum* and *Cladosporium herbarum* has been reported (*239*).

Withaferin A and its companion withanolide, withacnistin (**113**), showed cytotoxicity against KB cell cultures. Significant inhibitory activity against Sarcoma 180 (SA) tumor in mice and Walker intramuscular carcinosarcoma 256 (WM) in rats was also shown by withaferin A (*19*). It exhibited significant growth retardation of Ehrlich ascites carcinoma in mice; i.p. administration of a single dose of withaferin A (25–40 mg/kg) 24 hour after Ehrlich ascites implantation decreased the growth of the tumor which subsequently disappeared in 80% of the mice (*240, 241*). It acted as a mitotic poison and arrested the division of cultured human larynx carcinoma cells at metaphase (*242*). Besides withaferin A and withacnistin, several other structurally related withanolides have been reported to have cytotoxic activity. Compounds that deserve mention are 4β-hydroxywithanolide E (**151**), withanolide E (**150**), withanolide D (**68**) and the *trans*-diequatorial chlorohydrin of withaferin A (*3, 13, 148, 243–245*). Of the different withanolides tested for antineoplastic activity, 4β-hydroxywithanolide E (**151**) was the most promising on the basis of its activity on B-16 melanosarcoma and L-1210 leukemia (*148, 243*) but its activity was less than necessary for clinical investigation (*13*). Withangulatin A (**63**) from *Physalis angulata* has been reported to act on topoisomerase II to induce topoisomerase II-mediated DNA damage *in vitro* (*113, 246*).

IKEKAWA and his group (*247*) examined the inhibition of the growth of mouse leukemia L5178y cells *in vitro* by AB ring analogues of withaferin A and established a relationship between chemical structure and antitumor activity. The essential requirements in a withanolide for antitumor activity were considered to be epoxide and enone functions in the AB rings and an unsaturated lactone in the side chain. The inactivity of physalolactone (**147**), the *trans*-diequatorial chlorohydrin of 4β-hydroxy-withanolide E (NSC 339137) against P338 lymphocytic leukemia (*12*), can be explained on these grounds, but the reported cytotoxicity of withaferin A chlorohydrin (*245*) does not fit in with this postulate.

The antitumor activity of the physalin group of withasteroids has not been studied thoroughly, Physalin B (**175**) has been reported to be cytotoxic for 9 KB cells and physalin D (**176**) for B-16 melanocarcinoma, while 5α,6α-epoxyphysalin B was found to be more cytotoxic in 9 KB than physalin B (*14, 203*). Physalin A (**173**) showed moderate *in vitro* cytotoxicity against Hela cells. Physalin B and physalin F (**179**) were found to be more active than physalin A, while physalin E (**177**) and physalin L (**186**) were inactive. The absence of the cyclohexenone moiety in physalin L is considered to be responsible for its inactivity (*103*).

BUDHIRAJA *et al.* (*248*) demonstrated the antiinflammatory activity of 3β-hydroxy-2,3-dihydrowithanolide F (**137**) by testing against subacute models of inflammation and claimed its effect to be comparable to that of hydrocortisone. The compound has been reported to be superior to hydrocortisone in its activity as a hepatoprotective agent against CCl_4-induced liver damage (*249*). Physangulide (**108**) and 24,25-epoxywithan-olide D (**107**) have also been reported to produce antiinflammatory effects both in exudative and proliferative types of experimentally induced inflammation. The effects were comparable to those of hydrocortisone (*250*).

The withanolides have been shown to possess both immunosuppressive and immunostimulating properties (*3*). Inhibition of the growth of Ehrlich ascites carcinoma in mice followed by complete disappearance of tumor cells on withaferin A treatment and resistance of the cured mice to rechallenge with Ehrlich ascites tumor cells (*241*) indicate the immunoactivating property of withaferin A. The immunosuppressive activity of withaferin A is evident from its ability to inhibit adjuvant arthritis in rats and the graft versus host reaction in chicks (*251*). Similar activity was also shown by withanolide D (**68**). The immunosuppressive effect of Lycium substance A (**5**) was demonstrated by BAHR and HÄNSEL (*252*) who studied its ability to inhibit proliferation of murine spleen cell cultures. The glycowithanolides, sitoindoside IX (**61**) and sitoindoside X (**62**), were shown to exhibit adaptogenic and immunostimulatory activity. Both compounds produced significant anti-stress activity in albino mice and rats and augmented learning acquisition and memory retention in young and old rats (*180, 253*).

The insect repellent and insecticidal properties of nicandrenone (**189**) and the insect antifeedant property of withanolide E (**150**) and of a number of related natural and synthetically modified withanolides have been studied (*13, 84, 94, 254*). Further investigation of this property of withasteroids is expected to yield fruitful results.

There is an undiminishing interest in studying the various biological aspects of this group of steroidal lactones which is expected to continue into the future.

VI. Chemistry of Withasteroids

In this section, the structure determination of some important with-asteroids reported during the last five years will be discussed.

A. Withanolides

1. Minabeolides

The occurrence of withasteroids in marine organisms was brought to light by the reported isolation of minabeolides from the soft coral, *Minabea* sp., collected in the Truk Lagoon (*206*). These compounds were found to co-occur with their C_{27}-analogues with a cholestane framework.

Minabeolide-1, $C_{28}H_{38}O_3$ (M^+, m/z 422.2833) was obtained as a colourless oil and its withanolide nature was revealed from its ^{1}H-NMR and mass spectral data. The diagnostic mass peak at m/z 125 (ion *a*), the appearance of a ^{1}H-NMR signal at δ4.36 as a double triplet ($J = 12.3, 4$ Hz) and the presence of two tertiary methyls, two allylic methyls and a secondary methyl group, as disclosed by the ^{1}H-NMR spectrum suggested a typical withanolide structure which was corroborated by proton decoupling experiments and ^{1}H-^{13}C-NMR correlation data. Comparison of ^{13}C NMR data of minabeolide-1 with those of androsta-1,4-diene-3,17-dione and Lycium substance B (**6**) revealed that the AB rings of the compound were like those of the androstane derivative and that minabeolide-1 and Lycium substance B had identical CD rings and side chain. Minabeolide was thus assigned structure (**98**).

Minabeolide-1 differs from other withanolides by the absence of an oxygen function at C-1 but it may very well be formed from a precursor withanolide having an AB ring substitution pattern like that of physalo-lactone B (**55**). From a comparison of the UV, IR, ^{1}H- and ^{13}C-NMR data of minabeolide-2 with those of minabeolide-1, the structure of the former was derived as (99) which bears a –CH$_2$OAc grouping in place of 18-methyl group. Minabeolide-3 (100) was proved to be 1,2-dihydro-minabeolide-1 by detailed spectral analysis.

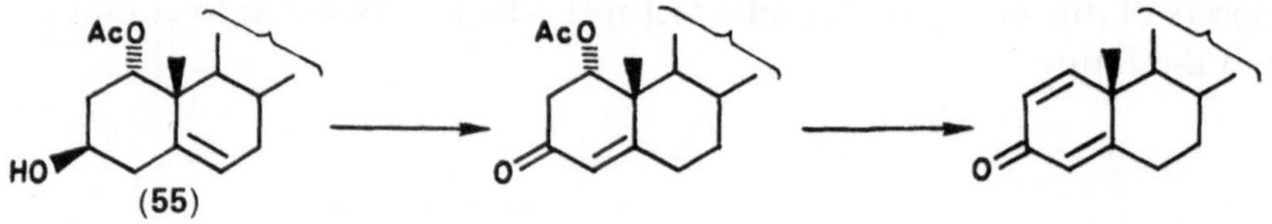

2. Withametelin, Isowithametelin, Secowithametelin and Withametelin B

EVANS et al. (32) who studied the withanolides of a number of Datura species and hybrids, reported Datura metel to be devoid of withasteroids but in spite of this report chemical examination of the plant was taken up almost simultaneously by three different groups, thus leading to isolation of a number of withanolides (33–45). Withametelin, the first known C-21 oxygenated withanolide with a novel bicyclic side chain was isolated by the reviewer and his group from the dried leaves of D. metel (33, 37). Withametelin, C$_{28}$H$_{36}$O$_4$ (M$^+$, m/z 436.2426), m.p. 210°, [α]$_D$ − 64.4° (CHCl$_3$) failed to exhibit the diagnostic spectral data of withanolides; its fragmentation mode on electron impact was rather uncharacteristic, its ^{1}H-NMR spectrum showed signals for only three methyl groups and the carbinyl hydrogen signal at δ4.65 appeared only as a broad singlet. Spectral data, however, suggested the presence of an enone system and an unsaturated δ-lactone both of which are common in withanolides. The presence of an exocyclic methylene and an oxymethylene in with-ametelin was disclosed by analysis of the ^{1}H-NMR spectrum and it was logically inferred that the two missing methyl groups were replaced by these two groups. Based on this finding and from a careful analysis of various spectral data, two alternative structures (44 and 44a) were advanced for withametelin. A correct choice between the two became possible from an analysis of the ^{13}C-shift correlated 2D spectrum (2D INADEQUATE) which clearly indicated that C-21 and C-24 rather than C-20 and C-28 are linked through an oxide bridge and that the expression 44 represents the structure of withametelin. Chemical evidence in support of this structure was secured by hydrogenolysis. While catalytic reduction of withametelin in methanol in the presence of perchloric acid gave a complex mixture, hydrogenation of a methanolic solution of the compound with 5% Pd-C yielded a hexahydro and a tetrahydro derivative (M$^+$, m/z 440). Unlike withametelin, this tetrahydro derivative showed the spectral features of typical withanolides; it exhibited the characteristic fragment ion peak at m/z 125 in its mass spectrum, the diagnostic double triplet at δ4.37 (J = 13.5, 3.6 Hz) and signals for two allylic methyls at δ1.88 and 1.93 in its ^{1}H-NMR spectrum indicating the presence of an α,β-unsaturated-δ-lactone with two vinyl methyl groups. The compound showed the presence of only one olefinic hydrogen (δ5.42 d), a hydroxymethyl group (δ3.74 dd; 4.00, br. d) for H-21 and a positive Cotton effect at 250 nm. Based on these observations, tetrahydrowithametelin was formulated as (268) which in turn confirmed the structure deduced for withametelin. The configuration at C-24 shown in 44 is based on the finding that a Dreiding model requires the 28-methyl and H-22 to be on the same side of the molecule.

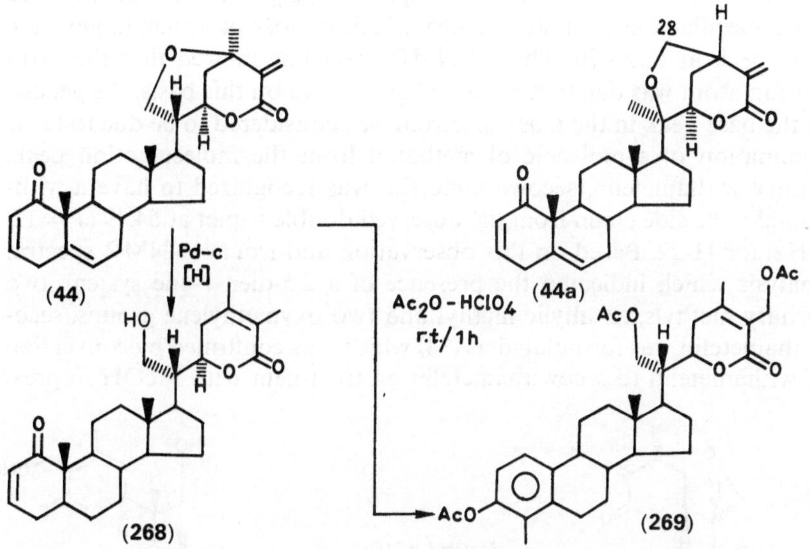

Withametelin underwent a dienone-phenol rearrangement with Ac$_2$O in presence of HClO$_4$ to a ring A aromatic withanolide with concomitant opening of the bicyclic side chain. The structure of the rearranged product was proposed as (269) from detailed spectral analysis. Daturilin isolated by Pakistani workers from the fresh leaves of *Datura metel* (34) was assigned the same gross structure as withametelin but with a 22S configuration on the basis of nOe experiments. There are reasons to believe that daturilin is identical with withametelin and that all the withanolides isolated from this source have the 22R configuration.

Isowithametelin, C$_{28}$H$_{36}$O$_4$, m.p. 280°, [α]$_D$ − 77.7° (CHCl$_3$), isolated from the mother liquor of withametelin, showed spectral features strikingly similar to those of withametelin. Careful comparison of their spectral data revealed that they differed only in the position of double bonds in the AB rings. The 3,5-dien-1-one system in isowithametelin was verified (*37*) by a two-step transformation of withametelin (**44**) to isowithametelin (**43**).

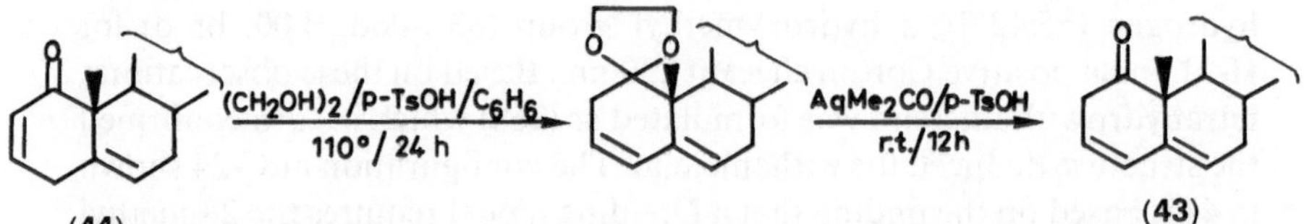

(**44**)   (**43**)

Secowithametelin, $C_{29}H_{40}O_5$, m.p. 190°, $[\alpha]_D + 28.53°$ (CHCl$_3$) had the molecular ion peak at m/z 468 which by loss of 32 a.m.u. gave the base peak at m/z 436. The ^{1}H-NMR spectrum showed that the extra carbon atom was due to a methoxyl group and on this basis, the genesis of the base peak in the mass spectrum was considered to be due to facile elimination of a molecule of methanol from the molecular ion peak. Unlike withametelin, secowithametelin was recognized to have a withanolide-like side chain from the observed double triplet at δ4.49 ($J = 12$, 4 Hz) for H-22. Based on this observation and from ^{1}H-NMR spectral analysis which indicated the presence of a 2,5-dien-1-one system, two tertiary methyls, an allylic methyl and two oxymethylene groups, secowithametelin was formulated as (**34**) which was confirmed by conversion of withametelin to secowithametelin on treatment with MeOH in pres-

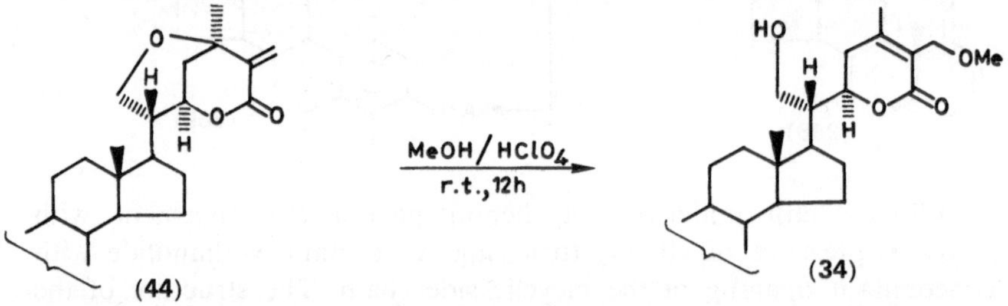

(**44**)   (**34**)

ence of HClO$_4$ (*38*). Secowithametelin has the same structure as daturametelin C, reported by NOHARA and his group (*42*).

Withametelin B, $C_{28}H_{36}O_5$ (M$^+$, m/z 452), m.p. 283–85° was isolated from the dried leaves of *Datura metel* collected in the suburbs of Calcutta which is believed to represent a chemotype different from the material collected in Varanasi. Withametelin B was recognised as a close relative of withametelin from spectral data and both the molecules were shown to have identical CD rings and side chain. The presence of a 2,4-dien-1-one chromophore in withametelin B was indicated by its uv maximum at 314 nm and confirmed by ^{1}H-NMR analysis. Unlike withametelin, withametelin B is acylable and gave a monoacetate (123-Ac) by acetylation with Ac_2O-Et_3N. Acetylation caused a downfield shift of the carbinyl hydrogen signal from δ4.59 to 5.50 and an upfield shift of the 19-methyl signal from δ1.42 to 1.33. The data suggested the presence of a 6β-hydroxyl group in the substance which was formulated as (123). Conclusive evidence for this structure came from the one-step transformation of withametelin (44) to withametelin B acetate (123-Ac) by treatment with $Hg(OAc)_2$ in the presence of acetic acid.

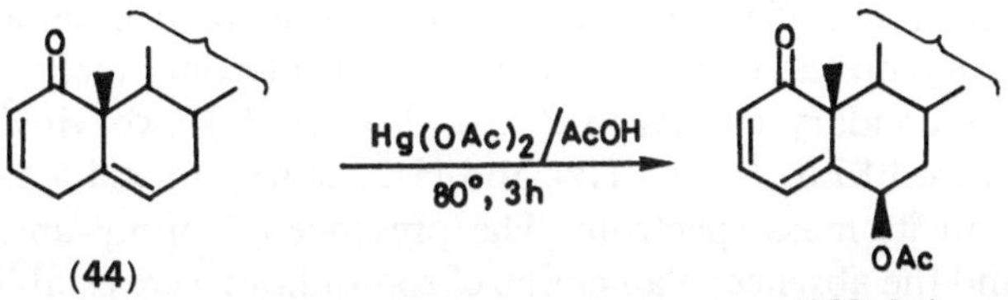

3. Daturataturins A and B

The withanolide glucosides, daturataturins A and B were isolated from the fresh aerial parts of *Datura tatura* L. (*52*). Daturataturin A, $C_{34}H_{48}O_{10}$, $[\alpha]_D - 38.1°$ (pyridine), showed a $[M + Na]^+$ peak at m/z 639 and a $[M-H_2O-Glc]^+$ peak at m/z 437 in its positive FAB-mass spectrum. The compound gave a pentaacetate indicating the presence of an acylable hydroxyl group in the withanolide part of the molecule. Analysis of the ^{1}H-NMR spectrum of the pentaacetate indicated the presence of a 2,5-dien-1-one system without any substituent at C-4 and an allylic oxymethylene whose chemical shift and splitting pattern (δ4.53 AB q, $J = 11$ Hz) suggested it to be C-27 linked with a glucose residue. Again the chemical shift of H-22 (δ4.41 dt, $J = 13.2, 3.3$ Hz) indicated a normal C-17 side chain without any hydroxyl group at C-20. The location of the acylable hydroxyl group in the withanolide part of the molecule was fixed at C-7 by finding that the H-7 signal was coupled to H-6 in the ^{1}H-^{1}H COSY spectrum. Based on these data and from ^{13}C-NMR evidence, daturataturin A was formulated as the 27-O-β-D-glucopyranoside of (22R)-7α,27-dihydroxy-1-oxowitha-2,5,24-trienolide (18). The aglycone part of this glucoside had been previously reported from *Withania somnifera* Dun. (*184*).

Daturataturin B, $C_{34}H_{52}O_{11}$, $[\alpha]_D - 17.7°$ (pyridine), was proved to have a side chain like that of viscosalactone A (81) and an AB ring system like that of pubescenin (106) with an additional 7α-hydroxyl group, from spectral comparison with reference compounds. Structure (47) assigned to daturataturin B agreed well with the ^{13}C-NMR data (52).

4. Withasomniferin A and 5-Dehydroxywithanolide R

These two withanolides were isolated from the aerial parts of *Withania somnifera* of Pakistani origin, with their structures deduced exclusively from comprehensive spectral data (175). Withasomniferin A, $C_{28}H_{38}O_5$(M^+, m/z 454.2730) showed spectral features commonly observed in withanolides. It showed signals for two tertiary methyls (δ0.76, 1.33 s), one secondary methyl (δ1.10, d, $J = 7.1$ Hz), two vinyl methyl groups (δ1.92 s, 1.87 d, $J = 1.1$ Hz), and H-22 (δ4.35 m) and a base peak at m/z 125 in its mass spectrum. The presence of an α,β-unsaturated δ-lactone and the absence of an enone chromophore were established by analysis of the IR and ^1H-NMR data. Withasomniferin A also showed ^1H-NMR signals for two mutually coupled carbinyl hydrogens (δ3.15, d, 3.52 dd) and an isolated olefinic hydrogen at δ5.53 (t) for a 4-en-6α,7α-epoxide function and on this basis the structure of withasomniferin A was proposed as (146) which was verified by the COSY 45° spectrum.

5-Deoxywithanolide R, $C_{28}H_{38}O_5$ (M^+, 454.2724) showed, in its mass spectrum, peaks at m/z 169, 141 and 125 indicating the presence of a hydroxyl group in the δ-lactone part of the molecule. From an analysis of various spectral data supplemented by information from COSY 45° spectrum, the compound was formulated as 5-deoxywithanolide R (10), but was stated to have the 22R configuration. The absolute configuration at C-22 of withanolide R (9) is the same as that of other withanolides but the presence of a hydroxyl group at C-23 changes the priority sequence and thus withanolide R has the 22S configuration (185). 5-Deoxywithanolide R therefore has also the 22S configuration.

5. Daturametelins D, E, F and G

All these withanolides were isolated from *Datura metel* by Japanese workers (42). Daturametelin D, $C_{29}H_{40}O_5$ (M^+, m/z 468), m.p. 199.5 − 201.5°, $[\alpha]_D - 90.6°$ was recognised to be a close relative of withametelin (44) by spectral analysis which revealed the identity of the carbocyclic part of the two molecules. Spectral data also disclosed that daturametelin D had a bicyclic side chain like withametelin but differed from the latter by the absence of an exocyclic methylene and the presence of a -CH$_2$-OMe grouping. The hydrogens of this oxymethylene had ^{1}H-NMR signals for the AB part of an ABX system (δ3.77, 1H, dd, $J = 9.8, 5.5$ Hz; 3.99, 1H, dd, $J = 9.8, 2.7$ Hz) and on this basis the structure of daturametelin D was advanced as (37), the correctness of which was verified by X-ray crystallographic analysis (41). The crystal structure revealed all R configurations at the chiral centres C-17, C-20, C-22, C-24 and C-25. Datumetelin, isolated by the Pakistani workers from the same plant was shown to have the same gross structure as daturametelin D but was assigned the unusual 22S, 24S configuration without conclusive evidence in support of this assignment (43, 44). The ^{1}H-NMR data for datumetelin resemble those of daturametelin D but the mp and rotations differ.

Daturametelin E, C$_{29}$H$_{42}$O$_9$S, [α]$_D$ − 25.8° (MeOH) and daturametelin F, C$_{28}$H$_{38}$O$_8$S [Negative FAB-MS: (M-H)$^+$, m/z 533] are two sulphur-containing withanolides of *Datura metel*. ^{1}H-NMR spectral analysis revealed that the carbocyclic parts of the two substances were identical and that the side chain of daturametelin E and daturametelin F was the same as that of secowithametelin (34) and withametelin (44), respectively. The ^{1}H-^{1}H COSY NMR spectrum of daturametelin F indicated the presence of a low-field methine (δ5.07 m) flanked by a ketomethylene and an allylic methylene in ring A. The low field hydrogen was thus proved to be H-3 geminal to an electro-negative group the exact nature of which was disclosed by hydrolysis. The polar nature of the compound and the IR band at 1226 cm^{-1} characteristic of an S-O function suggested the presence of a –OSO$_3$H group and in conformity with this assumption daturametelin F on treatment with pyridine-dioxane gave a product (270) in which the signal of the low field methine had moved upfield to δ3.86. The reaction mixture prior to the separation of (270) was found to respond positively to potassium rhodizonate reagent. These findings in conjunction with the observation that daturametelin F was transformed to withametelin (44) on attempted acetylation with pyridine and Ac$_2$O led to the assignment of the structure of daturametelin F as (42) and daturametelin E as (33).

(270)   (42)   (44)

(i) Pyridine-dioxane   (ii) Ac$_2$O/Pyridine

Daturametelin G, isolated as its acetate derivative, was proved to have the same hexacyclic structure as daturametelin D (37) from a comparison of the ¹H-NMR spectra. The absence of a methoxy signal and the presence of signals due to a per-O-acetyl-β-glucopyranosyl moiety in the ¹H-NMR spectrum of daturametelin G acetate settled the structure of daturametelin G as (38). The ¹³C-NMR spectrum also supported this structure as the signals of the carbons of the entire steroid skeleton of daturametelin D and daturametelin G acetate were identical.

6. Withacoagin

Withacoagin (127), $C_{28}H_{38}O_5$ (M⁺-H_2O, m/z 436.2611), m.p. 230 − 32°, $[\alpha]_D$ + 114° (CHCl₃) from the roots of *Withania coagulans* Dunal was recognised as a typical 20-hydroxywithanolide from its mass spectral peaks at m/z 169 and 125 (base peak) and from its ¹H-NMR signals which included those of five methyl groups including two vinyl methyls and one of H-22 appearing as a dd at δ4.21. The UV and IR spectral data indicated the presence of enone and α,β-unsaturated δ-lactone chromophores. The multiplicity of the enone hydrogens signals (δ5.89 dd and 6.58 ddd for H-2 and H-3) showed that C-4 was unsubstituted. Signals of the methylene hydrogens at C-4 appeared at δ2.36 dd (J = 19, 5.2 Hz) and 2.63 dt (J = 19, 2.3 Hz) suggesting the absence of hydrogen at C-5. Based on this observation and the presence of only three C-O signals in the ¹³C-NMR spectrum, a hydroxyl group was placed at this position. Unlike common withanolides, the ¹H-NMR spectrum of withacoagin contained in addition to the enone hydrogen signals an additional set of mutually coupled olefinic signals (δ5.57 dd, J = 9.9, 2.7 Hz; δ5.73 dd, J = 9.9, 1.7 Hz). Of the three possible positions of this *cis* double bond, the one between C-6 and C-7 was considered most likely by comparing the ¹³C-NMR spectrum with that of Lycium substance A (5) which contains a 5α-hydroxy,6α,7α-epoxy system. Conclusive evidence in support of the structure of withacoagin came from its epoxidation with *m*-chloroperbenzoic acid to Lycium substance A (5).

Withacoagin showed a negative Cotton effect at 333 nm and a positive Cotton effect at 250 nm which suggested respectively, a *trans*-fused A/B ring and a 22R configuration (*168*). Withacoagin is one of the two known Δ⁶-withanolides.

7. *Sominone and Sominolide*

These two withanolides were isolated from the dried whole plant of *Withania somnifera* growing in Pakistan (*176*). Sominone, $C_{28}H_{42}O_5$ (M^+, m/z 458.3051), $[\alpha]_D - 4.72°$ (CHCl$_3$), obtained as a colourless amorphous powder, was recognised as a withanolide with a $1\alpha,3\beta$-dihydroxy-5-ene system by virtue of its ^{1}H-NMR spectrum. The presence of two angular C-methyls, an allylic methyl and an allylic alcohol (AB quartet around $\delta 4.36$), also disclosed from its ^{1}H-NMR spectrum, and the mass spectral peak at m/z 141 suggested that sominone was a 27-hydroxywithanolide. The proposed structure (**58**) was verified by a 2D COSY 45° experiment and the ^{13}C-NMR spectrum. Sominone has the same structure as pubesenolide (**57**), $[\alpha]_D + 28.5°$ (CHCl$_3$) but the signs of their sp. rotation are different.

Sominolide, $C_{28}H_{38}O_5$ (M^+, m/z 470.2670), $[\alpha]_D - 2.66°$ (CHCl$_3$) was proved to have the same side chain as sominone (**58**) but differed from the latter in the AB ring substituents. Sominolide was shown to have a 4β-hydroxy-2-en-1-one system with a hydrogen in the 5α-position and a trisubstituted oxirane ($\delta 3.32$, brs) the site of which was fixed between C-14 and C-15 from mass spectral evidence. The α-orientation of the epoxide function was based on comparison of the ^{1}H-NMR spectrum with that of withanolide M (**27**), a 14,15α-epoxy-withanolide (**182**). The structure of sominolide (**92**) thus derived was verified by establishing proton-proton connectivities through a 2D COSY 45° spectrum.

8. *Withanolide Y*

Withanolide Y, $C_{28}H_{38}O_7$ (M^+-H_2O, m/z 468.2474), m.p. 270 − 73°, was isolated from a hybrid of *Withania somnifera* chemotype III (Israel) and Indian I (Delhi). The ^{1}H-NMR and mass spectra indicated the presence of a 2-en-1-one system and a 20-hydroxywithanolide structure but complete information on structure and stereochemistry was secured by X-ray single crystal analysis which showed that withanolide Y was (**129**) (*177*). Withanolide Y is isomeric with withanolide T (**11**); both substances contain an epoxy and three hydroxyl groups but while the hydroxyl groups in withanolide T are tertiary, withanolide Y has a secondary hydroxyl group at C-7. The carbinyl proton at C-7 of with-anolide Y showed a fairly large coupling constant ($J = 11$ Hz) with the proton of the hydroxyl group attached to the same carbon as shown by D_2O exchange as well as by decoupling. H-22 of withanolide Y appeared as a dd indicating the presence of a hydroxyl group at C-20 (this is supported by the chemical shift of the 21-methyl at δ1.27); its chemical shift – δ4.52 – suggested the presence of a hydroxyl group at C-17. The appearance of the 18-methyl signal at δ0.83 was typical of a normal β-side chain.

The 5α,6α-epoxy-7α-hydroxy system present in withanolide Y is rather unusual among withasteroids. While 5β,6β-epoxides are rather common, physalin J (183) is the only other withasteroid known to have a 5α,6α-epoxide without any hydroxyl group at C-7. It is believed that withanolide Y is derived from an intermediate 1-oxo-2,5-dien-7α-hydroxy derivative and that the α-epoxide formation is directed by the 7α-hydroxyl group (177).

9. Withaperuvin H

Withaperuvin H, $C_{30}H_{42}O_9S$ (FDMS, MH$^+$, m/z 579), m.p. 213 – 14°, $[\alpha]_D$ + 84.3° (pyridine) is a minor constituent of the roots of *Physalis peruviana* and a congener of withaperuvin (154), the major steroidal constituent of this plant part. Withaperuvin H showed a striking resemblance to withaperuvin in its spectral properties; the mass spectrum had important fragment ions at m/z 125, 152 and 169, and the ^{1}H-NMR spectrum had signals indicating the presence of five methyl groups, an enone system with an oxygen substituent at C-4 and a side chain identical with that of withaperuvin. Acetylation of withaperuvin H gave a monoacetate, a transformation which did not result in a shift in the signal of the C-4 carbinyl proton – a narrow triplet at δ4.83 – but caused a downfield shift in the signal of the carbinyl proton at δ5.24 (dd, $J = 9, 5$ Hz) to δ6.07. This indicated that the C-4-oxygen substituent is not a hydroxyl but an ether and that withaperuvin H contains a hemiacetal group ($δ_C$ 96.6 d). Irradiation of the hemiacetal hydrogen simplified the signals of a methylene group which appeared as geminally coupled doublets at δ2.75 and 3.06 ($J = 16$ Hz). The chemical shift of the hydrogens of this methylene and that of a methine at δ2.89 (dd, $J = 13.4$ Hz) suggested the corresponding carbons were attached to a sulphur atom, a suggestion which was supported by a carbon resonance at δ31.2 (t) and 47.9 (d). Based on these findings, withaperuvin H was formulated as (166) which was found to be in agreement with the ^{13}C-NMR data. It has been conjectured that cysteine-derived α-mercaptoacetaldehyde provides the extra two carbon atoms and the sulphur atom present in the molecule. Interestingly, unlike withaperuvin which shows an anomalous negative Cotton effect around 350 nm, withaperuvin H showed a positive Cotton effect at 350 nm suggesting *cis*-juncture of the AB rings (*141*).

10. Physagulin C

Physagulin C (**167**) was isolated from the methanolic extract of the fresh leaves and stems of *Physalis angulata* (*114*) and crystallised from n-butanol saturated with water as colourless needles, m.p. 242.5 − 43°, $[\alpha]_D$ + 105.8°. A typical withanolide structure with a 5β,6β-epoxy-4β-hydroxy-2-en-1-one system was disclosed by analysis of its NMR spectrum which also contained signals for an acetoxymethyl (δ2.09 s), a carbinyl hydrogen geminal to the acetoxy function (δ5.21 s) and a hydrogen of a trisubstituted oxirane (δ3.72 s). The ^{13}C-NMR spectrum not only substantiated the presence of these groupings but also indicated the presence of a tertiary hydroxyl group (δ_C 80.9 s). To establish the location and configuration of these functional groups, physagulin C was subjected to single crystal X-ray diffraction analysis and its structure was unambiguously settled as (20R,22R)-15α-acetoxy-5β,6β : 16β,17β-diepoxy-4β,14β-dihydroxy-1-oxowitha-2,24-dienolide (**167**). The compound is a close relative of physapubenolide (**60**) but differs in orientation of the side chain (*114*).

11. Jaborosalactol M and Jaborosalactone M

Jaborosalactol M (**159**) and its 2,3-dehydro derivative (**160**) were isolated from the whole plant of *Jaborosa bergii* Hieron as an inseparable mixture which on catalytic (10% Pd/C) hydrogenation yielded pure jaborosalactol M, m.p. 234–236° (*63*). Although the diagnostic double triplet at δ4.04 (*J* = 12, 4 Hz) spoke of its withanolide nature, the other spectroscopic evidence, unlike that of typical withanolides, showed the absence of olefinic carbons and lactone carbonyls and the presence of a hemiacetal group (δ4.99 d, *J* = 11 Hz; δ_C 91.4). Based on these findings and from the observed chemical shift of two low field methyls at δ1.40 and 1.41, an epoxylactol side chain in jaborosalactol M was considered likely and this was proved to be correct from its mass spectral peaks at m/z 143,127 and 109 as well as by chromic acid oxidation to a lactone (**161**) identical with jaborosalactone M. The chemical shift of the 18-methyl protons (δ1.10 s) of jaborosalactol M indicated the presence of a hydroxyl group either at C-14 or at C-17 but the presence of hydroxyl groups at both of these positions was shown by analysis of its ^{13}C-NMR spectrum. While the formation of a cyclic sulphite (271) on treatment of jaborosalactone M (161) with thionyl chloride indicated that both the hydroxyl groups were on the same side of the molecule, the pyridine-induced downfield shift of 18-methyl signal of jaborosalactol M by 0.24 ppm suggested that both were β-oriented. The hydroxyl group at C-26 and the epoxide function in the lactol ring were proved to have the α-configuration by comparing the ^{1}H-NMR spectrum with spectra of known compounds.

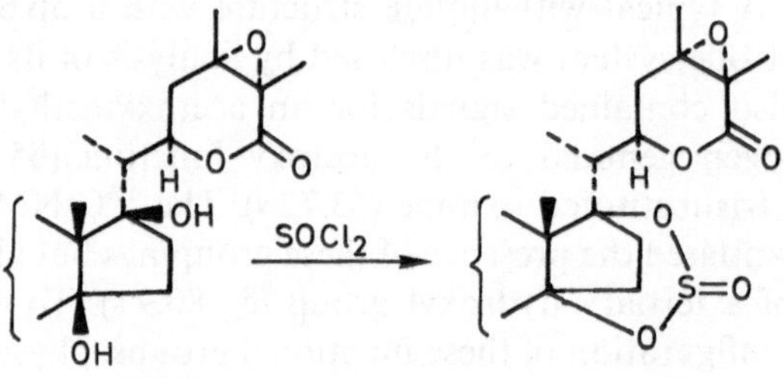

In contrast to the 14-hydroxyl groups of cardenolides, the 14-hydroxyl groups of withasteroids generally have the α-configuration. Thus jaborosalactol M and jaborosalactone M are unusual exceptions to the rule.

12. Jaboromagellone

Jaborosa magellanica (Griseb.) Dusen is a rich source of withasteroids and contains both withanolide and ixocarpalactone type compounds. Jaboromagellone (138), $C_{28}H_{40}O_7$ (M^+-H_2O, m/z 488.2813), m.p. 289°, $[\alpha]_D$ + 109° (MeOH) was recognised to be a typical 20-H withanolide from its mass spectral peaks at m/z 125 and 152 as well as from its UV, IR and NMR spectral data (73). The absence of an enone system and the presence of a lactone carbonyl and two keto carbonyl functions in the molecule were disclosed by its ^{13}C-NMR spectrum. The additional keto group was located at C-12 on the basis of the observation that the 21-methyl doublet appeared at a field (δ0.90) higher than that of the 18-methyl singlet (δ1.11 s). Further information secured from the 500 MHz ^{1}H-NMR spectrum with spin decoupling, the ^{13}C-NMR spectrum, the GASPE spectrum, a NOESY experiment and COLOC analysis permitted Freyer and coworkers (73) to assign chemical shifts to each of the hydrogens and formulate it as 5α,6β,17β-trihydroxy-1,12-dioxowitha-24-enolide (**138**). The β-stereochemistries of the hydroxyl groups at C-6 and C-17 were based on the pyridine-induced downfield shifts, of teh 19- and 18-methyl singlets. The compound showed positive Cotton effects at 240 and 280 nm in support of the 22*R* configuration and the *trans* geometry of the A/B rings, respectively.

B. Ring A Aromatic Withanolides

1. Jaborol

Jaborol, $C_{28}H_{36}O_6$ (M$^+$, m/z 468.2460), m.p. 135°, is the first known withanolide with an aromatic A ring, and was isolated from *Jaborosa magellanica* growing along the northern shores of the Straits of Magellan in Chile (*72*). Its phenolic nature was revealed by the bathochromic shift of the UV spectral maxima in the presence of alkali. Detailed spectral analysis indicated the presence of three contiguous aromatic hydrogens (δ6.68, dd, $J = 7.4, 1.3$ Hz; 6.99, dd, $J = 7.6$ Hz; 7.03, dd, $J = 7.6, 1.3$ Hz), a benzylic methyl (δ2.11 s) in place of the angular 19-methyl and a 12-keto function. That the only acylable hydroxyl group present was a phenolic hydroxyl became evident from the observation that acetylation

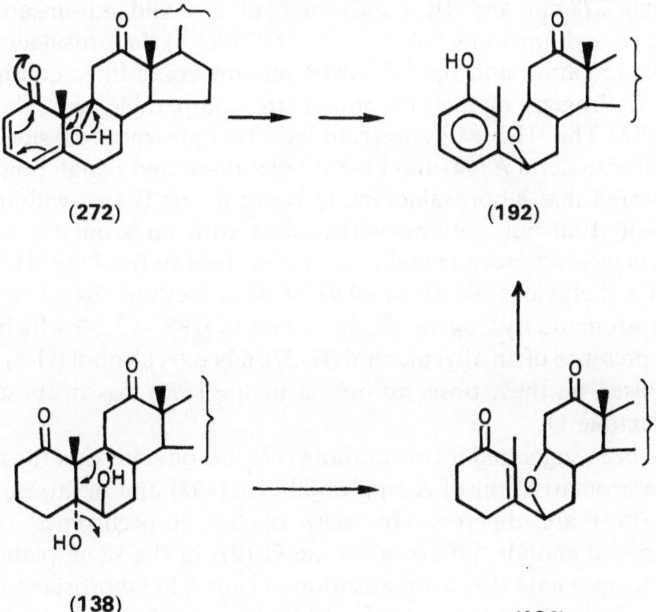

(272)   (192)

(138)   (194)

caused a downfield shift of signals of the aromatic hydrogens in the ^{1}H-NMR spectrum and of the benzylic methyl group. Presence of a normal withanolide side chain was apparent from the ^{1}H-NMR and mass spectral data and the presence of a tetrahydrofuran moiety with a benzylic ether system was deduced from analysis of the ^{1}H-NMR spectrum. These findings in conjunction with the information secured from the ^{13}C-NMR spectrum and nOe difference spectrometry led FAJARDO et al. (72) to deduce structure and stereochemistry (192) for (+)-jaborol which was confirmed by an X-ray analysis (75).

The genesis of jaborol, a 9,10-seco-, ring A aromatic withanolide was initially conceived (72) to occur by way of a 9-hydroxy-2,4-dien-1-one precursor (272) but the isolation of projaborol (194) from the same plant (73) changed the opinion of the authors. It is now believed that as shown above the immediate precursor of jaborol (192) is projaborol (194) which in turn is biosynthesised from jaboromagellone (138).

2. Jaborosalactone Q

Jaborosalactone Q, $C_{27}H_{36}O_5$ (M$^+$ − H$_2$O, m/z 422), m.p. 176–177° is the second known naturally occurring withanolide with an aromatic A ring. It was isolated from the whole plant of *Jaborosa leucotricha*, growing in Argentina (71). The compound had UV absorption maxima at 220 and 278 nm and IR bands of hydroxyl and α,β-unsaturated lactone carbonyl functions (3400, 2920, 1690 cm^{-1}). Jaborosalactone Q is phenolic in nature and the ^{1}H-NMR parameters of three contiguous aromatic hydrogens of the compound are comparable with those of jaborol (192). The ^{1}H-NMR spectrum indicated presence of a side chain like that of withaferin A (64) and absence of a 19-methyl signal. Logically, it was inferred that jaborosalactone Q is not a ring B-*seco* withanolide like jaborol (192) but a 19-norwithanolide with an aromatic A ring. Acetylation gave a triacetate and caused downfield shifts of the ^{1}H-NMR signals of a methylene (δ4.38 to δ4.91 br. s), a methine (δ4.75 to δ5.99 br. t) and aromatic hydrogens (δ6.06 − 7.06 to 6.98 − 7.24) which indicated the presence of an allyl alcohol (H-27), a benzyl alcohol (H-6) and a phenol. Based on these observations, structure (193) was proposed for jaborosalactone Q.

As has been suggested by the authors (71), it is obvious that the routes leading to aromatisation of A ring of jaborol (192) and of jaborosalactone Q (193) are different. In view of the co-occurrence of the 19-hydroxywithanolide, jaborosalactone O (97) in the same plant, it is tempting to speculate that aromatisation of ring A in jaborosalactone Q takes place by retroaldolisation of a 19-hydroxy-withanolide precursor.

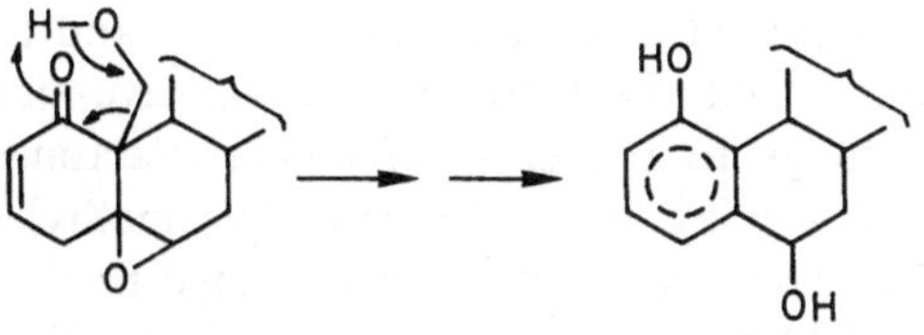

(193)

C. Ring D Aromatic Withanolide

1. Salpichrolide A

The novel group of withanolides containing an aromatic ring D was once believed to be confined within the monotypic genus, *Nicandra* and 'the story of such compounds seems to be closed' as commented by E. GLOTTER in his latest review article (*13*). A recent report of the isolation of salpichrolide A from *Salpichroa origanifolia* (*159*), however, belies this comment.

Salpichrolide A, $C_{28}H_{34}O_5$ (M$^+$, m/z 450), m.p. 179–180° was recognised as a close relative of nicandrenone (**189**) by analysis of the ^{1}H- and ^{13}C-NMR spectra. The proton signal at δ4.99 which appeared as a doublet ($J = 9.5$ Hz) but collapsed to a singlet on D_2O exchange, shifted downfield to δ6.01 on acetylation and disappeared on Jones oxidation indicated the presence of a lactol system. The chemical shifts of the 27- and 28-methyl groups (δ1.35 and 1.37 s) further suggested presence of an epoxylactol side chain. The stereochemistry at C-24, C-25 and C-26 was proved to be the same as that of nicandrenone by comparison of the ^{13}C-NMR spectra. The presence of a 2-en-1-one grouping, a 1,2,4-trisubstituted phenyl moiety and a 5,6-epoxide was also indicated by the NMR spectra. The rather large coupling constant ($J = 4.9$ Hz) of H-6 appearing as a doublet at δ3.24 was consistent with the calculated values for a 5α,6α-epoxide which was proved to be correct by appropriate chemical reactions. Salpichrolide A was thus formulated as (**191**).

D. Ixocarpalactones

1. Trechonolide A

Trechonolide A (**207**), $C_{28}H_{36}O_7$ (M$^+$, m/z 484.2436), m.p. 264–266°, was isolated from the leaves of *Trechonaetes laciniata* Miers (*160*). Its UV absorption maximum at 218 nm (ε 17400) indicated the presence of enone and α,β-unsaturated lactone chromophores and an IR band at 1751 cm^{-1} suggested that the lactone ring was five-membered. That trechonolide A belongs to the class of ixocarpalactones and not of perulactones was shown by the presence of five methyl signals in the ^{1}H-NMR spectrum. This was further substantiated by the mass peak at m/z 111.0464 ($C_6H_7O_2$), formed by fission of the 22,23 bond. The substance was proved to have two tertiary hydroxyl groups through formation of a carbamate ester (-OTAC) by treatment with trichloroace-tyl isocyanate. The appearance of H-22 as a dd at δ4.02 in the ^{1}H-NMR spectrum of the parent molecule and at δ3.99 in that of its carbamate ester indicated that the oxygen function at C-22 was an ether which was eventually proved to be part of a cyclic hemiketal attached to C-12. Addition of a few drops of trichloroacetyl isocyanate to an NMR tube containing trechonolide A in CDCl$_3$ solution and examination of the spectra at definite time intervals revealed the appearance of two lowfield signals at δ8.71 and 8.93 within 6 minutes (for NH of –OCONHCOCCl$_3$). With time, however, one of these lowfield signals began to disappear, a phenomenon which was accompanied by appear-ance of an additional olefinic hydrogen signal. The transformation was complete within 16 hours at which time only one amidic NH signal at δ8.29 and the new olefinic proton signal (H-11) at δ4.93 were discernible. The observation suggested the formation of a cyclic enol ether from the hemiketal carbamate by elimination of 12-OTAC. Based on these find-ings trechonolide A was formulated as (207) which was verified by an X-ray analysis. Trechonolide A is an ixocarpalactone with an α-oriented side chain; its precursor has a carbonyl function at C-12 and a hydroxyl group at C-22. The α-orientation of the side chain allows the C-22 hydroxyl to approach the C-12 carbonyl for formation of cyclic hemiketal.

A substance isolated subsequently from *Jaborosa magellanica* and named jaborosalactone M (*74*) is identical with trechonolide A. However, the name jaborosalactone M had been bestowed earlier (*63*) on (161) from *Jaborosa bergii*. To avoid confusion it has been suggested (*73*) that (161) be referred to as jaborosalactone M-bergii and (207) as jaborosa-lactone M-magellanica. This seems quite unnecessary as the name trechonolide A has priority even though the genus *Trechonaetes* seems to have been absorbed in *Jaborosa* (*76*).

2. Jaborosalactone P

Jaborosalactone P, $C_{28}H_{36}O_6$, m.p. 262–264° from *Jaborosa odoneliana* showed structural similarity with trechonolide A (207) by having two tertiary methyls ($\delta 1.15$, 1.20 s), one secondary methyl ($\delta 1.21$ d, $J = 7$ Hz) and two vinylic methyls on a γ-lactone ($\delta 1.91$, 2.27 s). The ^{1}H-NMR spectrum contained signals associated with a 2,5-dien-1-one moiety and the carbinyl hydrogen of a secondary alcohol which appeared as a dd at $\delta 4.22$ but collapsed to a clean doublet ($J = 12.5$ Hz) on D_2O exchange. On acetylation, this signal appeared as a doublet at $\delta 5.75$ indicating that it was coupled only with one vicinal hydrogen. The proton responsible for this signal was assigned to H-22 coupled to H-20 which required that C-23 was substituted. This was corroborated by the ^{13}C-NMR spectrum in which C-23 appeared as a singlet at $\delta 97.6$. The data indicated a spiro-γ-lactone structure (212) for jaborosalactone P which was confirmed by X-ray diffraction (76).

3. Taccalonolides A and B

The taccalonolides were isolated from the Chinese medicinal plant *Tacca plantaginea*, a member of the family Taccaceae (204). This is the first report of withasteroids from a non-solanaceous plant. While the molecular formula of taccalonolide A was $C_{36}H_{46}O_{14}$ the presence of four acetoxy groups indicated that the carbon skeleton contained 28 carbons like that of other withasteroids. The compound gave a positive steroidal colour test ($Ac_2O + H_2SO_4$), showed ^{1}H-NMR signals of five methyl groups in addition to those for the acetate functions and an IR

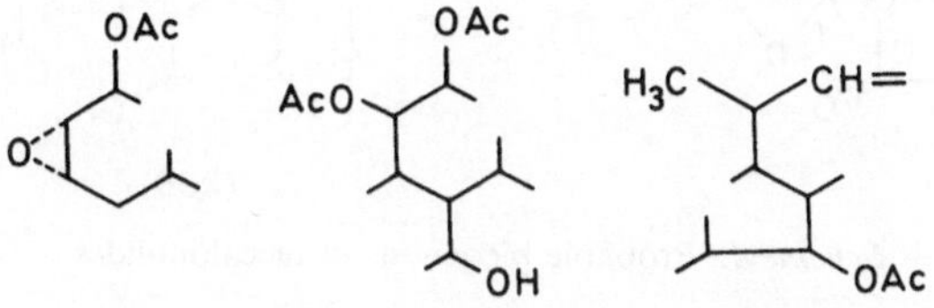

(x)   (y)   (z)

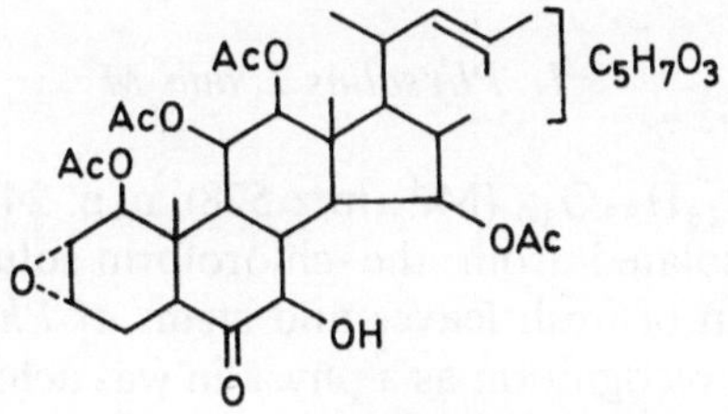

(273)

band (1807 cm^{-1}) characteristic of a five-membered enol lactone. The ^{13}C-NMR spectrum revealed the presence of a keto carbonyl, five ester carbonyls, nine methyls, one methylene, nine methines, three tetra-substituted carbons, one tetrasubstituted oxycarbon, five oxymethines and a –CH=C–O–system. Analysis and spin decoupling of the ^{1}H-NMR spectrum of taccalonolide A revealed the presence of the three moieties (x), (y) and (z) shown above and on the basis of this information the part structure of the compound was settled as (273).

The nature of the $C_5H_7O_3$ unit was deduced from the recognition of the presence of an enol lactone moiety, of an additional tertiary C-Me and of a methyl bound to a carbon bearing a hydroxyl group, thus leading to structure (209) for taccalonolide A. The stereochemistry shown in expression (209) was deduced by proton-proton nOe difference studies. The correctness of this structure was finally confirmed by X-ray crystallographic analysis. Taccalonolide B was proved to be 15-deacetyltacca-lonolide A (210). The taccalonolides are 16,24-cyclo-ixocarpalactones just as the physalins are 16,24-cyclowithanolides. The presence of an oxygen function at C-15 further suggests that this oxygen function was probably a carbonyl prior to cyclisation and was subsequently reduced to an alcohol (Scheme 4). Alternatively, the presence of a 15α,16α-epoxide and an unsaturated γ-lactone moiety in the precursor molecule may also give rise to taccalanolides.

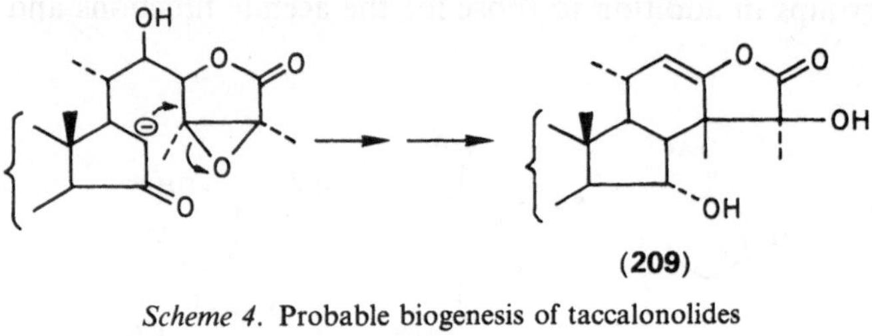

(209)

Scheme 4. Probable biogenesis of taccalonolides

E. Physalins

1. Physalins L and M

Physalin L, $C_{28}H_{32}O_{10}$ (M$^+$, m/z 528), m.p. 248–249°, [α]$_D$ – 118° (Me$_2$CO), was isolated from the chloroform-soluble fraction of the aqueous decoction of fresh leaves and stems of *Physalis alkekengi* var. *francheti* (*103*). Its recognition as a physalin was achieved by comparison of the ^{13}C-NMR spectrum with that of physalin A (173); resonances of 21 out of the 28 carbons of physalin L corresponded closely to those of (**173**). However, unlike typical physalins whose ¹H-NMR spectra exhibit signals of only three tertiary methyl groups, physalin L showed, in addition, a signal of a secondary methyl group. This methyl group was logically considered to be that of C-27 as C-25 is the only site where a secondary methyl group can be accommodated in a physalin skeleton. Like physalin A, physalin L had one secondary and two tertiary hydroxyl groups but differed from the other physalins in the absence of a cyclohexenone moiety and the presence of a conjugated diene having a secondary hydroxyl at the allylic position of the trisubstituted double bond. It is on this basis that physalin L was considered to have either of the following two substitution patterns in the A/B rings. Like physalin A, physalin L

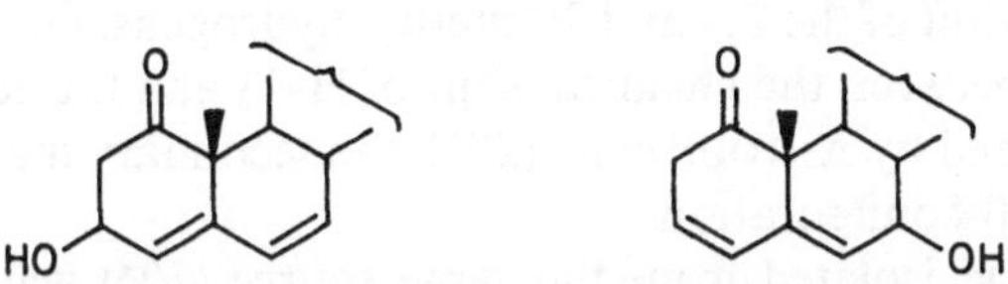

underwent acid-catalysed dehydration to yield a conjugated trienone (**274**) which on catalytic hydrogenation gave a product identical with deoxyhexahydrophysalin A (**275**), obtained by catalytic hydrogenation of

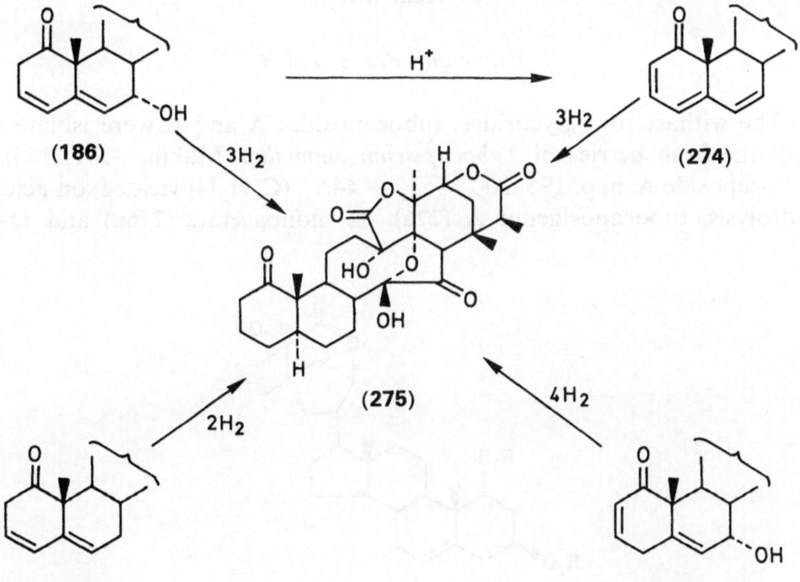

Chart 10. Conversion of Physalins L and M to common product

physalin A. The chemical correlations of physalin L with physalin A, summarised in Chart 10, was enough to advance a gross structure for physalin L but not adequate to settle the correct substitution pattern in the A/B rings. A 7-hydroxy-3,5-dien-1-one structure (186) was assigned to physalin L from the observation that the terminal olefinic hydrogen of the disubstituted double bond of the conjugated diene (δ5.88, brd, $J = 10$ Hz) was coupled with the hydrogens of the ketomethylene group (δ2.68 and 3.47). The structure assigned to physalin L was corroborated by DQF-COSY, NOESY and ^{13}C-NMR spectral analysis.

The hydroxyl group at C-7 was assigned the α-configuration on the basis of the small coupling constant ($J = 2$ Hz) between H-7 and H-8 while the stereochemistry at C-25 of physalin L (186) was deduced from the chemical shift of the 27- and 28-methyl hydrogens. On the basis of the relationship between the chemical shift of H-28 and the configuration at C-25 established by ANTOUN et al. (203), the secondary methyl group was assigned the β-configuration.

Physalin M isolated from the same source (106) was proved to be 7-deoxyphysalin L (187) by comprehensive spectral analysis and by its conversion to a common hydrogenation product (Chart 10).

F. Acnistins

1. Tubocaposides A and B

The withasteroid glycosides, tubocaposides A and B were isolated from the fresh berries of *Tubocapsicum anomalum* Makino (161, 162). Tubocaposide A, m.p. 195–200°, $[\alpha]_D - 44.2°$. (C_5H_5N) yielded, on acid hydrolysis, tubocaposigenin A (276), its monoacetate (276a) and D-

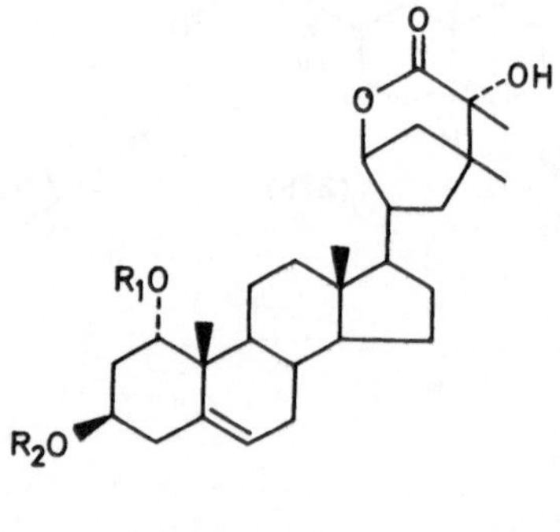

276: $R_1 = R_2 = H$, 276a: $R_1 = Ac$, $R_2 = H$, 276b: $R_1 = R_2 = Ac$

glucose. Comprehensive spectral analysis of tubocaposigenin A diacetate (**276b**) and comparison of its ^{1}H-NMR spectrum with that of the pentaacetate derivative of pubescenin (**106**) revealed the presence of a steroid skeleton with a 1α,3β-diacetoxy-5-ene system. The C-17 side chain of elemental composition $C_9H_{13}O_3$ was proved to be bicyclic like that of acnistin (VI) by ^{13}C-NMR spectral analysis leading to the structural proposal (**276b**) which was corroborated by an X-ray crystallographic analysis. Based on this evidence and the information secured by analysis of the ^{13}C-NMR spectrum of the parent glycoside, tubocaposide A was formulated as (**197**) with an acetoxy function at C-1 and a β-D-glucopyranosyl (1-6)-β-D-glucopyranosyl residue at C-3. The structure of tubocaposide B (**198**) was deduced in a similar manner.

TH-6 and TH-12 are two novel steroidal constituents isolated from the hydrolysate of the methanolic extract of the aerial parts of *Tubocapsicum anomalum (162)*. From detailed spectroscopic analysis and a single crystal X-ray analysis, the structure of TH-6 was settled as (**216**). TH-12 was assigned structure (**217**) by spectral comparison with TH-6. The structures of TH-6 and TH-12 clearly indicate that these are neither withasteroids nor natural products; their formation from a typical withanolide having a 16α,17α-*cis* diol system has been proposed by NOHARA and his group (*162*).

A compound like 27-deoxywithaferin A (**65**) having a 16α,17α-epoxy group may be a more likely starting material for the formation of TH-6 and TH-12. A concerted mechanism involving opening of the protonated epoxide, migration of the 18-methyl group to C-17 and formation of a 13,14-double bond by loss of H-14 which is outlined below appears to be a reasonable proposal. Methanolysis of the lactone ring and opening of the 5β,6β epoxide to a *trans*-diequatorial chlorohydrin and diol in the presence of a 4β-hydroxy group are well-known steps (*215*).

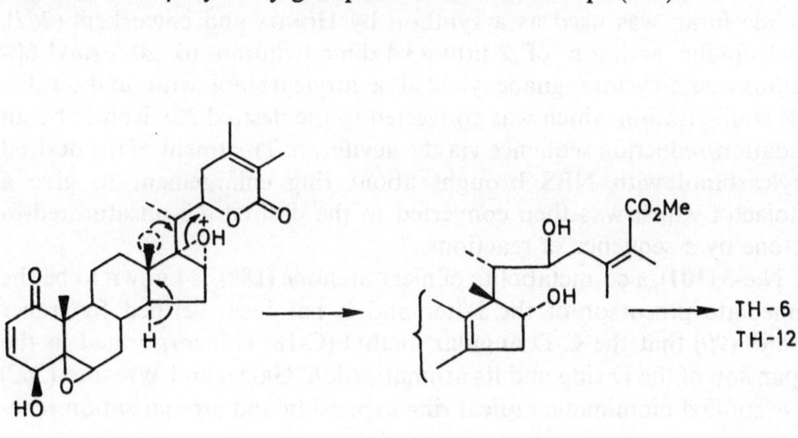

Addendum

Additions to the growing list of withasteroids have appeared after completion of this review and several recent publications on the subject deserve to be mentioned.

NOHARA and coworkers (*256, 257*) have reported the isolation of six new withanolides, physagulins A, B, D, E, F and G in addition to withaminimin (**93**), withangulatin A (**63**) and physagulin C (**167**) from *Physalis angulata* and determined their structures by spectroscopic methods. The structures of physagulins A, B, D and F were settled, respectively, as (20S,22R)-15α-acetoxy-5β,6β-epoxy-14α-hydroxy-1-oxowitha-2,16,24-trienolide, (20S,22R)-15α-acetoxy-5α-chloro-6β,14α-dihydroxy-1-oxowitha-2,16,24-trienolide, (20S,22R)-1α,27-dihydroxy-3-O-β-D-glucopyranosyl-witha-5,24-dienolide, and 15α-acetoxy-16β,17β-epoxy-5α,6β,14β-trihydroxy-1-oxowitha-2,24-dienolide. Physagulin E was characterised as 15α-acetoxy-28-O-β-D-glucopyranosyl-5α,6β,14β-trihydroxy-1-oxowitha-2,16,24-trienolide and physagulin G as its 16β,17β-epoxide.

Iochroma coccinium has been shown to be a rich source of withanolides. Besides withaferin A (**64**), withacnistin (**113**) and iochromolide (**59**), it yielded a number of new withanolides which were characterised as 24,25-dihydrowithacnistin, 16α-hydroxywithacnistin, 27-hydroxywithacnistin, 3-hydroxy-2,3-dihydrowithacnistin, 24,25-dihydroiochromolide and 27-hydroxyiochromolide (*258, 259*).

The phytoecdysteroids, 29-norcyasterone and 29-norsengosterone isolated from *Ajuga reptans* (Family Labiatae) (*260*) belong to the class of perulactones and should be regarded as withasteroids.

The synthesis of minabeolide-3 (**100**), a withanolide from a soft coral, has been reported. In the construction of the side chain of this withanolide furan was used as a synthon by HONDA and coworkers (*261*). Nucleophilic addition of 2-lithio-3,4-dimethylfuran to 20-formyl-6β-methoxy-3α,5-cyclopregnane yielded a furylcarbinol with undesirable 22R configuration which was converted to the desired 22S isomer by an oxidation/reduction sequence via the acylfuran. Treatment of the desired furylcarbinol with NBS brought about ring enlargement to give a ketolactol which was then converted to the desired α,β-unsaturated-δ-lactone by a sequence of reactions.

Nic-3 (**101**), a co-metabolite of nicandrenone (**189**), is known to be the immediate precursor of the latter and it has been verified by tracer studies (*90*) that the C/D angular methyl (C-18) is incorporated in the expansion of the D ring and its aromatisation. GREEN and WHITING (*262*) have studied biomimetic radical ring expansion and aromatisation reactions as a model for the biogenesis of nicadrenone type ring D aromatic steroids.

Withajardins, the structures of which have since been revised (*263*), constitute a new class of withasteroids. In the bicyclic side chain of withajardins, C-21 is linked with C-25. *Discopodium* has been reported (*264*) to be an additional withasteroid-yielding genus of Solanaceae.

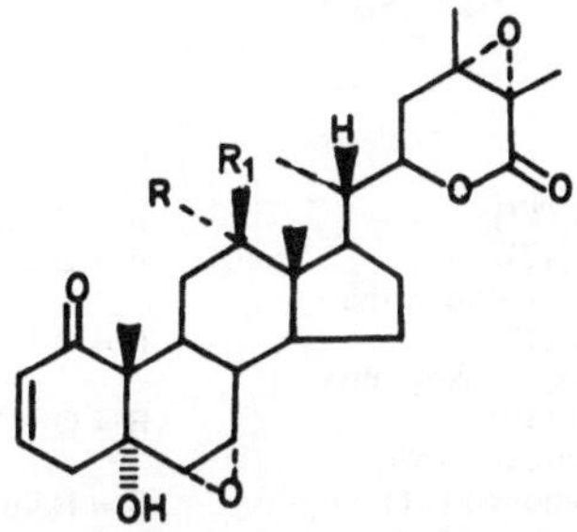

Daturalactone-1 (**1**)·   R = OH, R$_1$ = H
Daturalactone-2 (**2**):   R + R$_1$ = O
Daturalactone-4 (**3**)·   R = R$_1$ = H

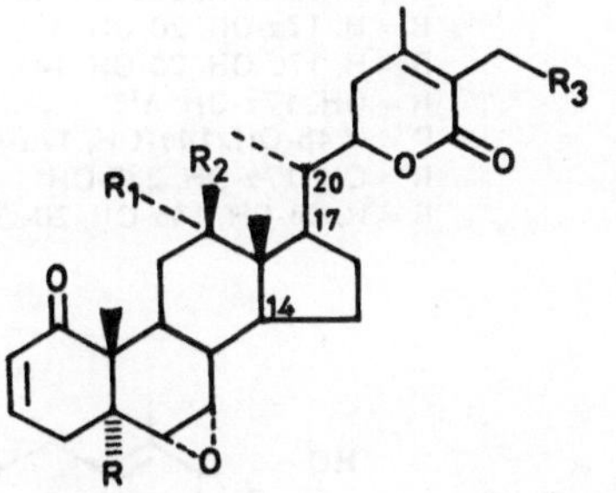

Daturalactone-3 (**4**):   R = R$_2$ = OH, R$_1$ = R$_3$ = H
Lycium substance A (**5**):   R = OH, R$_1$ = R$_2$ = R$_3$ = H, 20-OH
Lycium substance B (**6**):   R = OH, R$_1$ = R$_2$ = R$_3$ = H
Nicandrin B (**7**)   R = R$_1$ = OH, R$_2$ = R$_3$ = H

Withanicandrin (**8**):   R = OH, R$_1$ + R$_2$ = O, R$_3$ = H
Withanolide R (**9**):   R = OH, R$_1$ = R$_2$ = R$_3$ = H, 23β-OH
5-Deoxywithanolide R (**10**):   R = R$_1$ = R$_2$ = R$_3$ = H, 23β-OH
Withanolide T (**11**):   R = OH, R$_1$ = R$_2$ = R$_3$ = H, 17α-OH, 20-OH,
  = 20-Hydroxy withanone

Withanone (**12**):   R = OH, R$_1$ = R$_2$ = R$_3$ = H, 17α-OH
14α-Hydroxywithanone (**13**)·   R = OH, R$_1$ = R$_2$ = R$_3$ = H, 14α-OH, 17α-OH
14β-Hydroxywithanone (**14**).   R = OH, R$_1$ = R$_2$ = R$_3$ = H, 14β-OH, 17α-OH
Withastramonolide (**15**)·   R = R$_1$ = R$_3$ = OH, R$_2$ = H
27-Hydroxywithanolide B =
12-Deoxywithastramonolide (**16**)   R = R$_3$ = OH, R$_1$ = R$_2$ = H

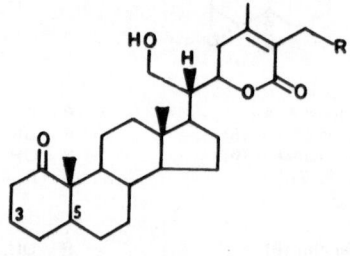

Daturametelin A (17):   R = OGlc
Daturataturin A (18):   R = OGlc, 7α-OH
7α-27-Dihydroxy-1-oxo-witha-
2,5,24-trienolide (19):   R = OH, 7α-OH
17α-27-Dihydroxy-1-oxo-witha-
2,5,24-trienolide (20):   R = OH, 17α-OH
4β, 7β, 20-Trihydroxy-1-oxo-
witha-2,5,24-trienolide (21):   R = H,4β-OH, 7β-OH, 20-OH
4β, 17α, 27-Trihydroxy-1-oxo-
witha-2,5,24-trienolide (22):   R = OH, 4β-OH, 17α-OH
Withanolide G (23):   R = H, 14α-OH, 20-OH
Withanolide H (24):   R = OH, 14α-OH, 20-OH
Withanolide J (25):   R = H, 14α-OH, 17α-OH, 20-OH
Withanolide L (26):   R = H, 17α-OH, 20-OH, Δ¹⁴
Withanolide M (27):   R = H, 17α-OH, 20-OH, 14α, 15α-epoxy
Withanolide N (28):   R = OH, 17α-OH, Δ¹⁴
Withanolide O (29):   R = H,4β-OH, 14α-OH, 17α-OH
Withanolide Q (30):   R = OH, 17α-OH, 23β-OH
Withanolide U (31):   R = H, 4β-OH, 14α-OH, 20-OH

Daturametelin B (32):   R = OGlc, Δ²·⁵
Daturametelin E (33):   R = OMe, 3β-OSO₃H, Δ⁵
Secowithametelin (34):   R = OMe, Δ²·⁵
Withametelin C (35):   R = H, 5α-OH, 6β-OH
Withametelin D (36):   R = H, 5α-OH, 6β-OH, Δ²

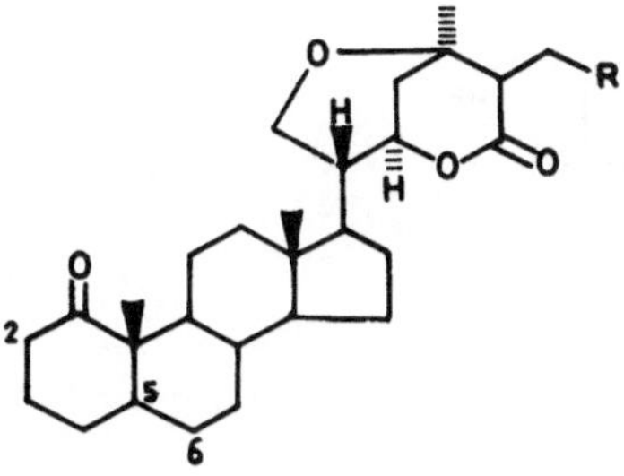

Daturametelin D (37)   R = OMe, $\Delta^{2.5}$
= Datumetelin
Daturametelin G (38):   R = Oglc, $\Delta^{2.5}$
Daturilinol (39):   R = OH (22S, 24S), $\Delta^{2.5}$
Withafastuosin A (40):   R = OH, 5β, 6β-epoxide
Withafastuosin B (41):   R = OH, 5β, 6β-epoxide, Δ^2

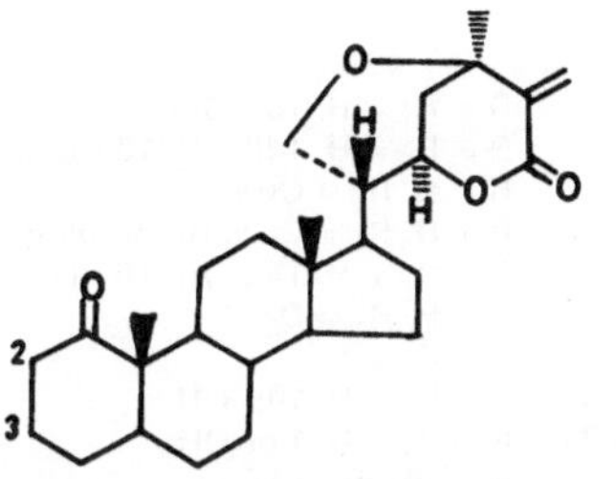

Daturametelin F (42):   3β-OSO$_3$H, Δ^5
Isowithametelin (43): $\Delta^{3.5}$
Withametelin (44)· $\Delta^{2.5}$
= Daturilin
Withametelin F (45).   5β, 6β-epoxy, Δ^2
Withametelin G (46):   5α-OH, 6β-OH, Δ^2

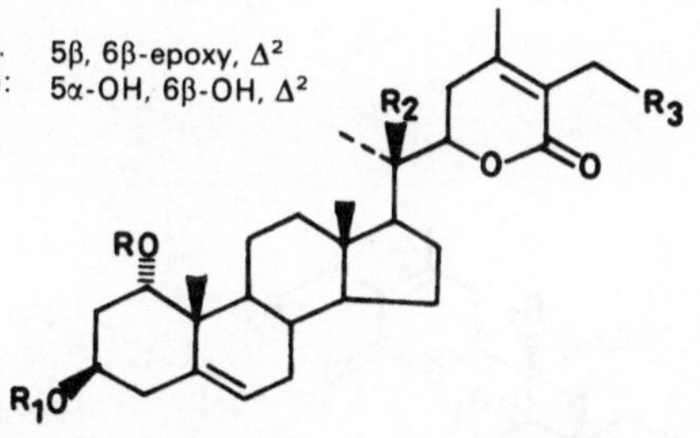

Daturataturin B (47).   R = R$_2$ = H, R$_1$ = Glc, R$_3$ = OH, 7α-OH
Dunawithagenin (48):   R = R$_1$ = R$_3$ = H, R$_2$ = OH
Dunawithanin A (49):   R = Ac, R$_1$ = Glc-Glc, R$_2$ = OH, R$_3$ = H
   Glc (2′,3′)
Dunawithanin B (50)·   R = Ac, R$_1$ = Glc-Glc, R$_2$ = OH, R$_3$ = H
Dunawithanin C (51)·   R = Ac, R$_1$ = Glc-Xyl, R$_2$ = OH, R$_3$ = H
   Xyl
Dunawithanin D (52)·   R = Ac, R$_1$ = Glc-Xyl, R$_2$ = OH, R$_3$ = H
   Glc
Dunawithanin E (53):   R = Ac, R$_1$ = Glc-Glc, R$_2$ = OH, R$_3$ = H
   Glc (3′,4′)
Dunawithanin F (54)   R = Ac, R$_1$ = Glc-Glc, R$_2$ = OH, R$_3$ = H, 12β-OH
   Glc (3′,4′)
Physalolactone B (55).   R = Ac, R$_2$ = OH, R$_1$ = R$_3$ = H
Physalolactone B glucoside (56):   R = Ac, R$_1$ = Glc, R$_2$ = OH, R$_3$ = H
Pubesenolide (57)·   R = R$_1$ = R$_2$ = H,   R$_3$ = OH
Sominone (58): = (57?)   R = R$_1$ = R$_2$ = H, R$_3$ = OH

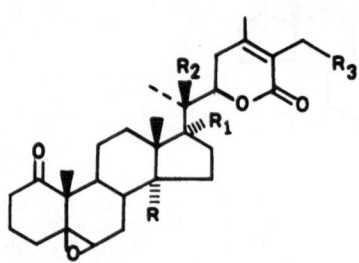

Iochromolide (**59**):   R = R₁ = H, 16α-OAc
Physapubenolide (**60**):   R = R₁ = H, 14β-OH, 15α-OAc
Sitoindoside IX (**61**):   R = H, R₁ = OGlc
Sitoindoside X (**62**):   R = H, R₁ = OGlc (6′-palmitoyl)
Withangulatin A (**63**):   R = R₁ = H, 14α-OH, 15α-OAc, Δ¹⁶
Withaferin A (**64**):   R = H, R₁ = OH
27-Deoxywithaferin A (**65**):   R = R₁ = H
14α-Hydroxy-27-deoxywithaferin A (**66**):   R = R₁ = H, 14α-OH
17α-Hydroxy-27-deoxywithaferin A (**67**):   R = R₁ = H, 17α-OH
Withanolide D (**68**):   R = OH, R₁ = H
7β-Acetoxywithanolide D (**69**):   R = OH, R₁ = H, 7β-OAc
7β-Hydroxywithanolide D (**70**)·   R = OH, R₁ = H, 7β-OH
14α-Hydroxywithanolide D (**71**)·   R = OH, R₁ = H, 14α-OH
17α-Hydroxywithanolide D (**72**):   R = OH, R₁ = H, 17α-OH
27-Hydroxywithanolide D (**73**)   R = R₁ = OH

Dihydrojaborosalactone A (**74**)   R = R₁ = R₂ = H, R₃ = OH
2,3-Dihydrowithaferin A (**75**):   R = R₁ = R₂ = H, R₃ = OH, 4β-OH
2,3-Dihydro-27-desoxywithaferin A (**76**):   R = R₁ = R₂ = R₃ = H, 4β-OH
2,3-Dihydrowithanolide D (**77**)   R = R₁ = R₃ = H, R₂ = OH, 4β-OH
Isowithanolide E (**78**):   R = R₁ = R₂ = OH, R₃ = H, Δ²
Jaborosalactone A (**79**):   R = R₁ = R₂ = H, R₃ = OH, Δ²
Jaborosalactone L (**80**):   R = R₂ = R₃ = H, R₁ = OH, Δ²
Viscosalactone A (**81**):   R = R₁ = R₂ = H, R₃ = OH, 2β, 3β-epoxy, 4β-OH
Viscosalactone B (**82**):   R = R₁ = R₂ = H, R₃ = OH, 3β-OH, 4β-OH
4α-27-Dihydroxy-5β, 6β-epoxy-1-
oxo-witha-2,24-dienolide (**83**):   R = R₁ = R₂ = H, R₃ = OH, 4α-OH, Δ²
7α, 17α, Dihydroxy-5β, 6β-epoxy-1-
oxo-witha-2,24-dienolide (**84**):   R = R₂ = R₃ = H, R₁ = OH, 7α-OH, Δ²

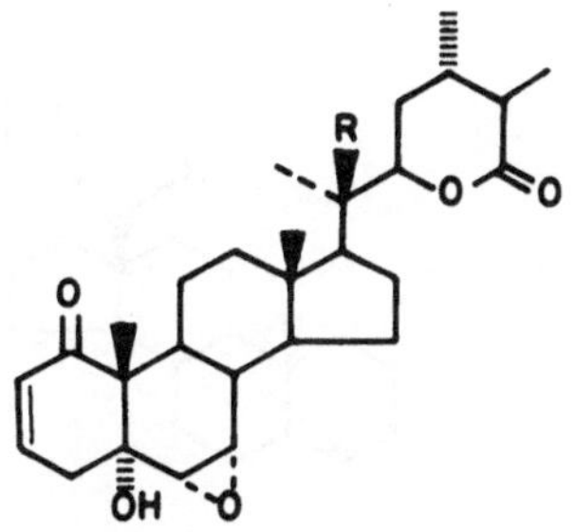

Ixocarpanolıde (**85**)·   R = OH
14α-Hydroxyixocarpanolıde (**86**):   R = OH, 14α-OH
Vamonolıde (**87**):   R = H, 14α-OH

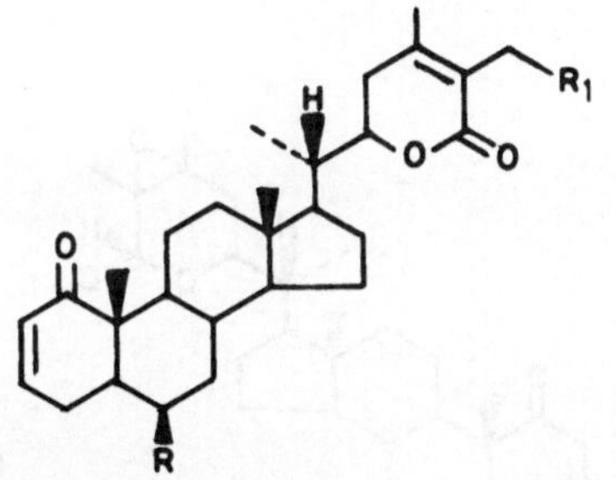

Jaborosalactone B (**88**): R.= R₁ = OH, Δ⁴
Jaborosalactone D (**89**): R = R₁ = OH, 5α-OH
= Acnıstoferin
Jaborosalactone E (**90**): R = R₁ = OH, 5α-Cl
Jaborosalactone F (**91**)· R = R₁ = OH, 5α-OH, 12α-OH
Somınolide (**92**):   R = H, R₁ = OH, 4β-OH, 14α, 15α-epoxy
Withamınimin (**93**).   R = OH, R₁ = H, 5α-OH, 14α-OH,
                                         15α-OAc, Δ¹⁶

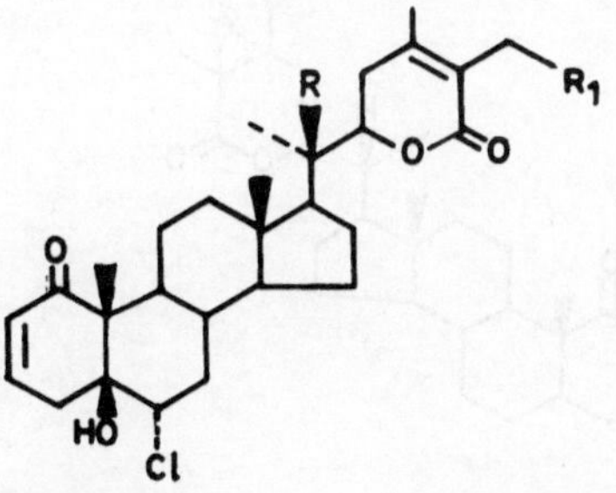

Jaborosalactone C (**94**):   R = H, R₁ = OH
4β-Hydroxy-   R = H, R₁ = OH, 4β-OH
jaborosalactone C (**95**)
Withanolide D-
chlorohydrin (**96**)   R = OH, R₁ = H, 4β-OH

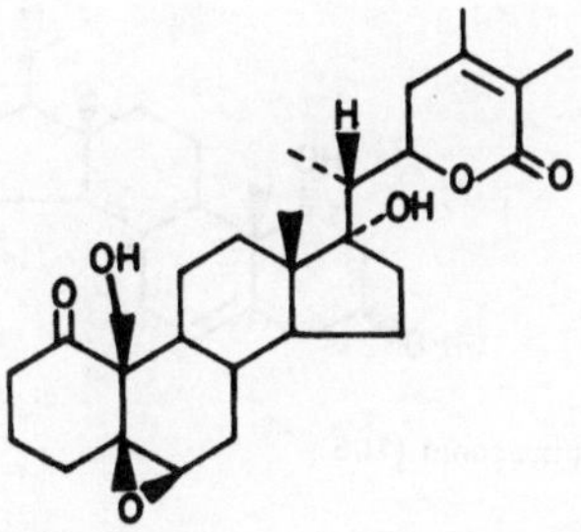

Jaborosalactone O (**97**)

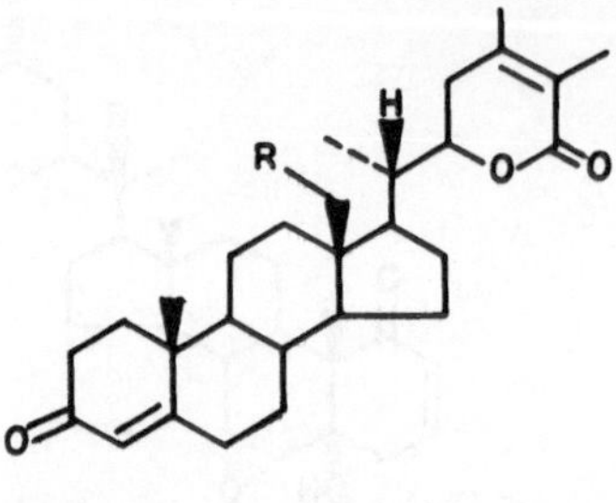

Minabeolide 1 (**98**):   R = H, Δ^1
Minabeolide 2 (**99**):   R = OAc, Δ^1
Minabeolide 3 (**100**):   R = H

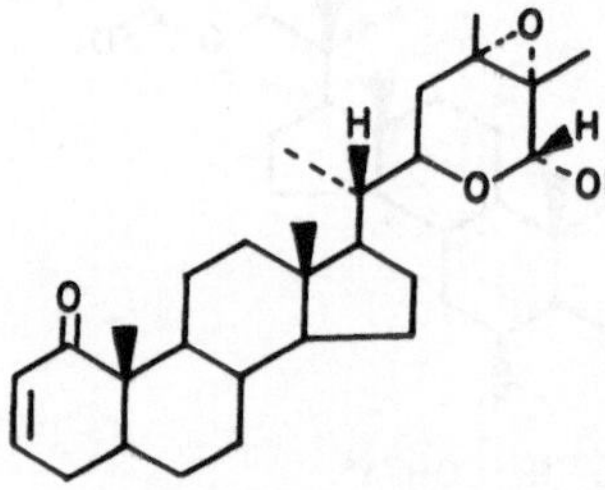

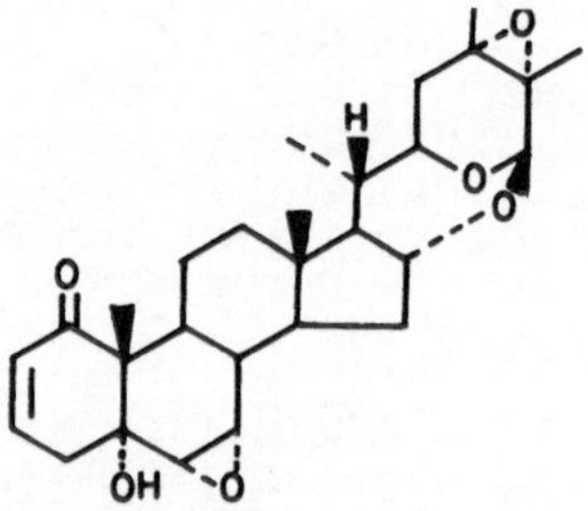

Nicalbin B (**105**)

Nic-3 (**101**). 5α-OH, 6α, 7α-epoxy
Nic-7 (**102**): 5α-OH, 6α, 7α-epoxy, 12-Oxo
Nicalbin A (**103**): 5α-OH, 6α, 7α-epoxy, 16α-OH
Physapubescin (**104**): 4β-OH, 5β, 6β-epoxy, 15α-OAc, (26R, 80%)

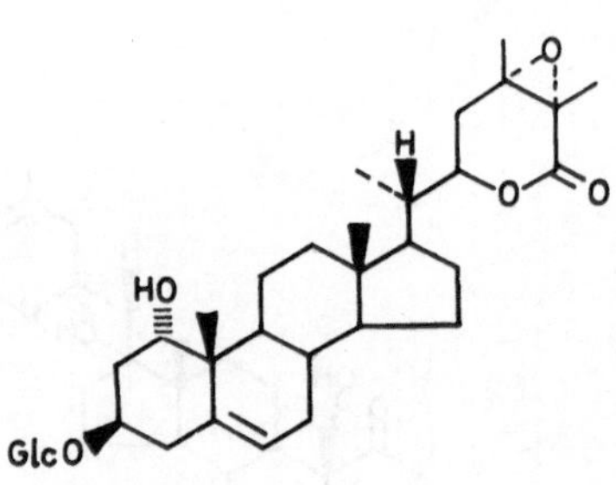

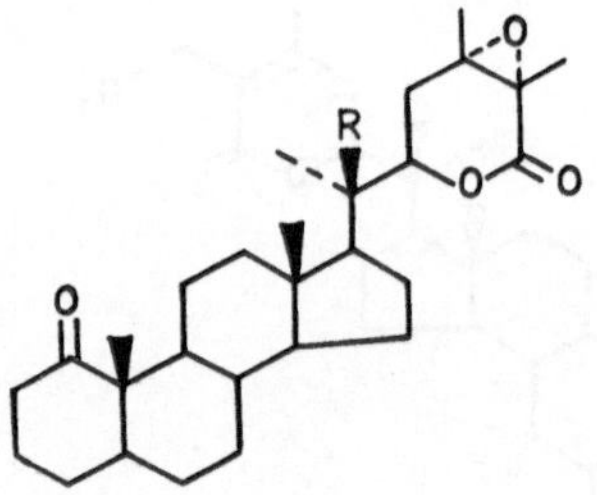

Pubescenin (**106**)

Physangulide (**108**)· 3β-OH, 4β-OH, 5β, 6β-epoxy,
                            R = OH
Epoxywithanolide D (**107**). 4β-OH, 5β, 6β-epoxy, Δ^2,
                            R = OH
Pubescenol (**109**). 4α-OH, 7α-OH, R = H

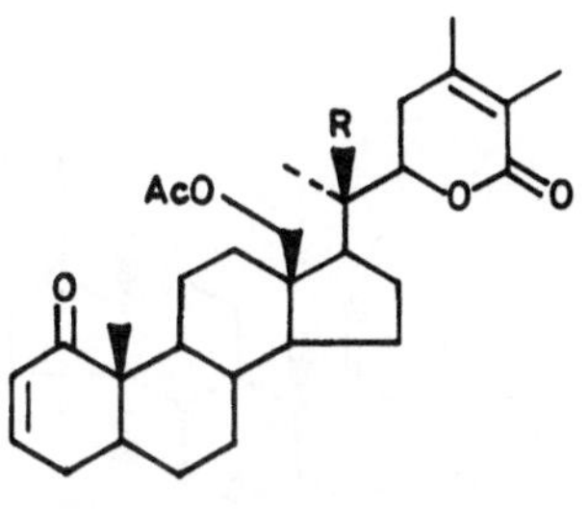

18-Acetoxy withanolide D (**110**): R = OH, 4β-OH, 5β, 6β-epoxy
18-Acetoxy-4,5,6,-deoxy-5-withenolide D (**111**): R = OH, Δ^5
18-Acetoxy-5,6-deoxy-5-withenolide D (**112**): R = OH, 4β-OH, Δ^5
Withacnistin (**113**): R = H, 4β-OH, 5β, 6β-epoxy

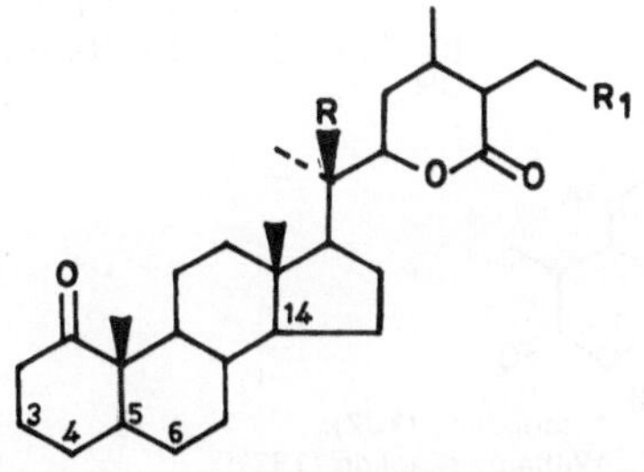

5,6-Deoxywithaferin A (**114**): R = H, R_1 = OH. 4β-OH, $\Delta^{2.5.24}$
2,3-Dihydro-5,6-deoxywithaferin A (**115**): R = H, R_1 = OH, 4β-OH, $\Delta^{5.24}$
24,25-Dihydro-27-deoxywithaferin A (**116**): R = R_1 = H, 4β-OH, 5β, 6β-epoxy, Δ^2
2,3,24,25-Tetrahydro-27-deoxywithaferin A (**117**): R = R_1 = H, 4β-OH, 5β, 6β-epoxy
Withaferin A chlorohydrin (**118**): R = H, R_1 = OH, 4β-OH, 5β-OH, 6α-Cl
4-Dehydrowithanolide D (**119**): R = OH, R_1 = H, 4-oxo, 5β, 6β-epoxy, $\Delta^{2.24}$
4-Dehydro-24,25-dihydrowithanolide D (**120**): R = OH, R_1 = H, 4-oxo, 5β, 6β-epoxy, Δ^2
24,25-Dihydrowithanolide D (**121**): R = OH, R_1 = H, 4β-OH, 5β, 6β-epoxy, Δ^2
3β-Hydroxy-2,3-dihydrowithanolide H (**122**): R = R_1 = OH, 3β-OH, 14α-OH, $\Delta^{5.24}$

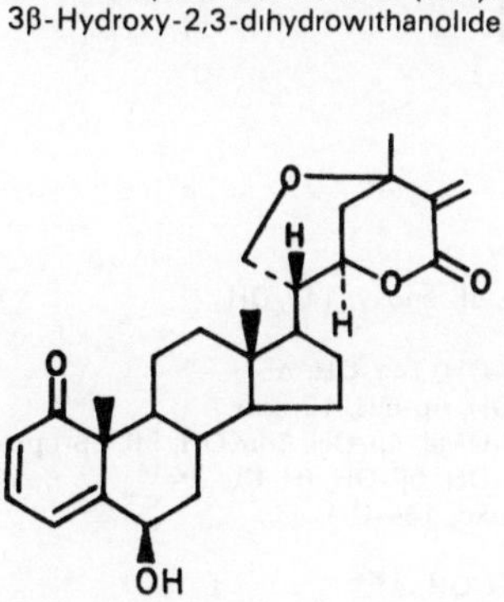

Withametelin B (**123**)

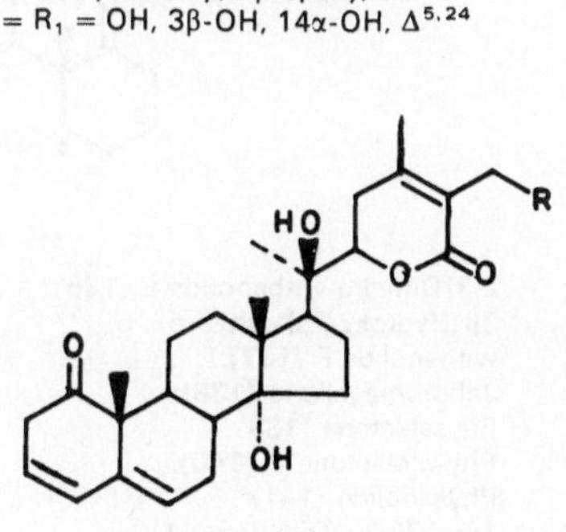

Withanolide I (**124**). R = H
27-Hydroxywithanolide I (**125**) R = OH
Withanolide K (**126**) R = H, 17α-OH

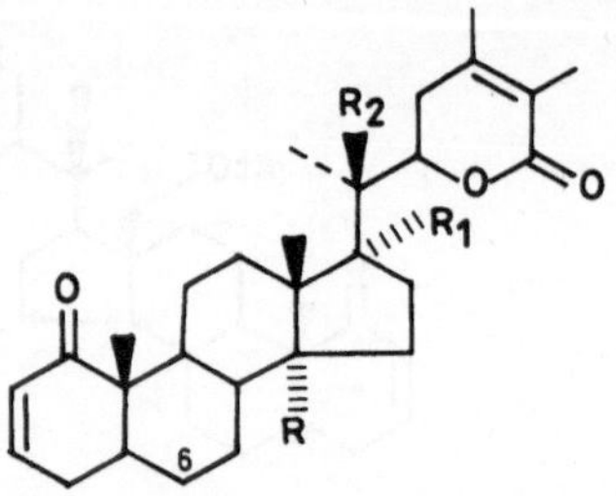

Withacoagin (127).   R = R₁ = H, R₂ = OH 5α-OH, Δ⁶
20-Deoxy-17-hydroxy-
withacoagin (128)·   R = R₂ = H, R₁ = OH, 5α-OH, Δ⁶
Withanolide Y (129).   R = H, R₁ = R₂ = OH, 5α, 6α-epoxy, 7α-OH
5α, 14α, 17α-Trihydroxy-6β,
7β-epoxy-1-oxo-witha-2,24-
dienolide (130):   R = R₁ = OH, R₂ = H, 5α-OH, 6β, 7β-epoxy
Withametelin E (131)·   R = R₁ = R₂ = H, 5α-OH, 6α, 7α-epoxy, 12β-OH, 27-
                                      OH

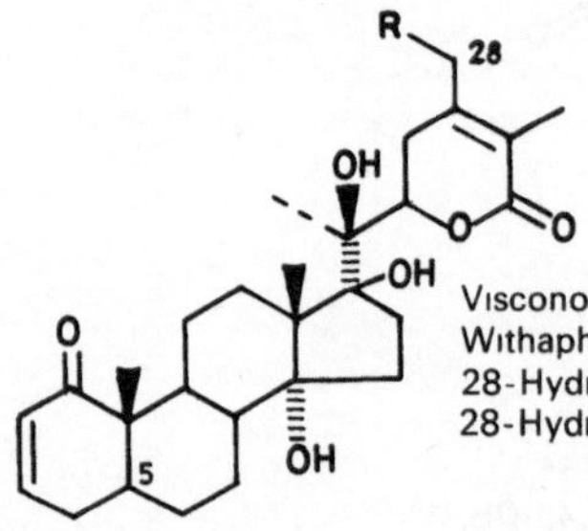

Visconolide (132):   R = OH, 4β-OH, 5β, 6β-epoxy
Withaphysanolide (133):   R = H, 4β-OH, 5α-OH
28-Hydroxy withaphysanolide (134):   R = OH, 4β-OH, 5α-OH
28-Hydroxy withaperuvin C (135)   R = OH, 6β-OH, Δ⁴

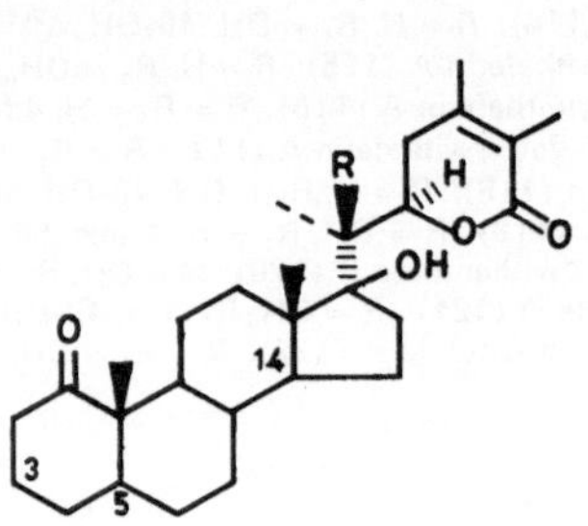

2,3-Dihydro withanolide E (136):   R = OH, 5β, 6β-epoxy, 14α-OH
3β-Hydroxy-2,3-dihydro
withanolide F (137):   R = OH, 3β-OH, 14α-OH, Δ⁵
Jaboromagellone (138):   R = H, 5α-OH, 6β-OH, 12-oxo
Physalactone (139):   R = OH, 3β-OMe, 4β-OH, 14α-OH, 5β, 6β-epoxy
Physalolactone C (140):   R = OH, 4β-OH, 5β-OH, 6α-CL, Δ²,¹⁴
Physanolide (141):   R = OH, 4-oxo, 14α-OH, Δ⁵
14α,17β,20-Trihydroxy-1-
oxo-witha-3,5,24-trienolide (142):   R = OH, 14α-OH, Δ³,⁵
Withanolide P (143):   R = H, 14α-OH, Δ²,⁵
Withaperuvin B (144):   R = OH, 4β-OH, 5β-OH, 6α-OH, Δ²,¹⁴
Withaperuvin G (145):   R = OH, 2β, 3β-epoxy,5β, 6β-epoxy, 14α-OH
Withasomniferin A (146):   R = H, 6α,7α-epoxy, Δ⁴

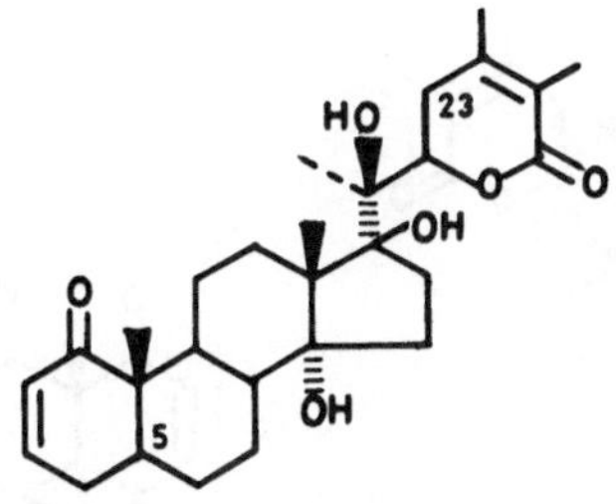

Physalolactone (**147**):   4β-OH, 5β-OH, 6α-Cl
4-Deoxyphysalolactone (**148**):   5β-OH, 6α-Cl
23-Hydroxyphysalolactone (**149**): 4β-OH, 5β-OH, 6α-Cl, 23α-OH
Withanolide E (**150**):   5β, 6β-epoxy
4β-Hydroxy withanolide E (**151**): 4β-OH, 5β, 6β-epoxy
Withanolide F (**152**): Δ^5
Withanolide S (**153**):   5α-OH, 6β-OH
Withaperuvin (**154**):   4β-OH, 5β-OH, 6α-OH
Withaperuvin C (**155**):   6β-OH, Δ^4
Withaperuvin E (**156**):   4-oxo, 5β, 6β-epoxy
Withanolide C (**156a**):   5α-Cl, 6β-OH

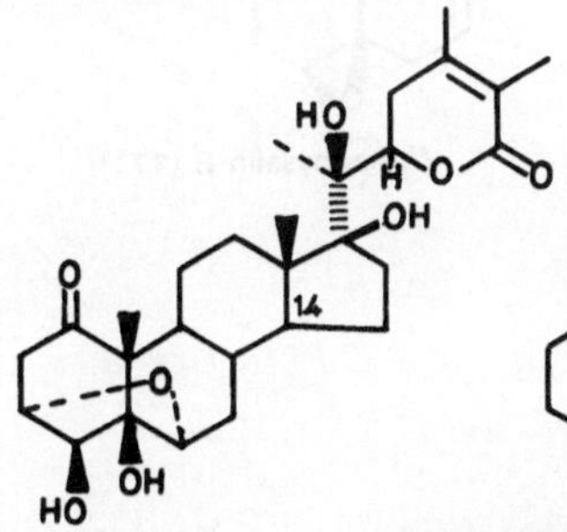

Withaperuvin D (**157**):   14α-OH
Withaperuvin F (**158**): Δ^{14}

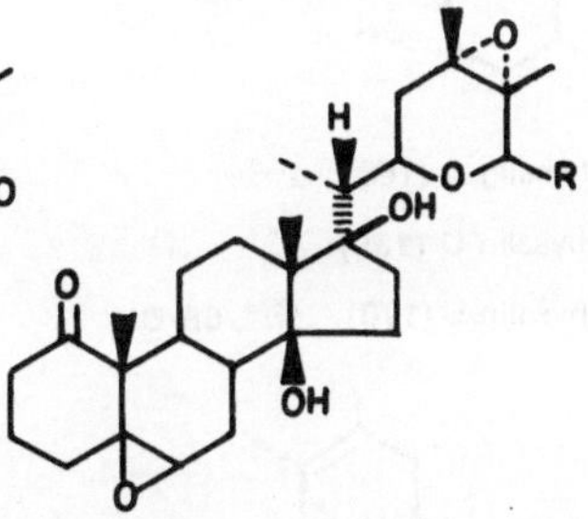

Jaborosalactol M (**159**):   R = α-OH, β-H
2,3-Dehydro-jaborosalactol M (**160**):   R = α-OH, β-H, Δ^2
Jaborosalactone M (**161**).   R = O
2,3-Dehydrojaborosalactone M (**162**)·   R = O, Δ^2

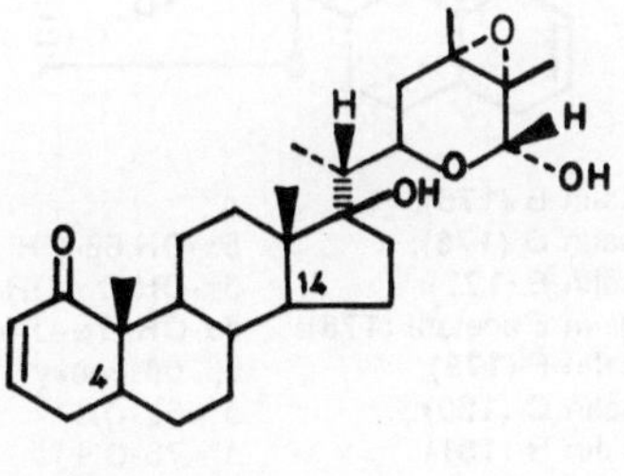

Jaborosalactol N (**163**):   6β-OH, 14β-OH, Δ^4
Nic-2 (**164**):   5α-OH, 6α, 7α-epoxy

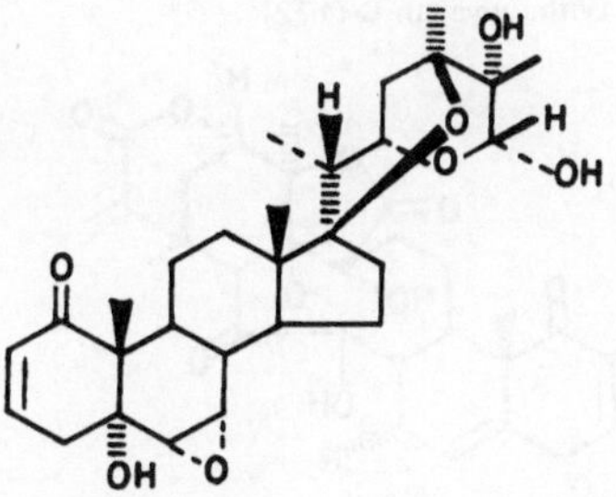

Nic-11 (**165**)

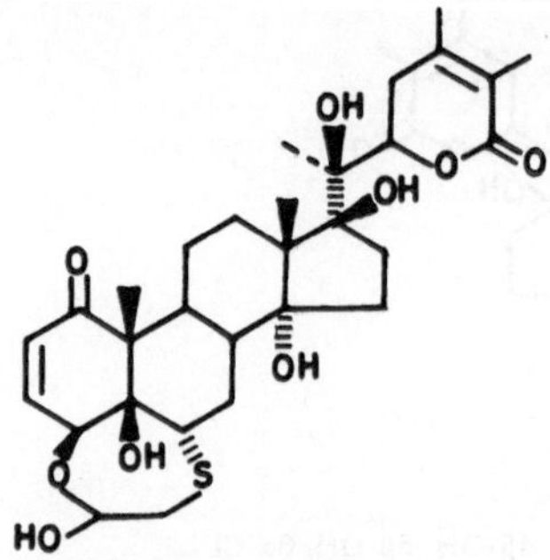

Withaperuvin H (**166**)

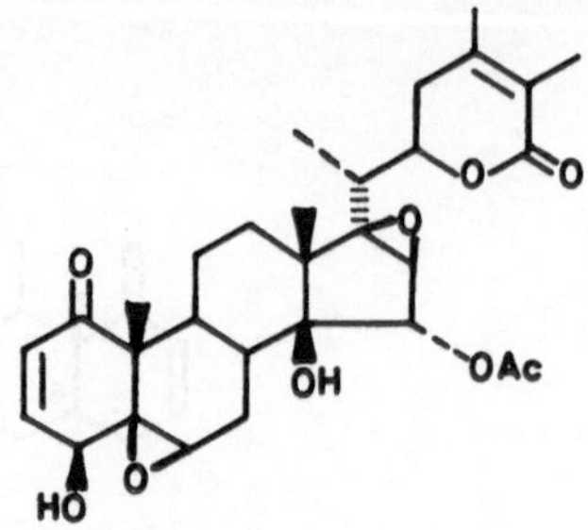

Physagulin C (**167**)

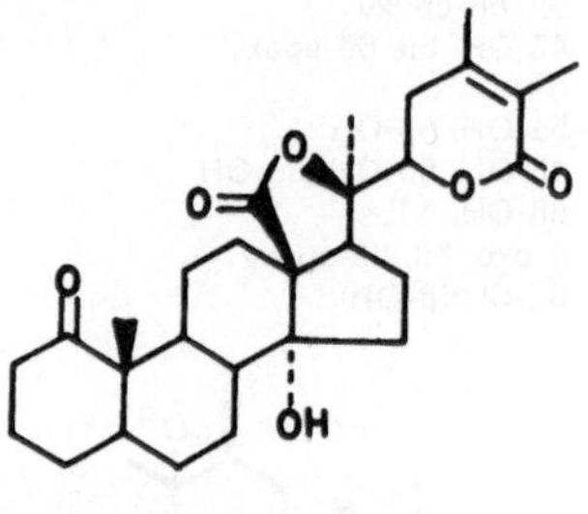

Withaphysalin A (**168**): $\Delta^{2,5}$

Withaphysalin D (**169**): $\Delta^{3,5}$

Withaphysalin E (**170**): $\Delta^{2,4}$, 6β-OH

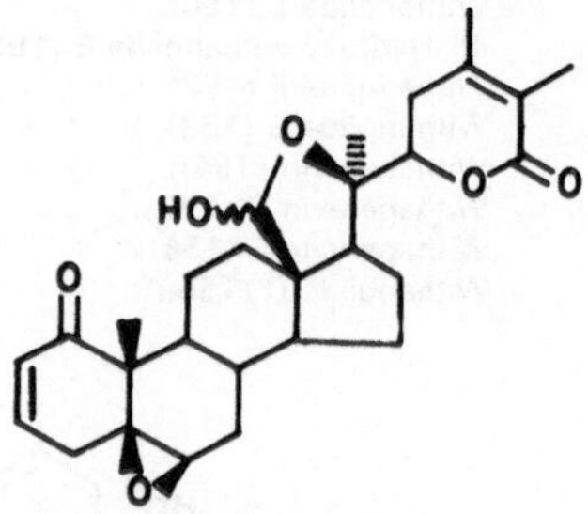

Withaphysalin B (**171**).

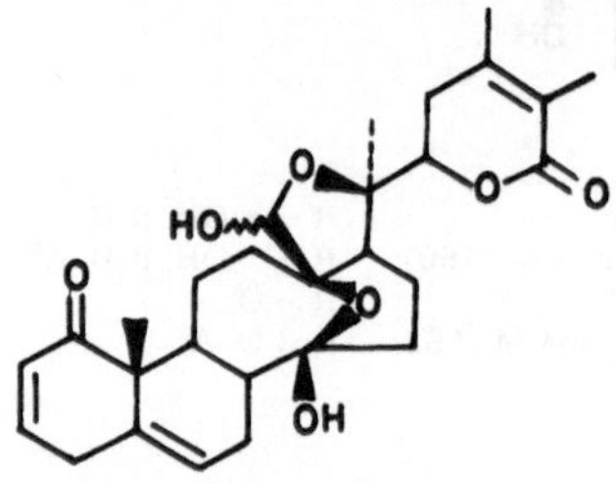

Withaphysalin C (**172**)

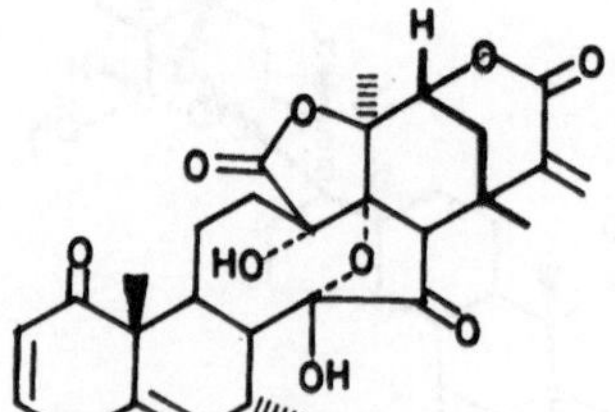

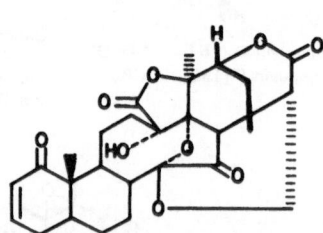

Physalin A (**173**): R = OH
Physalin C (**174**): R = H

Physalin B (**175**):	Δ^5
Physalin D (**176**):	5α-OH,6β-OH
Physalin E (**177**):	5α-OH, 7α-OH
Physalin E acetate (**178**):	5α-OH, 7α-OAc
Physalin F (**179**):	5β, 6β-epoxy
Physalin G (**180**):	Δ^4, 6α-OH
Physalin H (**181**):	Δ^5, 7β-OH
Physalin I (**182**):	5α-OMe, 6β-OH
Physalin J (**183**):	5α, 6α-epoxy
Physalin K (**184**):	4α, 5α-epoxy,6α-OH
Physalin N (**184a**).	Δ^5, 7α-OH

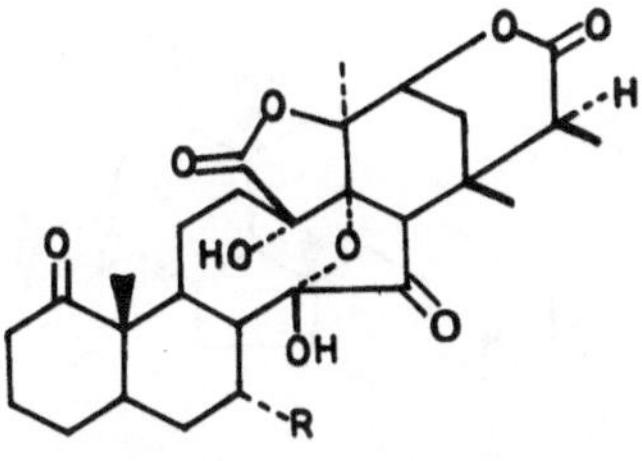

25-Epi-25,27-dihydrophysalin C (**185**):   R = H, $\Delta^{2.5}$
Physalin L (**186**):   R = OH $\Delta^{3.5}$
Physalin M (**187**):   R = H $\Delta^{3.5}$
Physalin O (**188**):   R = OH, $\Delta^{2.5}$

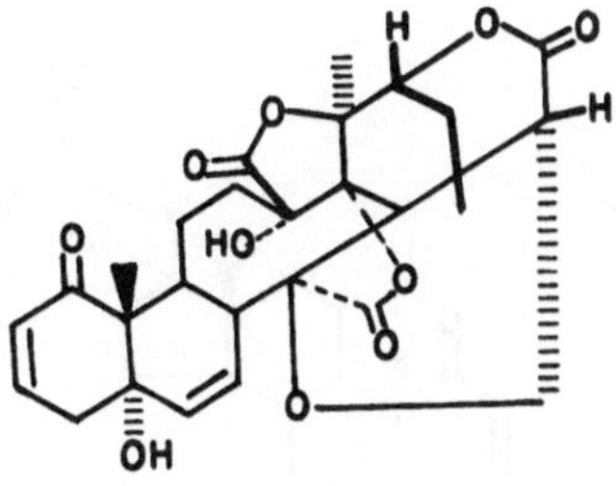

Physalin P (**188a**)

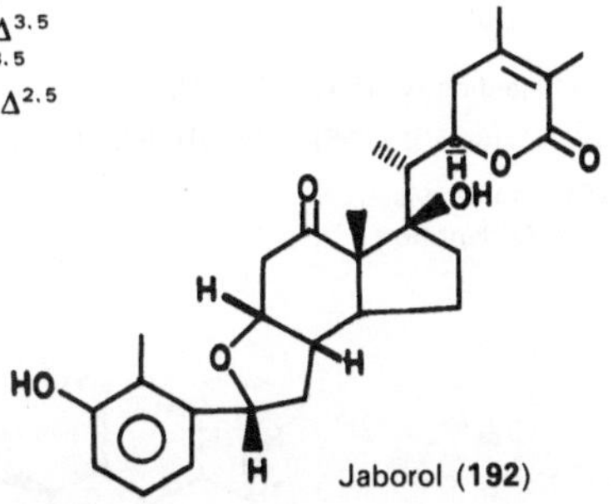

Jaborol (**192**)

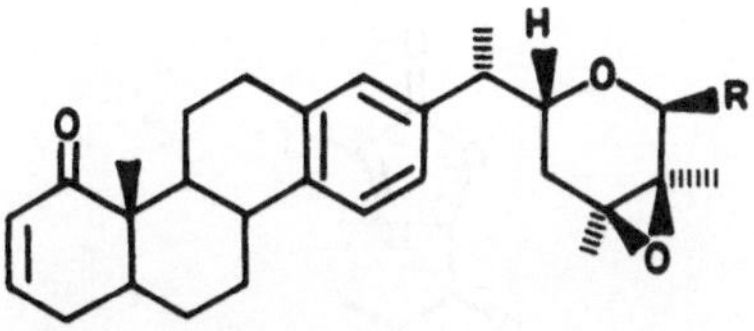

Nicandrenone (**189**):   R = β-OH, α-H, 5α-OH, 6α, 7α-epoxy
Nicandrenolactone (**190**): R = O, 5α-OH, 6α, 7α-epoxy
Salpichrolide A (**191**):   R = β-OH, α-H, 5α, 6α-epoxy

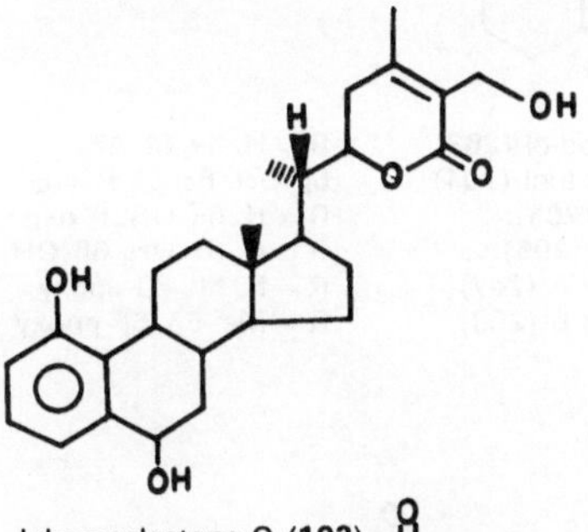

Projaborol (**194**)

Jaborosalactone Q (**193**)

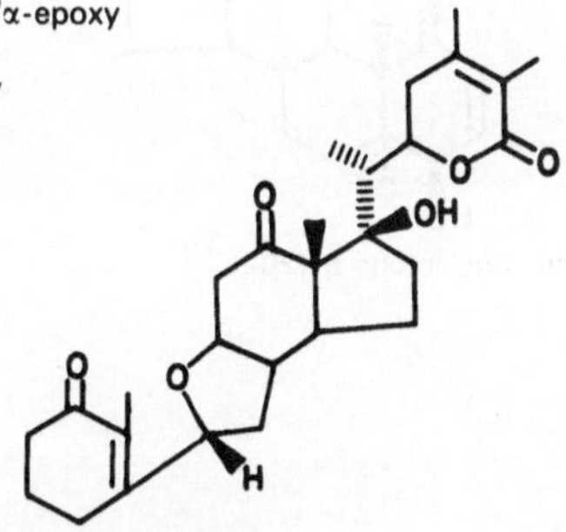

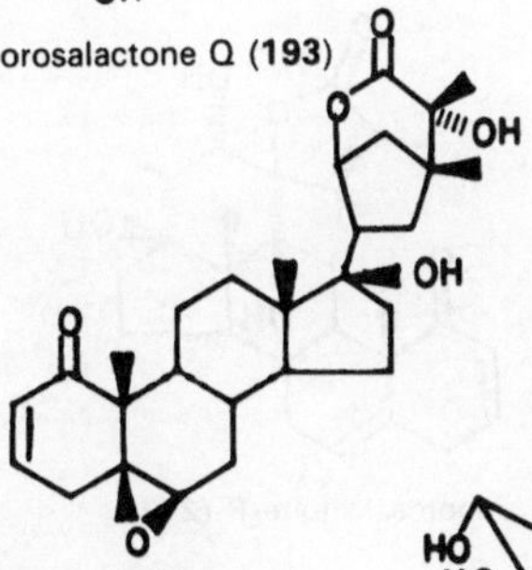

Acnistin A (**195**)
Acnistin E (**196**): 4β-OH

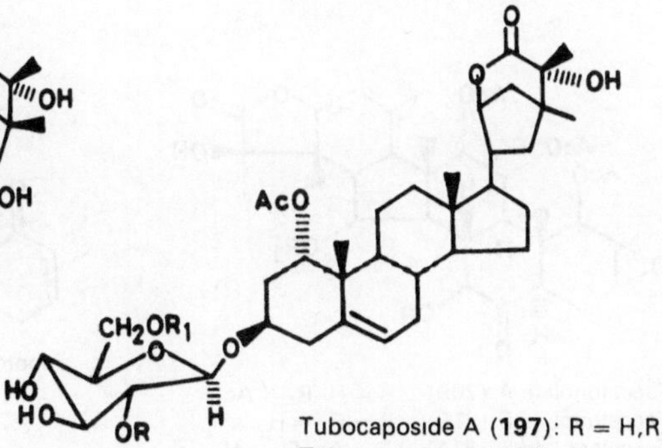

Tubocaposide A (**197**): R = H,R$_1$ = Gl
Tubocaposide B (**198**). R = R$_1$ = Gl

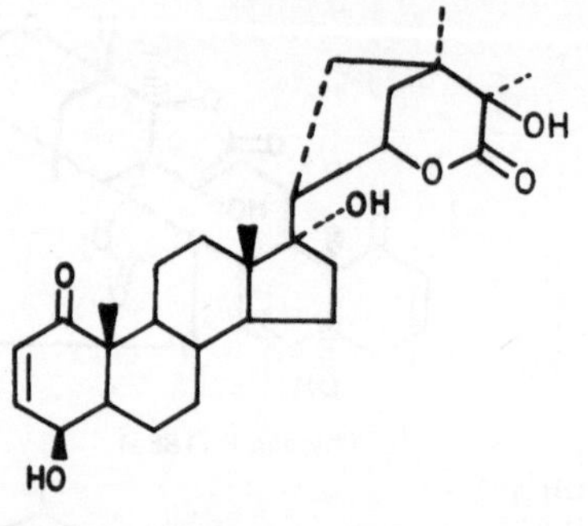

Withajardine A (**199**)· 14α-OH, Δ⁵

Withajardine B (**200**)· 14β-OH, 5β, 6β-epoxy

(Structure revised,
   see Addendum)

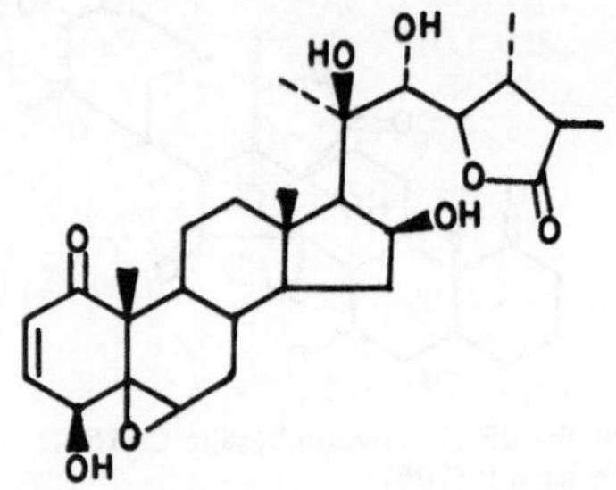

Ixocarpalactone A (**201**)

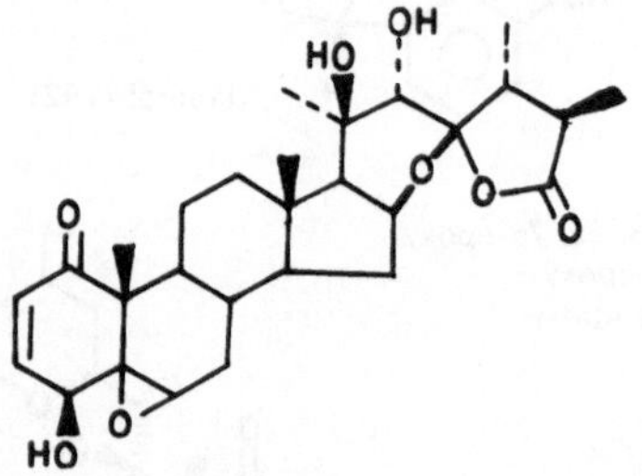

Ixocarpalactone B (**202**)

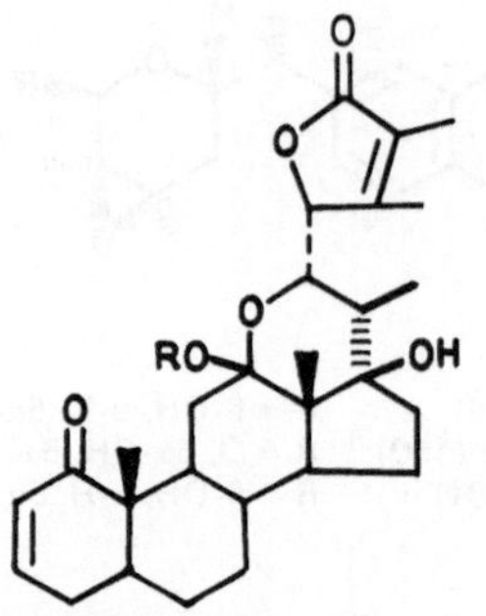

Jaborochlorodiol (**203**)·	R = H, 6α-Cl, Δ⁴
Jaborochlorotriol (**204**)	5β-OH, 6α-Cl, R = H
Jaborolone (**205**):	R = H, 5α-OH, 6-oxo
Jaborotetrol (**206**):	R = H, 5α-OH, 6β-OH
Trechonolide A (**207**):	R = H, 5β, 6β-epoxy
Trechonolide B (**208**):	R = Me, 5β, 6β-epoxy

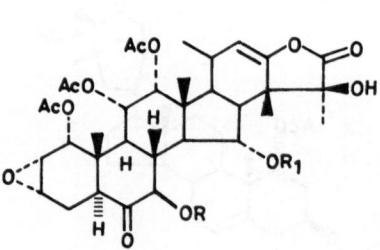

Taccalonolide A (**209**)· R = H, R₁ = Ac
Taccalonolide B (**210**): R = R₁ = H
Taccalonolide D (**211**)   R = Ac, R₁ = H

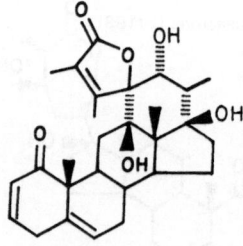

Jaborosalactone-P (**212**)

References pp. 93–106

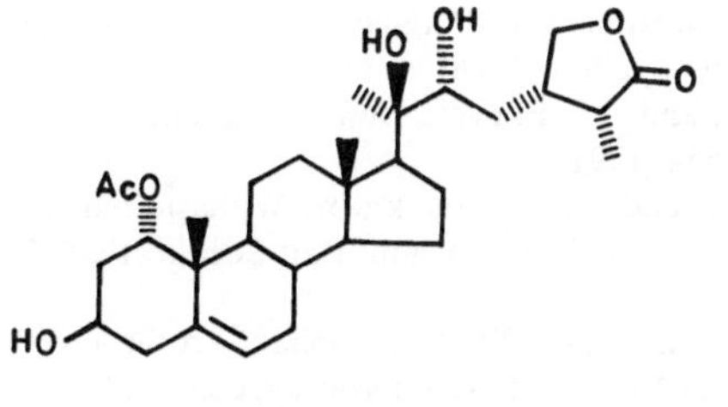

Perulactone (**213**)

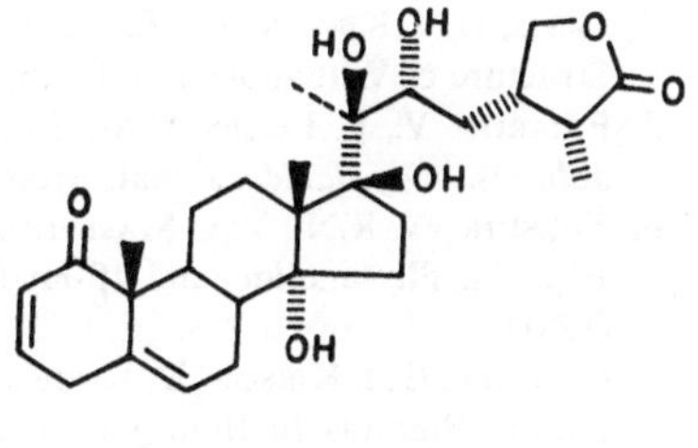

Perulactone B (**214**)

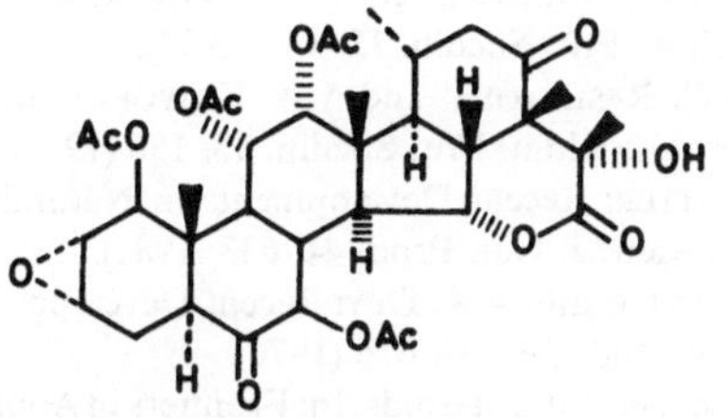

Taccalonolide C (**215**)

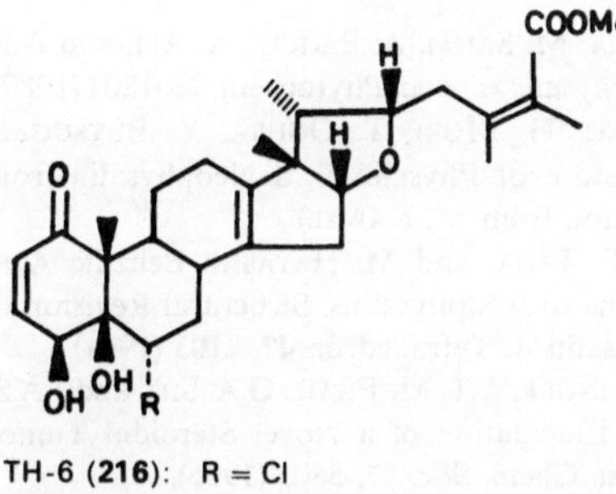

TH-6 (**216**): R = Cl
TH-12 (**217**): R = OH

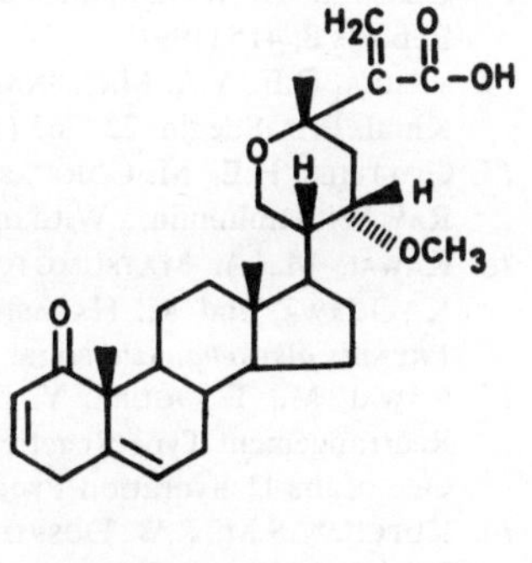

Dutumelin (**218**)

Acknowledgement

The work was supported partly by a grant from CSIR, New Delhi.

References

1. KURUP, P.A.: Antibiotic Principle of the Leaves of *Withania somnifera*. Current Sci. (India) **25**, 57 (1956).

2. LAVIE, D., E. GLOTTER, and Y. SHVO: Constituents of *Withania somnifera* Dun. The Structure of Withaferin A. J. Chem. Soc. **1965**, 7517.

3. BUDHIRAJA, R.D., and S. SUDHIR: Review of Biological Activity of Withanolides. J. Scient. Ind. Res. **46**, 488 (1987).

4. LAVIE, D., I. KIRSON, and E. GLOTTER: Constituents of *Withania somnifera* Dun. The Structure of Withanolide D. Israel J. Chem. **6**, 671 (1968).
5. FAJARDO, V., F. PODESTA, M. SHAMMA, and A.J. FREYER: New Withanolides from *Jaborosa magellanica*. J. Nat. Prod. **54**, 554 (1991).
6. TURSUNOVA, R.N., V.A. MASLENNIKOVA, and N.K. ABUBAKIROV: Withasteroids of *Physalis*. Physanolide, and 4β-Hydroxy-withanolide E. Khim. Prir. Soedin. **17**, 187 (1981).
7. GLOTTER, E., I. KIRSON, D. LAVIE, and A. ABRAHAM: The Withanolides – A Group of Natural Steroids. In: Bioorganic Chemistry Vol. 2 (E.E. VAN TAMELEN, ed.), pp. 57–95 New York: Academic Press, 1978.
8. TURSUNOVA, R.N., A. MASLENNIKOVA, and N.K. ABUBAKIROV: Withanolides in the Vegetable Kingdom. Khim. Prir. Soedin. **13**, 147 (1977).
9. KAMERNITSKII, A.V., I.G. RESHETOVA, and V.A. KRIVORUCHKO: Withanolides – A New Type of Phytosteroids. Khim. Prir. Soedin. **13**, 156 (1977).
10. KIRSON, I., and E. GLOTTER: Recent Developments in Naturally Occurring Ergostane-type Steroids. A Review. J. Nat. Prod. **44**, 633 (1981).
11. KUNDU, A.B., A. MUKHERJEE, and A.K. DEY: Recent Developments in the Chemistry of Withanolides. J. Scient. Ind. Res. **35**, 616 (1976).
12. RAY, A.B.: Recent Progress in Withasteroids. In: Frontiers in Applied Chemistry. (A.K. BISWAS, ed.). New Delhi: Narosa, 1989.
13. GLOTTER, E.: Withanolides and Related Ergostane-type Steroids. Natural Product Reports **8**, 415 (1991).
14. VASINA, O.E., V.A. MASLENNIKOVA, and N.K. ABUBAKIROV: *Physalis* Withasteroids. Khim. Prir. Soedin. **22**, 263 (1986).
15. GOTTLIEB, H.E., M. COJOCARU, S.C. SINHA, M. SAHAI, A. BAGCHI, A. ALI, and A.B. RAY: Withaminimin, a Withanolide from *Physalis minima*. Phytochem. **26**, 1801 (1987).
16. KAWAI, M., A. MATSUMOTO, B. MAKINO, H. MORI, T. OGURA, Y. BUTSUGAN, K. OGAWA, and M. HAYASHI: The Structure of Physalin P, a Neophysalin from *Physalis alkekengi*. (Personal communication from M. KAWAI.)
17. KAWAI, M., T. OGURA, Y. BUTSUGAN, T. TAGA, and M. HAYASHI: Benzilic Acid Rearrangement Type Reaction of Physalins to Neophysalins. Structural Revision of One of the Dehydration Products of Physalin A. Tetrahedron **47**, 2103 (1991).
18. KUPCHAN, S.M., R.W. DOSKOTCH, P. BOLLINGER, A.T. MCPHAIL, G.A. SIM, and J.A.S. RENAULD: The Isolation and Structural Elucidation of a Novel Steroidal Tumor Inhibitor from *Acnistus arborescens*. J. Am. Chem. Soc. **87**, 5805 (1965).
19. KUPCHAN, S.M., W.K. ANDERSON, P. BOLLINGER, R.W. DOSKOTCH, R.M. SMITH, J.A.S. RENAULD, R.H.K. SCHNOES, A.L. BURLINGAME, and D.R. SMITH: Tumor Inhibitors: Active Principles of *Acnistus arborescens*. Isolation and Structural and Spectral Studies of Withaferin A and Withacnistin. J. Org. Chem. **34**, 3858 (1969).
20. BARATA, L., W.B. MORS, I. KIRSON, and D. LAVIE: New Withanolide from *Acnistus arborescens* from the State of Guanabara, Brazil. An. Acad. Brasil. Cienc. **42**, 401 (1970).
21. CONTA, A.G., S.M. ALBONICO, and H.R. JULIANI: Isolation of Withaferin from Some Argentinian Species of *Acnistus* and *Dunalia*. Ann. Asoc. Quim. Argent. **59**, 373 (1971).
22. KIRSON, I., D. LAVIE, S.M. ALBONICO, and H.R. JULIANI: The Withanolides of *Acnistus australis*. Tetrahedron **26**, 5062 (1970).
23. BUKOVITS, G.J., and E.G. GROS: Acnistoferin, A New Withanolide from *Acnistus breviflorus*. Phytochem. **18**, 1237 (1979).

24. BUKOVITS, G.J., and E.G. GROS: Phytochemical Studies on *Acnistus breviflorus* Withanolides. An. Asoc. Quim. Argent. **69**, 7 (1981).
25. BURTON, G., A.S. VELEIRO, and E.G. GROS: Reversed Phase Chromatographic Separation of Withanolides from *Acnistus breviflorus*. J. Chromatogr. **315**, 435 (1984).
26. MONTEGUDO, E.S., A.S. VELEIRO, G. BURTON, and E.G. GROS: Biosynthesis of Withanolides in *Acnistus breviflorus*. Chemical Degradation of ^{14}C-Labeled Jaborosalactone A, and Withaferin A. Z. Naturforsch. **B42**, 1471 (1987).
27. a. NITTALA, S.S., and D. LAVIE: Withanolides of *Acnistus breviflorus*. Phytochem. **20**, 2735 (1981). b. VELEIRO, A.S., G. BURTON, E.G. GROS, and G. EDUARDO: 2,3-Dihydrojaborosalactone A, a Withanolide of *Acnistus breviflorus*. Phytochem. **24**, 1799 (1985).
28. VELEIRO, A.S., G. BURTON, and E.G. GROS: Biosynthesis of Withanolides in *Acnistus breviflorus*. Phytochem. **24**, 2263 (1985).
29. VELEIRO, A.S., G. BURTON, and E.G. GROS: Biosynthesis of Withanolides in *Acnistus breviflorus*; Biogenetic Relationships Among Main Withanolides. Phytochem. **24**, 2573 (1985).
30. USUBILLAGA, A., G. DECASTELLANO, V. ZABEL, and W.H. WATSON: Acnistins, a New Class of Steroidal Lactones from *Acnistus ramiflorum* Miers, X-Ray Structure of Acnistin E. J. Chem. Soc. Chem. Commun. **1980**, 854.
31. MANICKAM, M., S.C. SINHA, A. SINHA-BAGCHI, M. GUPTA, and A.B. RAY: Withanolides of *Datura fastuosa* Leaves. Phytochem. **34**, 868 (1993).
32. EVANS, W.C., R.J. GROUT, and M.L.K. MENSAH: Withanolides of *Datura* spp. and Hybrids. Phytochem. **23**, 1717 (1984).
33. OSHIMA, Y., A. BAGCHI, H. HIKINO, S.C. SINHA, M. SAHAI, and A.B. RAY: Withametelin, a Hexacyclic Withanolide of *Datura metel*. Tetrahedron Letters **28**, 2025 (1987).
34. SIDDIQUI, S., N. SULTANA, S.S. AHMAD, and S.I. HAIDER: A Novel Withanolide from *Datura metel*. Phytochem. **26**, 2641 (1987).
35. SIDDIQUI, S., S.S. AHMAD, and T. MAHMOOD: Datumelin – A New Withanolide from *Datura metel* L. Pakistan J. Sci. Ind. Res. **30**, 567 (1987).
36. SHINGU, K., T. KAJIMOTO, Y. FURUSAWA, and T. NOHARA: The Structure of Daturametelin A and B. Chem. Pharm. Bull. **35**, 4359 (1987).
37. SINHA, S.C., S. KUNDU, R. MAURYA, A.B. RAY, Y. OSHIMA, and H. HIKINO: Structures of Withametelin and Isowithametelin, Withanolides of *Datura metel* Leaves. Tetrahedron **45**, 2165 (1989).
38. KUNDU, S., S.C. SINHA, A. BAGCHI, and A.B. RAY: Secowithametelin, a Withanolide of *Datura metel* Leaves. Phytochem. **28**, 1769 (1989).
39. GUPTA, M., A. BAGCHI, and A.B. RAY: Additional Withanolides of *Datura metel*. J. Nat. Prod. **54**, 599 (1991).
40. GUPTA, M., M. MANICKAM, S.C. SINHA, A. SINHA-BAGCHI, and A.B. RAY: Withanolides of *Datura metel*. Phytochem. **31**, 2423 (1992).
41. SHINGU, K., Y. FURUSAWA, N. MARUBAYASHI, I. UEDA, S. YAHARA, and T. NOHARA: The Structure of Daturametelin D. Chem. Pharm. Bull. **38**, 2866 (1990).
42. SHINGU, K., Y. FURUSAWA, and T. NOHARA: New Withanolides, Daturametelins C, D, E, F and G-Ac from *Datura metel* L. Chem. Pharm. Bull. **37**, 2132 (1989).
43. MAHMOOD, T., S.S. AHMAD, and A. FAZAL: A New Withanolide Datumetelin from the Leaves of *Datura metel*. J. Indian Chem. Soc. **65**, 526 (1988).
44. MAHMOOD, T., S.S. AHMAD, and A. FAZAL: A New Withanolide Datumetelin from the Leaves of *Datura metel*. Planta Med. **54**, 468 (1988).

45. MAHMOOD, T., S.S. AHMAD, and S. SIDDIQUI: Daturilinol – A New Withanolide from the Leaves of *Datura metel*. Heterocycles **27**, 101 (1988).
46. JAHROMI, M.A.F., M. MANICKAM, Y. OSHIMA, S. HATAKEYAMA, M. GUPTA, and A.B. RAY: Withametelins F and G, Two New Withanolides of *Datura metel*. J. Chem. Res(s). **1993**, 234.
47. a. DHAR, K.L., and M.L. RAINA: A Novel Withanolide from *Datura quercifolia*. Phytochem. **12**, 476 (1973). b. DHAR, K.L., and A.K. KALLA: A 12-Oxowithanolide from *Datura quercifolia*. Phytochem. **15**, 339 (1976).
48. QURISHI, M.A., K.L. DHAR, and C.K. ATAL: A Novel Withanolide from *Datura quercifolia*. Phytochem. **18**, 283 (1979).
49. QURISHI, M.A., A.K. KALLA, and K.L. DHAR: A Novel Withanolide from *Datura quercifolia*. Phytochem. **18**, 1756 (1979).
50. KALLA, A.K., M.L. RAINA, K.L. DHAR, M.A. QURISHI, and G. SNATZKE: Revised Structures of Daturalactone and 12-Oxowithanolide. Phytochem. **18**, 637 (1979).
51. TURSUNOVA, R.N., V.A. MASLENNIKOVA, and N.K. ABUBAKIROV: Withanolides of *Datura stramonium* II. Withastramonolide. Khim. Prir. Soedin. **14**, 91 (1978).
52. SHINGU, K., S. YAHARA, and T. NOHARA: New Withanolides, Daturataturins A and B from *Datura tatura*. Chem. Pharm. Bull. **38**, 3485 (1990).
53. ECHEVERRI, F., J. VEGA, F. TORRES, C. PELAEZ, G. CARDONA, W. QUINONES, A.G. GONZALEZ, J.G. LUIS, A. RAVELO: Structure of Withajardin B, a Withanolide from *Deprea orinocensis*. Spectroscopy **7**, 345 (1989).
54. ECHEVERRI, F., J. VEGA, F. TORRES, A.G. GONZALEZ, R. ANGEL, and J.G. LUIS: Isolation and Structure Determination of Withajardin from the Leaves of *Deprea procumbens*. Spectroscopy **6**, 137 (1988).
55. ADAM, G., N.Q. CHIEN, and N.H. KHOI: Dunawithanines A and B, the First Withanolide Glycosides from *Dunalia australis*. Phytochem. **23**, 2293 (1984).
56. ADAM, G., N.Q. CHIEN, and N.H. KHOI: Dunawithanine A and B, the First Plant Withanolide Glycosides from *Dunalia australis*. Proc. 11th Int. Conf. Chem. Biotechnol. of Biol. Act. Nat. Prod. **3**, 191 (1981).
57. ADAM, G., and M. HESSE: New C_{28} Steroid Lactone of the Withaferin Type from *Dunalia australis*. Tetrahedron Letters **17**, 1199 (1971).
58. LISCHEWSKI, M., N.T.B. HANG, A. PORZEL, G. ADAM, G. MASSIOT, and C. LAVAUD: Withanolides from *Dunalia australis*. Phytochem. **30**, 4184 (1991).
59. LISCHEWSKI, M., N.T.B. HANG, A. PORZEL, G. ADAM, G. MASSIOT, and C. LAVAUD: Withanolide Glycosides from *Dunalia australis*. Phytochem. **31**, 939 (1992).
60. RECK, G., N.H. KHOI, and G. ADAM: (20R,22R)-1α,3β,20-Trihydroxy-witha-5,24-dienolide Monohydrate (Dunawithagenine Monohydrate). Cryst. Struct. Comm. **11**, 355 (1982).
61. ALFONSO, D., I. KAPETANDIS, and G. BERNARDINELLI: Iochromolide: A New Acetylated Withanolide from *Iochroma coccineum*. J. Nat. Prod. **54**, 1576 (1991).
62. RAFFAUF, R.F., M.J. SHEMLUCK, and P.W. LE QUESNE: The Withanolides of *Iochroma fuchsioides*. J. Nat. Prod. **54**, 1601 (1991).
63. MONTEAGUDO, E.S., G. BURTON, C.M. GONZALEZ, J.C. OBERTI, and E.G. GROS: 14β,17β-Dihydroxywithanolides from *Jaborosa bergii*. Phytochem. **27**, 3925 (1988).
64. TSCHESCHE, R., H. SCHWANG, and G. LEGLER: Inhaltstoffe aus *Jaborosa integrifolia* Lam. Isolierung und chemische Charakterisierung der Jaborosalactone A und B. Tetrahedron **22**, 1121 (1966).
65. TSCHESCHE, R., H. SCHWANG, H.W. FEHLHABER, and G. SNATZKE: Inhaltstoffe aus *Jaborosa integrifolia* Lam. Strukturermittlung der Jaborosalactone A und B. Tetrahedron **22**, 1129 (1966).

66. TSCHESCHE, R., M. BAUMGARTH, and P. WEIZEL: Weitere Inhaltstoffe aus *Jaborosa integrifolia* Lam. – Zur Struktur der Jaborosalactone C, D und E. Tetrahedron **24**, 5169 (1968).

67. TSCHESCHE, R., K. ANNEN, and P. WEIZEL: Die Struktur des Jaborosalactons F. Tetrahedron **28**, 1909 (1972).

68. TSCHESCHE, R., K. ANNEN, and P. WEIZEL: Zur Konfiguration der Jaborosalactone, ein neuer Abbauweg für Withanolide. Chem. Ber. **104**, 3556 (1971).

69. LAVIE, D., H.E. GOTTLIEB, M.J. PESTCHANKER, and O.S. GIORDANO: Jaborosalactone L, a Withanolide from *Jaborosa leucotricha*. Phytochem. **25**, 1765 (1986).

70. MONTEAGUDO, E.S., G. BURTON, E.G. GROS, C.M. GONZALEZ, and J.C. OBERTI: A 19-Hydroxywithanolide from *Jaborosa leucotricha*. Phytochem. **28**, 2514 (1989).

71. VELEIRO, A.S., C.E. TROCCA, G. BURTON, and J.C. OBERTI: A Phenolic Withanolide from *Jaborosa leucotricha*. Phytochem. **31**, 2250 (1992).

72. FAJARDO, V., A.J. FREYER, R.D. MINARD, and M. SHAMMA: (+)-Jaborol, an Unusual Phenolic Withanolide from *Jaborosa magellanica*. Tetrahedron **43**, 3875 (1987).

73. FAJARDO, V., F. PODESTA, M. SHAMMA, and A.J. FREYER: New Withanolides from *Jaborosa magellanica*. J. Nat. Prod. **54**, 554 (1991).

74. PARVEZ, M., V. FAJARDO, and M. SHAMMA: (+)-Jaborosalactone M, a Hemiketal Withanolide from *Jaborosa magellanica*. Acta Cryst. **C44**, 553 (1988).

75. PARVEZ, M., V. FAJARDO, and M. SHAMMA: (+)-Jaborol, a B-Secowithanolide from *Jaborosa magellanica*. Acta Cryst. **C44**, 556 (1988).

76. MONTEAGUDO, E.S., J.C. OBERTI, E.G. GROS, and G. BURTON: A Spiranic Withanolide from *Jaborosa odonelliana*. Phytochem. **29**, 933 (1990).

77. HÄNSEL, R., J.T. HUANG, and D. ROSENBERG: Two Withanolides from *Lycium chinense*. Arch. Pharm. **308**, 653 (1975).

78. HÄNSEL, R., and J.T. HUANG: *Lycium chinense*. Semiquantitative Determination of Withanolides. Arch. Pharm. **310**, 35 (1977).

79. CHRISTEN, P., and I. KAPETANIDIS: Isolation of a Withanolide from the Leaves of *Lycium halimifolium*. Pharm. Acta Helv. **63**, 263 (1988).

80. ADAM, G., P.D. SETHI, and S.S. SUBRAMANIAN: Investigations on Nicandrenone – A Withanolide-like Aromatic Plant Steroid from *Nicandra physaloides*. Pharmazie **31**, 647 (1976).

81. ANDREWS-SMITH, W., H.K. GILL, R.W. SMITH, and D.A. WHITING: Stages in the Biosynthesis of the Epoxy-Lactol Side Chain of Nic-1, Insect-Antifeedant Steroid of *Nicandra physaloides*. J. Chem. Soc. Perkin I **1991**, 291.

82. BAGCHI, A., M. SAHAI, and A.B. RAY: C_{28}-Steroidal Lactones of the Seeds of *Nicandra physaloides*. J. Indian Chem. Soc. **61**, 173 (1984).

83. BAGCHI, A., P. NEOGI, M. SAHAI, A.B. RAY, Y. OSHIMA, and H. HIKINO: Withaperuvin E and Nicandrin B, Withanolides from *Physalis peruviana* and *Nicandra physaloides*. Phytochem. **23**, 853 (1984).

84. BATES, R.B., and D.J. ECKERT: Nicandrenone, an Insecticidal Plant Steroid Derivative with Ring D Aromatic. J. Am. Chem. Soc. **94**, 8258 (1972).

85. BATES, R.B., and S.R. MOREHEAD: Structure of Nic-2, a Major Steroidal Constituent of the Insect-Repellant Plant *Nicandra physaloides*. J. Chem. Soc. Chem. Commun. **1974**, 125.

86. BEGLEY, M.J., L. CROMBIE, P.J. HAM, and D.A. WHITING: Constitution of Four Novel Methyl Steroid Relatives (Ring-D Aromatic) from the Insect Repellant Plant *Nicandra physaloides*; X-ray Analysis of Nic-10. J. Chem. Soc. Chem. Commun. **1972**, 1250.

87. BEGLEY, M.J., L. CROMBIE, P.J. HAM, and D.A. WHITING: Structures of Three Oxygenated 24-Methyl Steroids (Nic-3, -7 and -11) from the Insect Repellant Plant

Nicandra physaloides (Solanaceae); X-Ray Analysis of Nic-3 Acetate and Nic-11 Ethyl Ether. J. Chem. Soc. Perkin I **1976**, 296.

88. BEGLEY, M.J., L. CROMBIE, P.J. HAM, and D.A. WHITING: A New Class of Natural Steroids, with Ring D Aromatic, from *Nicandra physaloides* (Solanaceae). X-Ray Analysis of Nic-10, and the Structures of Nic-1 (Nicandrenone), -12, and -17. J. Chem. Soc. Perkin I **1976**, 304.

89. GIZYCKI, F.V., and G. KOTITSCHKE: Nicandrin, a Bitter Principle Isolated from *N. physaloides*. Arch. Pharm. **284**, 129 (1951).

90. GILL, H.K., R.W. SMITH, and D.A. WHITING: Biosynthesis of the Insect Antifeedant Steroid Nic-1: Origins of the Aromatic Ring-D. J. Chem. Soc. Chem. Commun. **1986**, 1457.

91. GLOTTER, E., P. KRINSKY, and I. KIRSON: Reactions of Nic-1 (Nicandrenone), a Naturally Occurring Ring D-Aromatic Steroid. J. Chem. Soc. Perkin I **1976**, 669.

92. GLOTTER, E., I. KIRSON, A. ABRAHAM, and P. KRINSKY: Nic-1-lactone, a Minor Steroidal Constituent of *Nicandra physaloides*. (Solanaceae). Phytochem. **15**, 1317 (1976).

93. GUNASEKARA, S.P., G.A. CORDELL, and N.R. FARNSWORTH: Plant Anticancer Agents: Constituents of *Nicandra physaloides*. Planta Med. **43**, 389 (1981).

94. NALBANDOV, O., R.T. YAMAMOTO, and G.S. FRAENKEL: Nicandrenone, a New Compound with Insecticidal Properties, Isolated from *Nicandra physaloides*. J. Agric. Food Chem. **12**, 55 (1964).

95. SUBRAMANIAN, S.S., P.D. SETHI, and G. ADAM: Nicandrenone, a New Compound with Insecticidal Properties, Isolated from *Nicandra physaloides*. Indian J. Pharm. **35**, 123 (1973).

96. BEGLEY, M.J., L. CROMBIE, P.J. HAM, and D.A. WHITING: X-Ray Analysis of Nic-11 from *Nicandra physaloides*, a 17-Spiro-bicyclo Methylsteroid. J. Chem. Soc. Chem. Commun. **1973**, 821.

97. KIRSON, I., D. LAVIE, S.S. SUBRAMANIAN, P.D. SETHI, and E. GLOTTER: Withanicandrin, a Ring C Substituted Withanolide from *Nicandra physaloides*. J. Chem. Soc. Perkin I **1972**, 2109.

98. GILL, H.K., R.W. SMITH, and D.A. WHITING: Biosynthesis of the Nicandrenoids: Stages in the Oxidative Elaboration of the Side Chain and the Fate of the Diastereotopic 25-Methyl Groups of 24-Methylene Cholesterol. J. Chem. Soc. Chem. Commun. **1986**, 1459.

99. KIRSON, I., H.E. GOTTLIEB, M. GREENBERG, and E. GLOTTER: Nicalbins A and B, Two Novel Ergostane-Type Steroids from *Nicandra physaloides* var. *albiflora*. J. Chem. Res. **1980**, (S) 69, (M) 1031.

100. MASLENNIKOVA, V.A., R.N. TURSUNOVA, and N.K. ABUBAKIROV: Withanolides of Physalis. Physalactone. Khim. Prir. Soedin **12**, 531 (1977).

101. KAWAI, M., T. TAGA, K. OSAKI, and T. MATSUURA: Bitter Principles of *Physalis alkekengi* var. *francheti*, X-Ray Analysis of 5-α-Acetoxy-6β-bromohexahydrophysalin A. Tetrahedron Letters **1969**, 1087.

102. KAWAI, M., and T. MATSUURA: The Structure of Physalin C, a Bitter Principle of *Physalis alkekengi* var. *francheti*. Tetrahedron **26**, 1743 (1970).

103. KAWAI, M., T. MATSUURA, S. KYUNO, H. MATSUKI, M. TAKENAKA, T. KATSUOKA, Y. BUTSUGAN, and K. SAITO: A New Physalin from *Physalis alkekengi*. Structure of Physalin L. Phytochem. **26**, 3313 (1987).

104. KAWAI, M., T. OGURA, A. MATSUMOTO, Y. BUTSUGAN, and M. HAYASHI: Isolation of (25S)-25,27-Dihydrophysalin A from *Physalis alkekengi* var. *francheti*. Chem. Express **4**, 97 (1989).

105. KAWAI, M., T. OGURA, B. MAKINO, A. MATSUMOTO, H. YAMAMURA, Y. BUTSUGAN, and M. HAYASHI: Physalins N and O from *Physalis alkekengi*. Phytochem. **31**, 4299 (1992).

106. KAWAI, M., T. OGURA, M. NAKANISHI, T. MATSUURA, Y. BUTSUGAN, Y. MORI, K. HARADA, and M. SUZUKI: Structure of Physalin M Isolated from *Physalis alkekengi* var. *francheti*. Bull. Chem. Soc. Jpn. **61**, 2696 (1988).

107. MATSUURA, T., and M. KAWAI: Bitter Principles of *Physalis alkekengi* var. *francheti*. Structure of Physalin B. Tetrahedron Letters **1969**, 1765.

108. MATSUURA, T., M. KAWAI, R. NAKASHIMA, and Y. BUTSUGAN: Bitter Principles of *Physalis alkekengi* var. *francheti*. Structure of Physalin A. Tetrahedron Letters **1969**, 1083.

109. MATSUURA, T., M. KAWAI, R. NAKASHIMA, and Y. BUTSUGAN: Structure of Physalins A and B, 13,14-Seco-16,24-cyclo-steroids from *Physalis alkekengi* var. *francheti*. J. Chem. Soc. (C) **1970**, 664.

110. ROW, L.R., N.S. SARMA, T. MATSUURA, and R. NAKASHIMA: Physalins E and H, New Physalins from *Physalis angulata* and *P. lancifolia*. Phytochem. **17**, 1641 (1978).

111. ROW, L.R., N.S. SARMA, K.S. REDDY, T. MATSUURA, and R. NAKASHIMA: The Structure of Physalins F and J from *Physalis angulata* and *P. lancifolia*. Phytochem. **17**, 1647 (1978).

112. ROW, L.R., K.S. REDDY, N.S. SARMA, T. MATSUURA, and R. NAKASHIMA: New Physalins from *Physalis angulata* and *Physalis lancifolia*. Structure and Reactions of Physalins D, I, G and K. Phytochem. **19**, 1175 (1980).

113. CHEN, C.M., Z.T. CHEN, C.H. HSIEH, W.S. LI, and S.Y. WEN: Withangulatin A, a New Withanolide from *Physalis angulata*. Heterocycles **31**, 1371 (1990).

114. SHINGU, K., N. MARUBAYASHI, I. UEDA, S. YAHARA, and T. NOHARA: Physagulin C, a New Withanolide from *Physalis angulata* L. Chem. Pharm. Bull. Japan **39**, 1591 (1991).

115. VASINA, O.E., V.A. MASLENNIKOVA, D. ABDULLAEV, and N.K. ABUBAKIROV: Withasteroids of *Physalis*. 14-α-Hydroxy-ixocarpanolide. Khim. Prir. Soedin. **22**, 596 (1986).

116. VASINA, O.E., N.D. ABDULLAEV, and N.K. ABUBAKIROV: Isolation of Vamonolide from Epigeal Part of *Physalis angulata* L. Chem. Nat. Compd. **23**, 712 (1987).

117. VASINA, O.E., N.D. ABDULLAEV, and N.K. ABUBAKIROV: Withasteroids of *Physalis*. Physangulide, the First Natural 22S-Withasteroid. Khim. Prir. Soedin. **26**, 366 (1990) (Chem. Abstr. 58881b, 1991).

118. KIRSON, I., and E. GLOTTER: Unpublished results reported in J. Chem. Res. (M) 2134 (1980).

119. KIRSON, I., A. COHEN, M. GREENBERG, H.E. GOTTLIEB, P. VARENNE, and A. ABRAHAM: Ixocarpalactones A and B, Two Unusual Naturally Occurring Steroids of the Ergostane Type. J. Chem. Res. **1979**, (S) 103, (M) 1178.

120. ABDULLAEV, N.D., O.E. VASINA, V.A. MASLENNIKOVA, and N.K. ABUBAKIROV: Withasteroids of *Physalis*. Study of Proton and Carbon-13 NMR Spectra of the Withasteroids, Ixocarpalactone A and Ixocarpanolide. Khim. Prir. Soedin. **17**, 326 (1986).

121. GLOTTER, E., I. KIRSON, A. ABRAHAM, P.D. SETHI, and S.S. SUBRAMANIAN: Steroidal Constituents of *Physalis minima*. J. Chem. Soc. Perkin I **1975**, 1370.

122. KIRSON, I., Z. ZARETSKII, and E. GLOTTER: Withaphysalin C, a Naturally Occurring 13,14-Secosteroid. J. Chem. Soc. Perkin I **1976**, 1244.

123. MULCHANDANI, N.B., S.S. IYER, and L.P. BADHEKA: Physalin D, a New 13,14-Seco-16,24-cyclo-steroid from *Physalis minima*. Planta Medica **37**, 268 (1979).

124. ALLURI, R.R., R.J. MILLER, W.H. SHELVER, and S.K.W. KHALIL: Dihydroxyphysalin B, a New Physalin from *Physalis minima* Leaves. Lloydia **39**, 405 (1976).

125. PURUSHOTHAMAN, K.K., and V. SARADA: A New Physalin from *Physalis minima*. Indian J. Chem. **20B**, 89 (1981).

126. SHIPAHIMALANI, A.T., V.A. BAPAT, P.S. RAO, and M.S. CHADHA: Biosynthetic Potential of Cultured Tissues and Regenerated Plants of *Physalis minima*. J. Nat. Prod. **44**, 114 (1981).

127. SAHAI, M., and I. KIRSON: Withaphysalin D, a New Withaphysalin from *Physalis minima* Linn. var. *indica*. J. Nat. Prod. **47**, 527 (1984).

128. SINHA, S.C., and A.B. RAY: Chemical Constituents of *Physalis minima* var. *indica*. J. Indian Chem. Soc. **65**, 740 (1988).

129. SINHA, S.C., A.B. RAY, Y. OSHIMA, A. BAGCHI, and H. HIKINO: Withaphysalin E, a Withanolide of *Physalis minima* var. *indica*. Phytochem. **26**, 2115 (1987).

130. GUPTA, M., A. BAGCHI, S.C. SINHA, M. SAHAI, and A.B. RAY: Fruit Constituents of *Physalis minima* var. *indica* and One-step Conversion of Physalin B to 6-Epiphysalin G Acetate. J. Indian Chem. Soc. **67**, 597 (1990).

131. ALI, A., M. SAHAI, A.B. RAY, and D.J. SLATKIN: Physalolactone C, a New Withanolide from *Physalis peruviana*. J. Nat. Prod. **47**, 648 (1984).

132. BHATTACHARYA, T.C., T.N.C. VEDANTAM, S.S. SUBRAMANIAN, and I. KIRSON: Withanolide S from *Physalis peruviana*. Indian J. Pharm. Sci. **40**, 177 (1978).

133. EGUCHI, T., Y. FUJIMOTO, K. KAKINUMA, N. IKEKAWA, M. SAHAI, M.P. VERMA, and Y.K. GUPTA: 23-Hydroxyphysalolactone, a New Withanolide with 23-Hydroxyl Group from *Physalis peruviana* (Solanaceae). Chem. Pharm. Bull. **36**, 2897 (1988).

134. FROLOW, F., A.B. RAY, M. SAHAI, E. GLOTTER, H.E. GOTTLIEB, and I. KIRSON: Withaperuvin and 4-Deoxyphysalolactone – Two Ergostane-Type Steroids from *Physalis peruviana* (Solanaceae). J. Chem. Soc. Perkin I **1981**, 1029.

135. GLOTTER, E., A. ABRAHAM, G. GÜNZBERG, and I. KIRSON: Naturally Occurring Steroidal Lactones with 17α-Oriented Side Chain, Structure of Withanolide E and Related Compounds. J. Chem. Soc. Perkin I **1977**, 341.

136. GOTTLIEB, H.E., I. KIRSON, E. GLOTTER, A.B. RAY, M. SAHAI, and A. ALI: Perulactone, a New Ergostane-Type Steroid from *Physalis peruviana* (Solanaceae). J. Chem. Soc. Perkin I **1980**, 2700.

137. KIRSON, I., A. ABRAHAM, P.D. SETHI, S.S. SUBRAMANIAN, and E. GLOTTER: 4β-Hydroxywithanolide E, a New Natural Steroid with a 17α-Oriented Side Chain. Phytochem. **15**, 340 (1976).

138. KIRSON, I., E. GLOTTER, A.B. RAY, A. ALI, H.E. GOTTLIEB, and M. SAHAI: Physalolactone B-3-O-β-D-Glucopyranoside, the First Glucoside in the Withanolide Series. J. Chem. Res (S) **1983**, 120.

139. NEOGI, P., M. SAHAI, and A.B. RAY: Withaperuvins F and G, Two Withanolides of *Physalis peruviana* Roots. Phytochem. **26**, 243 (1987).

140. NEOGI, P.: Chemistry of Minor Steroidal Lactones of *Physalis peruviana* Roots. Ph.D. Thesis, Banaras Hindu University, 1985.

141. OSHIMA, Y., H. HIKINO, M. SAHAI, and A.B. RAY: Withaperuvin H. a Withanolide of *Physalis peruviana* Roots. J. Chem. Soc. Chem. Commun. **1989**, 628.

142. RAY, A.B., M. SAHAI, and B.C. DAS: Physalolactone, a New Withanolide from *Physalis peruviana*. J. Indian Chem. Soc. **55**, 1175 (1978).

143. RAY, A.B., M. SAHAI, P.L. SCHIFF, Jr., J.E. KNAPP, and D.J. SLATKIN: Physalolactone B, a Novel Withanolide of *Physalis peruviana*. Chem & Ind. **1981**, 62.

144. SAHAI, M., P. NEOGI, A.B. RAY, Y. OSHIMA, and H. HIKINO: Structures of Withaperuvins B and C, Withanolides of *Physalis peruviana*. Heterocycles **19**, 37 (1982).

145. SAHAI, M., H.E. GOTTLIEB, A.B. RAY, A. ALI, E. GLOTTER, and I. KIRSON: Perulactone B, a New Ergostane-Type Steroid with an α-Oriented Side Chain. J. Chem. Res. **1982**, (S) 346.

146. SAHAI, M., A. ALI, A.B. RAY, D.J. SLATKIN, and I. KIRSON: Withaperuvin D, a Novel Withanolide of *Physalis peruviana*. J. Chem. Res. **1983**, (S) 152.

147. SAHAI, M., and P. NEOGI: Chemical Constituents of *Physalis peruviana*. J. Indian Chem. Soc. **61**, 171 (1984).

148. SAKURAI, K., H. ISHII, S. KOBAYASHI, and T. IWAO: Isolation of 4β-Hydroxywithanolide E, a New Withanolide from *Physalis peruviana* (Solanaceae). Chem. Pharm. Bull. **24**, 1403 (1976).

149. SUBRAMANIAN, S.S., and P.D. SETHI: Bitter Principles of *P. minima* and *P. peruviana*. Curr. Sci. **40**, 85 (1971).

150. GLOTTER, E., M. SAHAI, I. KIRSON, and H.E. GOTTLIEB: Physapubenolide and Pubescenin, Two New Ergostane-Type Steroids from *Physalis pubescens* L. (Solanaceae). J. Chem. Soc. Perkin I **1985**, 2241.

151. REDDY, K.S., L.R. ROW, and T. MATSUURA: Pubescenol, a New Withanolide from *Physalis pubescence*. J. Chem. Soc. Perkin I **1985**, 419.

152. KIRSON, I., H.E., GOTTLIEB, and E. GLOTTER: Physapubescin, a New Ergostane-Type Steroid from *Physalis pubescens* L. (Solanaceae). J. Chem. Res. **1980**, (S) 125 (M) 2134.

153. ROW, L.R., K.S. REDDY, K. DHAVEJI, and T. MATSUURA: Pubescenol, a Withanolide of *Physalis pubescence*. Phytochem. **23**, 427 (1984).

154. SAHAI, M.: Pubesenolide, a New Withanolide from *Physalis pubescens* J. Nat. Prod. **48**, 474 (1985).

155. PELLETIER, S.W., G. GEBEYEHU, J. NOWACKI, and N.V. MODY: Viscosalactone A and Viscosalactone B, Two New Steroidal Lactones from *Physalis viscosa*. Heterocycles, **15**, 317 (1981).

156. ABDULLAEV, N.D., O.E. VASINA, V.A. MASLENNIKOVA, and N.K. ABUBAKIROV: Withasteroids of *Physalis*. A Study of the ^{1}H- and ^{13}C-NMR Spectra of the Withasteroids Visconolide and 28-Hydroxy Withaperuvin C. Khim. Prir. Soedin. **21**, 657 (1985).

157. ABDULLAEV, N.D., V.A. MASLENNIKOVA, R.N. TURSUNOVA, N.K. ABUBAKIROV, and M.R. YAGUDAEV: Physalis Withasteroids. 28-Hydroxywithaphysanolide. Khim. Prir. Soedin. **20**, 197 (1984).

158. MASLENNIKOVA, V.A., R.N. TURSUNOVA, K.L. SEITANIDI, and N.K. ABUBAKIROV: Physalis Withanolides. Withaphysanolide. Khim. Prir. Soedin. **16**, 214 (1980).

159. VELEIRO, A.S., J.C. OBERTI, and G. BURTON: A Ring-D Aromatic Withanolide from *Salpichroa origanifolia*. Phytochem. **31**, 935 (1992).

160. LAVIE, D., R. BESSALLE, M.J. PESTCHANKER, H.E. GOTTLIEB, F. FROLOW, and O.S. GIORDANO: Trechonolide A, a New Withanolide Type from *Trechonaetes laciniata*. Phytochem. **26**, 1791 (1987).

161. YOSHIDA, K., K. SHINGU, S. YAHARA, and T. NOHARA: A New Class of Ergostane Glycosides from *Tubocapsicum anomalum*. Tetrahedron Letters **29**, 673 (1988).

162. SHINGU, K., N. MARUBAYASHI, I. UEDA, S. YAHARA, and T. NOHARA: Two New Ergostane Derivatives from *Tubocapsicum anomalum* (Solanaceae). Chem. Pharm. Bull. **38**, 1107 (1990).

163. GONZALEZ, A.G., J.L. BRETON, and J.M. TRUJILLO: Withanolides of *Withania aristata* and *Withania frutescens*. An. Quim. **68**, 107 (1972).

164. GONZALEZ, A.G., J.L. BRETON, and J.M. TRUJILLO: *Withania* steroids. Five Steroidal Lactones of *Withania aristata*. An. Quim. **70**, 64 (1974).

165. SUBRAMANIAN, S.S., and P.D. SETHI: Withaferin A from the Roots of *Withania coagulans*. Current Sci. (India) **38**, 267 (1969).

166. Subramanian, S.S., P.D. Sethi, E. Glotter, I. Kirson, and D. Lavie: 5,20α(R)-Dihydroxy-6α,7α-epoxy-1-oxo-(5α)-witha-2,24-dienolide, a New Steroidal Lactone from *Withania coagulans*. Phytochem. **10**, 685 (1971).

167. Ramaiah, P.A., D. Lavie, R.D. Budhiraja, S. Sudhir, and K.N. Garg: Spectroscopic Studies on a Withanolide from *Withania coagulans*. Phytochem. **23**, 143 (1984).

168. Neogi, P., M. Kawai, Y. Butsugan, Y. Mori, and M. Suzuki: Withacoagin, a New Withanolide from *Withania coagulans* Roots. Bull. Chem. Soc. Jpn. **61**, 4479 (1988).

169. Velde, V.V., D. Lavie, R.D. Budhiraja, S. Sudhir, and K.N. Garg: Potential Biogenetic Precursors of Withanolides from *Withania coagulans*. Phytochem. **22**, 2253 (1983).

170. Gonzalez, A.G., J.L. Breton, and J.M. Trujillo: Steroidal Lactones of *Withania frutescens*. An. Quim. **70**, 69 (1974).

171. Modawi, B., G.M. Iskander, M.A. Karim, C.K. Fair, and E.O. Schlemper: Crystal and Molecular Structure of a New Withanolide Isomeric with Withaferin A Isolated from *Withania obtusifolia* Dun. J. Prakt. Chemie **328**, 291 (1986).

172. Abraham, A., I. Kirson, E. Glotter, and D. Lavie, A Chemotaxonomic Study of *Withania somnifera*. Phytochem. **7**, 957 (1968).

173. Abraham, A., I. Kirson, D. Lavie, and E. Glotter: The Withanolides of *Withania somnifera* Chemotypes I and II. Phytochem. **14**, 189 (1975).

174. Ascher, K.R.S., M. Eliyahu, E. Glotter, I. Kirson, and A. Abraham: Distribution of Chemotypes of *Withania somnifera* in Some Areas of Israel: Feeding Studies with *Littoralis* Larvae and Chemical Examination of Withanolide Content. Phytoparasitica **12**, 147 (1984).

175. Atta-Ur-Rahman, S.A. Jamal, M.I. Choudhary, and E. Asif: Two New Withanolides from *Withania somnifera*. Phytochem. **30**, 3824 (1991).

176. Atta-Ur-Rahman, S.A. Jamal, and M.I. Choudhary: Two New Withanolides from *Withania somnifera*. Heterocycles **34**, 689 (1992).

177. Bessalle, R., D. Lavie and F. Frolow: Withanolide Y, a Withanolide from a Hybrid of *Withania somnifera*. Phytochem. **26**, 1797 (1987).

178. Dhalla, N.S., M.S. Shastry, and C.L. Malhotra: Chemical Studies on the Leaves of *Withania somnifera*. J. Pharm. Sci. **50**, 876 (1961).

179. Eastwood, F.W., I. Kirson, D. Lavie, and A. Abraham: New Withanolides from a Cross of a South African Chemotype by Chemotype II (Israel) in *Withania somnifera*. Phytochem. **19**, 1503 (1980).

180. Ghosal, S., R. Kaur, and R.S. Srivastava: Sitoindosides IX and X, New Glycowithanolides from *Withania somnifera*. Indian J. Nat. Prod. **4**, 12 (1988).

181. Glotter, E., R. Waitman, and D. Lavie: Constituents of *Withania somnifera*: A New Steroidal Lactone, 27-Deoxy-14-hydroxy-withaferin A. J. Chem. Soc. (C) **1966**, 1765.

182. Glotter, E., I. Kirson, A. Abraham, and D. Lavie: Constituents of *Withania somnifera* Dun. The Withanolides of Chemotype III. Tetrahedron **29**, 1353 (1973).

183. Kirson, I., E. Glotter, A. Abraham, and D. Lavie: Constituents of *Withania somnifera* Dun. The Structure of Three New Withanolides. Tetrahedron **26**, 2209 (1970).

184. Kirson, I., E. Glotter, D. Lavie, and A. Abraham: Constituents of *Withania somnifera* Dun. The Withanolides of an Indian Chemotype. J. Chem. Soc. (C) **1971**, 2032.

185. Kirson, I., A. Cohen, and A. Abraham: Withanolides Q and R, Two New 23-Hydroxy Steroidal Lactones. J. Chem. Soc. Perkin I **1975**, 2136.

186. KIRSON, I., A. ABRAHAM, and D. LAVIE: Chemical Analysis of Hybrids of *Withania somnifera* L. (Dun.), Chemotypes III (Israel) by Indian I (Delhi). Israel J. Chem. **16**, 20 (1977).

187. KIRSON, I., and H.E. GOTTLIEB: 14α-Hydroxy Steroids from *Withania somnifera* (L.) Dun. (Solanaceae). J. Chem. Res. **1980**, (S) 338 (M) 4275.

188. LAVIE, D., E. GLOTTER, and Y. SHVO: Constituents of *Withania somnifera* Dun.: The Structure of Withaferin A. J. Chem. Soc. **1965**, 7517.

189. LAVIE, D., E. GLOTTER, and Y. SHVO: Constituents of *Withania somnifera* Dun.: The Side Chain of Withaferin A. J. Org. Chem. **30**, 1774 (1965).

190. LAVIE, D., G.S. GREENFIELD, and E. GLOTTER: Constituents of *Withania somnifera* Dun.: The Stereochemistry of Withaferin A. J. Chem. Soc. (C) **1966**, 1753.

191. LAVIE, D., Y. KASHMAN, and E. GLOTTER: Constituents of *Withania somnifera* Dun.: Studies on Some Model Steroidal Epoxides. Tetrahedron **22**, 1103 (1966).

192. LAVIE, D., I. KIRSON, E. GLOTTER, D. RABINOVITCH, and Z. SHAKKED: Crystal and Molecular Structure of Withanolide E, a New Steroidal Lactone with a 17α-Side Chain. J. Chem. Soc. Chem. Commun. **1972**, 877.

193. LOCKLEY, W.J.S., D.P. ROBERTS, H.H. REES, and T.W. GOODWIN: 24-Methylcholesta-5,24(25)-dien-3β-ol, a New Sterol from *Withania somnifera*. Tetrahedron Letters **43**, 3773 (1974).

194. MENSSEN, H.G., and G. STAPEL: C_{28}-Steroidal Lactone from the Roots of *Withania somnifera*. Planta Med. **24**, 8 (1973).

195. NITTALA, S.S., F. FROLOW, and D. LAVIE: Novel Occurrence of 14β-Hydroxy Group on a Withanolide Skeleton. X-Ray Crystal and Molecular Structure of 14β-Hydroxy-withanone. J. Chem. Soc. Chem. Commun. **1981**, 178.

196. NITTALA, S.S., V.V. VELDE, F. FROLOW, and D. LAVIE: Chlorinated Withanolides from *Withania somnifera* and *Acnistus breviflorus*. Phytochem. **20**, 2547 (1981).

197. NITTALA, S.S., and D. LAVIE: Chemistry and Genetics of Withanolides in *Withania somnifera* Hybrids. Phytochem. **20**, 2741 (1981).

198. VELDE, V.V., and D. LAVIE: A Δ^{16}-Withanolide in *Withania somnifera* as a Possible Precursor for α-Side Chain. Phytochem. **21**, 731 (1982).

199. YARDEN, A., and D. LAVIE: Constituents of *Withania somnifera*. The Functional Groups of Withaferin. J. Chem. Soc. **1962**, 2925.

200. VELDE, V.V., and D. LAVIE: New Withanolides of Biogenetic Interest from *Withania somnifera*. Phytochem. **20**, 1359 (1981).

201. LAVIE, D., I. KIRSON, E. GLOTTER, and G. SNATZKE: Conformational Studies on Certain Six-Membered Ring Lactones. Tetrahedron **26**, 2221 (1970).

202. NITTALA, S.S., H.E. GOTTLIEB, and I. KIRSON: A New Chlorine Containing With-anolide from *Withania somnifera*. 12th IUPAC Symposium. Chemistry of Natural Products, 1980. Tenerife, Spain, Abstract p. 152.

203. ANTOUN, M.D., D. ABRAMSON, R.L. TYSON, C. CHANG, J.L. MCLAUGHLIN, G. PECK, and J.M. CASSADY: Potential Antitumor Agents. Physalin B and 25,26-Epidihydro-physalin C from *Witheringia coccoloboides*. J. Nat. Prod. **44**, 579 (1981).

204. CHEN, Z.L., B.D. WANG, and M.Q. CHEN: Steroidal Bitter Principles from *Tacca plantaginea*. Structures of Taccalonolide A and B. Tetrahedron Letters **28**, 1673 (1987).

205. CHEN, Z.L., B.D. WANG, and J.H. SHEN: Taccalonolide C and D. Two Pentacyclic Steroids of *Tacca plantaginea*. Phytochem. **27**, 2999 (1988).

206. KSEBATI, M.B., and F.J. SCHMITZ: Minabeolides: A Group of Withanolides from a Soft Coral, *Minabea* sp. J. Org. Chem. **53**, 3926 (1988).

207. SRIVASTAVA, C., I.R. SIDDIQUI, J. SINGH, and H.P. TIWARI: An Antifeedant and Insecticidal Steroid and a New Hydroxyketone from *Cassia siamea* Bark. J. Indian Chem. Soc. **69**, 111 (1992).

208. GOTTLIEB, H.E., and I. KIRSON: ^{13}C-NMR Spectroscopy of the Withanolides and Other Highly Oxygenated C_{28}-Steroids. Org. Magn. Res. **16**, 20 (1981).

209. PELLETIER, S.W., N.V. MODY, J. NOWACKI, and J. BHATTACHARYA: Carbon-13 Nuclear Magnetic Spectral Analysis of Naturally Occurring Withanolides and Their Derivatives. J. Nat. Prod. **42**, 512 (1979).

210. KUNDU, S.: Ph.D. Thesis. Banaras Hindu University, 1989.

211. SNATZKE, G., H. SCHWANG, and P. WELZEL: Some Newer Physical Methods in Structural Chemistry, p. 159 (R. BONNET and J.G. DAVIS eds.). London: United Trade Press, 1967.

212. RABINOVICH, D., Z. SHAKKED, I. KIRSON, G. GÜNZBERG, and E. GLOTTER: An Unusual D-Homo-Rearrangement of a 14α,17β,20α-Trihydroxywithanolide. Crystal and Molecular Structure of 5-Hydroxy-14α,17α-epoxy-17β-methyl-5β-D-homo-androstan-1,17a-dione. J. Chem. Soc. Chem. Commun. **1976**, 461.

213. KIRSON, I., G. GÜNZBERG, H.E. GOTTLIEB, and E. GLOTTER: Acid-Catalysed Dehydration of Withanolide E, a 14α,17β,20α_F-Trihydroxy Steroid, a Revision. J. Chem. Soc. Perkin I **1980**, 531.

214. LAVIE, D., Y. KASHMAN, E. GLOTTER, and D.N. DANIELI: Constituents of *Withania somnifera*. Rearrangements in Withaferin A. J. Chem. Soc. (C) **1966**, 1757.

215. NITTALA, S., and D. LAVIE: Studies on the 5β,6β-Epoxide Opening in Withanolides. J. Chem. Soc. Perkin I **1982**, 2835.

216. ISHIGURO, M., A. KAJIKAWA, T. HARUYAMA, M. MORISAKI, and N. IKEKAWA: Synthetic Studies of Withanolides. Synthesis of AB Ring Moiety of Withaferin A. Tetrahedron Letters **1974**, 1471.

217. ISHIGURO, M., A. KAJIKAWA, T. HARUYAMA, Y. OGURA, M. OKUBAYASHI, M. MORISAKI, and N. IKEKAWA: Synthetic Studies of Withanolides. Synthesis of 5,6β-Epoxy-4β-hydroxy-5β-cholest-2-en-1-one and Related Compounds. J. Chem. Soc. Perkin I **1975**, 2295.

218. WEISSENBERG, M., E. GLOTTER, and D. LAVIE: The Synthesis of the Rings A/B in Withaferin A and Other Withanolides. Tetrahedron Letters **1974**, 3063.

219. WEISSENBERG, M., D. LAVIE, and E. GLOTTER: 1-Oxo-Steroids. Model Studies for the Synthesis of the Withanolides. J. Chem. Soc. Perkin I **1977**, 795.

220. KAJIKAWA, A., M. MORISAKI, and N. IKEKAWA: Exclusive γ-Coupling in the Aldol Reaction of α,β-Unsaturated Esters. Tetrahedron Letters **47**, 4135 (1975).

221. GLOTTER, E., M. ZVIELY, and I. KIRSON: Synthesis of (22R)-3β-Acetoxyergosta-5,24-dien-22,26-olide, the Parent Withanolide Steroidal Lactone. J. Chem. Res. **1982**, (S) 32 (M) 373.

222. WEIHE, G.R., and T.C. MCMORRIS: Stereoselective Synthesis of 23-Deoxyantheridiol. J. Org. Chem. **43**, 3942 (1978).

223. HIRAYAMA, M., K. GAMOH, and N. IKEKAWA: Synthesis of Jaborosalactone A, B and D. J. Am. Chem. Soc. **114**, 3735 (1982).

224. HIRAYAMA, M., K. GAMOH, and N. IKEKAWA: Synthetic Studies of Withanolides, Stereoselective Synthesis of Withaferin-A and 27-Deoxywithaferin-A. Tetrahedron Letters **23**, 4725 (1982).

225. GAMOH, K., M. HIRAYAMA, and N. IKEKAWA: Stereocontrolled Synthesis of Withanolide D and Related Compounds. J. Chem. Soc. Perkin I **1984**, 449.

226. PEREZ-MEDRANO, A., and P.A. GRIECO: Synthesis of the Highly Oxygenated Ergostane Type Steroid (+)-Withanolide E. J. Am. Chem. Soc. **113**, 1057 (1991).

227. LOCKLEY, W.J.S., H.H. REES, and T.W. GOODWIN: Biosynthesis of Steroidal With-anolides in *Withania somnifera*. Phytochem. **15**, 937 (1976).
228. LEETE, E., H. GREGORY, and E.G. GROS: Biosynthesis of Plant Steroids. The Origin of the Butenolide Ring of Digitoxigenin. J. Am. Chem. Soc. **87**, 3475 (1965).
229. SEO, S., A. UOMORI, Y. YOSHIMURA, and K. TAKEDA: Stereospecificity in the Biosynthesis of Phytosterol Side Chains: ^{13}C-NMR Signal Assignments of C-26 and C-27. J. Am. Chem. Soc. **105**, 6343 (1985).
230. VELEIRO, A.S., G. BURTON, and E.G. GROS: Biosynthesis of Withanolides in *Acnistus breviflorus*: Biogenetic Relationships Among the Main Withanolides. Phytochem. **24**, 2573 (1985).
231. MULCHANDANI, N.B., S.S. IYER, and L.P. BADHEKA: Biosynthesis of Physalin D. BARC-764 27 (1974).
232. CHAKRABORTI, S.K., B.K. DE, and T. BANDYOPADHYAY: Variations in the Antitumor Constituents of *Withania somnifera*. Experientia **30**, 852 (1974).
233. BESSALLE, R., and D. LAVIE: Withanolide C, a Chlorinated Withanolide from *Withania somnifera*. Phytochem. **31**, 3648 (1992).
234. NES, W.D., and E. HEFTMANN: A Comparison of Triterpenoids with Steroids as Membrane Components. J. Nat. Prod. **44**, 377 (1981).
235. SHARMA, K., and P.C. DANDIYA: *Withania somnifera* Dunal – Present Status. Indian Drugs **29**, 247 (1992).
236. CHATTERJEE, S., and S.K. CHAKRABORTI: Antimicrobial Activity of Some Antineoplas-tic and Other Withanolides. Antonie van Leeuwenhoek **46**, 59 (1980).
237. SINHA, S.C., A. ALI, A. BAGCHI, M. SAHAI, and A.B. RAY: Physalindicanols, New Biogenetic Precursors of C_{28}-Steroidal Lactones from *Physalis minima* var. *indica*. Planta Med. **53**, 55 (1987).
238. SHARMA, P.V.: Dravya Vijnana, Part II, pp. 763–765. Varanasi: Choukhamba Bharati Academy, 1984.
239. SETHI, P.D., and R.L. KHOSA: Antifungal Activity of Some C_{28}-Steroidal Lactones. Current Sci. (India) **44**, 867 (1975).
240. SHOHAT, B., S. GITTER, A. ABRAHAM, and D. LAVIE: Antitumor Activity of Withaferin A. Cancer. Chemother. Rep. **51**, 271 (1969).
241. SHOHAT, B., S. GITTER, and D. LAVIE: Effect of Withaferin A on Ehrlich Ascites Tumor Cells: Cytological Observations. Int. J. Cancer **5**, 244 (1970).
242. SHOHAT, B.: Antimitotic Properties of Withaferin A in Tissue Culture. Harefuah **83**, 582 (1972); C.A. **79**, 139h (1973).
243. CASSADY, J.M., C.J. CHANG, and J.L. McLAUGHLIN: Recent Advances in the Isolation and Structural Elucidation of Antineoplastic Agents (J.L. BEAL and E. REINHARD, eds.), pp. 93–124. Stuttgart: Hippokrates Verlag, 1981.
244. DAS, H., S.K. DUTTA, B. BHATTACHARYA, and S.K. CHAKRABORTI: Antineoplastic Agents from Plants; Antitumor Activity of Withanolide D. Indian J. Cancer Chemo-ther. **7**, 59 (1985).
245. GONZALEZ, A.G., Y. DARIAS, D.A. MARTIN HERRERA, and M.C. SUAREZ: Cyto-toxic Activity of Natural Withanolides from Spanish Withanias. Fitoterapia **53**, 85 (1982).
246. JUANG, J.K., H.W. HUANG, C.M. CHEN, and H.J. LIU: A New Compound With-angulatin A Promotes Type II DNA Topoisomerase-Mediated DNA Damage. Biochem. Biophys. Res. Commun. **159**, 1128 (1989).
247. YOSHIDA, M., A. HOSHI, K. KURETANI, M. ISHIGURO, and N. IKEKAWA: Studies on Steroids. Relationship Between Chemical Structure and Antitumor Activity of With-aferin A Analogs. J. Pharm. Dyn. **2**, 92 (1979).

248. BUDHIRAJA, R.D., S. SUDHIR, and K.N. GARG: Antiinflammatory Activity of 3β-Hydroxy-2,3-dihydrowithanolide F. Planta Medica **1984**, 134.

249. BUDHIRAJA, R.D., K.N. GARG, S. SUDHIR, and B. ARORA: Protective Effect of 3β-hydroxy-2,3-dihydrowithanolide F Against Carbon Tetrachloride-Induced Hepatotoxicity. Planta Medica **1986**, 28.

250. SYROV, V.N., Z.A. KHUSBAKTOVE, and O.E. VASINA: Antiinflammatory Properties of Withanolides. Khim. Farm. Zh. **23**, 610 (1989); C.A. **111**, 166955a (1989).

251. FUGNER, A.: Hemmung immunologisch bedingter Entzündungen durch das Pflanzensteroid Withaferin A. Arzneim. Forsch. **23**, 932 (1973).

252. BAHR, V., and R. HÄNSEL: Immunomodulating Properties of 5,20α-(R)-Dihydroxy-6α,7α-epoxy-1-oxo-(5α)-witha-2,24-dienolide and Solasodine. Planta Medica **44**, 32 (1982).

253. GHOSAL, S., J. LAL, R. SRIVASTAVA, S.K. BHATTACHARYA, S.N. UPADHYAY, A.K. JAISWAL, and U. CHATTOPADHYAY: Immunomodulatory and CNS Effects of Sitoindosides IX and X, Two New Glycowithanolides from *Withania somnifera*. Phytotherapy Research **3**, 201 (1989).

254. ASCHER, K.R.S., N.E. NEMMY, M. ELIYAHU, I. KIRSON, A. ABRAHAM, and E. GLOTTER: Insect Antifeedant Properties of Withanolides and Related Steroids from Solanaceae. Experientia **36**, 998 (1980).

255. KEINAN, E., M. SAHAI, and I. KIRSON: Reductive Elimination of Vicinal Oxygen Functions with Palladium (0). Applications in the Withanolide Series. J. Org. Chem. **48**, 2550 (1983).

256. SHINGU, K., S. YAHARA, T. NOHARA, and H. OKABE: Three New Withanolides, Physagulins A, B and D from *Physalis angulata* L. Chem. Pharm. Bull. **40**, 2088 (1992).

257. SHINGU, K., S. YAHARA, H. OKABE, and T. NOHARA: Three New Withanolides, Physagulins E, F and G from *Physalis angulata* L. Chem. Pharm. Bull. **40**, 2448 (1992).

258. ALFONSO, D., and G. BERNARDINELLI: New Withanolides from *Iochroma coccinium*. Planta Medica **57**, A67 (1991).

259. ALFONSO, D., G. BERNARDINELLI, and I. KAPETANIDIS: Four New Withanolides from *Iochroma coccinium*. Planta Medica **58**, A712 (1992).

260. TOMAS, J., F. CAMPS, E. CLAVERIA, J. COLL, E. MELE, and J. MESSEGUER: Composition and Location of Phytoecdysteroids in *Ajuga reptans in vivo* and *in vitro* Cultures. Phytochem. **31**, 1585 (1992).

261. TSUBUKI, M., K. KANAI, K. KEINO, N. KAKINUMA, and T. HONDA: A Facile Construction of Withanolide Side Chain: Synthesis of Minabeolide-3. J. Org. Chem. **57**, 2930 (1992).

262. GREEN, S.P., and D.A. WHITING: Biomimetic Radical Ring Expansion and Aromatisation; a Model for the Biogenesis of Natural Ring-D Aromatic Steroids. J. Chem. Soc. Chem. Commun. **1992**, 1754.

263. LUIS, J.G., F. ECHEVERRI, W. QUINONES, A.G. GONZALEZ, F. TORRES, G. CARDONA, R. ARCHBOLD, and A. PERALES: Withajardins, Withanolides with a New Type of Skeleton, Structure of Withajardins A, B, C and D, Absolute Configuration of Withajardin C. *Tetrahedron* **50**, 1217 (1994).

264. HABTEMARIAM, S., A.I. GRAY, and P.G. WATERMAN: 16-Oxygenated Withanolides from the Leaves of *Discopodium penninervium*. Phytochem. **34**, 807 (1993).

(*Received March 8, 1993*)

Clerodane Diterpenes in *Labiatae**

L. Rodríguez-Hahn, B. Esquivel, and J. Cárdenas,
Instituto de Química, Universidad Nacional Autónoma de México, D.F.,
México

Contents

I. Introduction

The chemical constituents of the *Labiatae* family have been studied mainly for their economic value which is closely related to their high content of essential oils. These consist mostly of volatile mono- and sesquiterpenes. There is much less information on the diterpenoid content of these plants although the latter has been considered (*1*) potentially more useful for taxonomic purposes.

Phytochemical studies of members of the genera *Ajuga, Teucrium, Scutellaria, Salvia* and to a lesser extent *Leonurus* and *Stachys* (Labiatae)

* Dedicated to Professor Sir Derek Barton

have led to the isolation of a number of *neo*-clerodane diterpenoids. Interest in these compounds has been stimulated by their activity as antitumoral, antimicrobial, antifungal and insect antifeedant agents.

Neo-clerodane diterpenoids have also been isolated from plants of several other families (*2*) such as Compositae (*Baccharis, Olearia, Conyza*), Euphorbiaceae (*Croton*), Sapindaceae (*Dodonea*) and Verbenaceae (*Clerodendron, Caryopteris*) among others. A list of clerodanes from all sources has been published (*3*). The genera *Clerodendron* and *Caryopteris* of the Verbenaceae deserve special mention. The botanical relationship between the genera *Ajuga, Scutellaria* and *Teucrium* of the Labiatae and some genera of the Verbenaceae has been recently analyzed and a reclassification has been proposed (*4*).

In this article we will review the chemistry of the *neo*-clerodane diterpenoids isolated from plants of the Labiatae belonging to the genera *Ajuga, Scutellaria, Teucrium, Salvia, Leonurus* and *Stachys* (*5*). The chemistry of *neo*-clerodanes isolated from other sources will be considered when there is an important structural relationship with *neo*-clerodane diterpenoids found in species of the Labiatae.

Elucidation of the structure and relative stereochemistry of clerodin, the first member of the *neo*-clerodane diterpenes, was the opening chapter in the chemistry of these important secondary metabolites. Clerodin was first isolated from *Clerodendron infortunatum* (Verbenaceae). Its structure and relative stereochemistry **1** (R=H) were established by chemical transformations and spectroscopic means (*6*) which included an X-ray analysis (*7*) of the bromolactone derivative **2** (R=H) prepared from clerodin. Attention should be drawn to the functionality of the A/B ring system, as this substitution pattern has been found in *neo*-clerodane diterpenes isolated from *Teucrium* and *Ajuga* species (Labiatae) as well as from *Clerodendron* and *Caryopteris* species (Verbenaceae).

The furo-furan function in the side chain of clerodin was found to be very sensitive to acid. Solution in acetic acid at room temperature converted clerodin to the corresponding hemiacetal acetate. Treatment of this adduct with aqueous acetic acid produced the corresponding hemiacetal (*6*). Knowledge of the chemical behaviour of the furo-furan moiety is important, as some *Ajuga* diterpenoids described as natural products might well be artifacts formed as a result of the methods used for their isolation. Treatment of clerodin with methanol at room temperature, gave the 15-methoxy derivative (*8*).

The tentative conclusion that the relative stereochemistry **1** also represented the absolute configuration of clerodin was based (*6*) on the ORD curves of two clerodin derivatives, A and B, which were enantiomeric to 6-keto-*trans*-A/B steroids such as C and seemed to be confirmed by the Bijvoet anomalous dispersion method as applied to bromolactone **2** (7). The biogenesis of clerodin was therefore thought to be similar to that of columbin (9) which involves two 1,2-methyl migrations from a normal labdane skeleton. However subsequent work showed that those conclusions were in error.

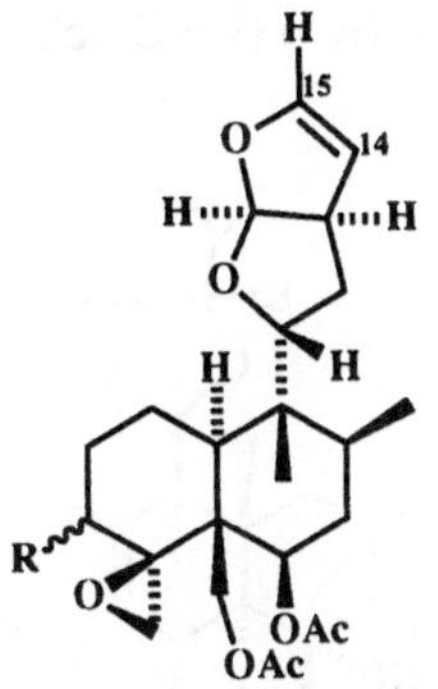

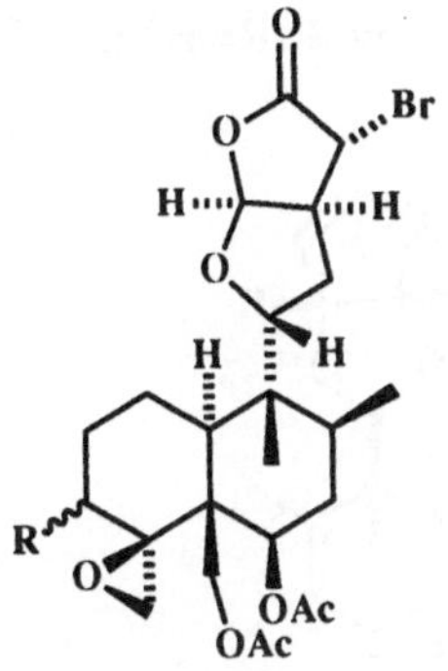

The absolute configuration of clerodendrin A, a clerodane diterpenoid isolated from *Clerodendron tricotomum* Thunb., was unequivocally established as **3a** by correlation with R(−)-2-hydroxy-2-methylbutyric acid and by the Bijvoet method (10) applied to a 3-p-bromobenzoate chlorohydrin derivative of clerodendrin (**3b**) (11). Hence clerodendrin appeared to be antipodal to clerodin (**1**) in all the common chiral centres. Subsequent reexamination of the chiroptical data found for clerodin (**1** R=H), caryoptin (presumably **1**, R=β-OAc) and 3-epi-caryoptin (presumably **1**, R=α-OAc), clerodane diterpenoids isolated from plants of the same genera (*Clerodendron* and *Caryopteris*, Verbenaceae), as well as molecular rotation difference data between dihydro or lactone derivatives and the corresponding unsaturated compounds led to the conclusion

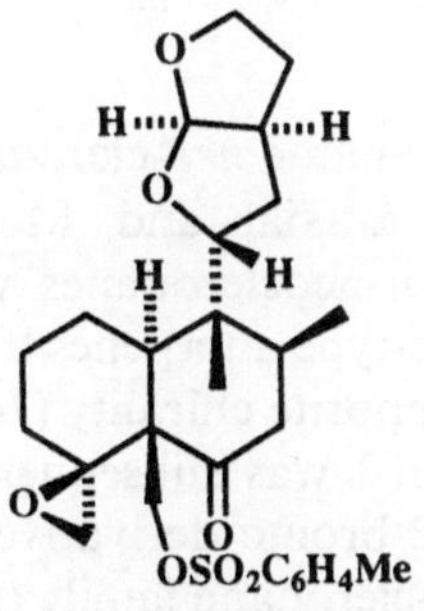

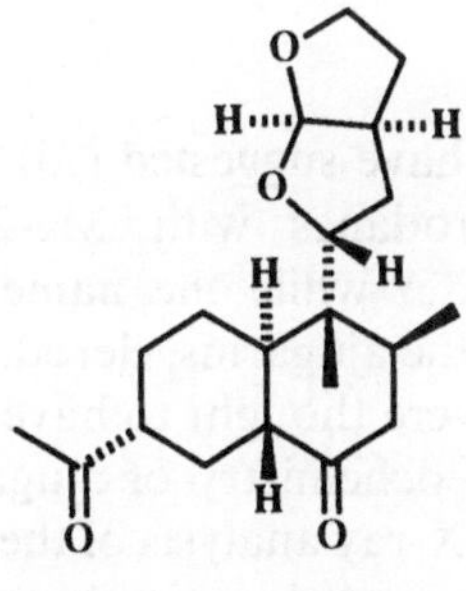

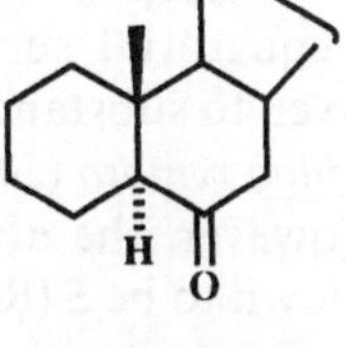

(*12*) that the absolute stereochemistries of clerodin, caryoptin and 3-epi-caryoptin were actually the reverse of those postulated previously and that these clerodane diterpenes have the same absolute configuration in all the corresponding chiral centres as clerodendrin (**3a**). A new X-ray study of clerodin bromolactone and an X-ray analysis of 3-epi-caryoptin confirmed the revised absolute stereochemistry (*13*). Clerodin must be, therefore, represented by **4** (R=H) and 3-epi-caryoptin by **4** (R=OAc).

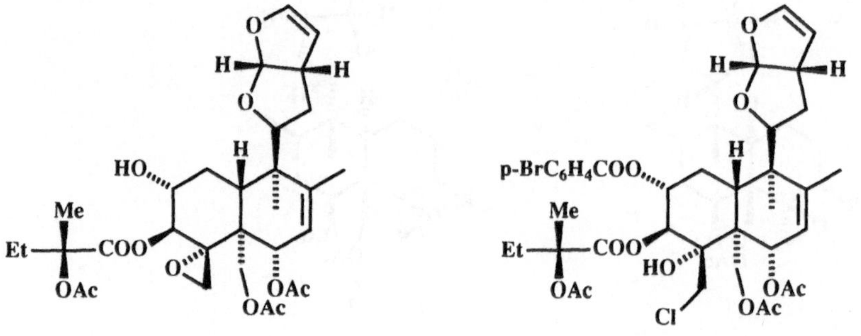

3a   3b

Rogers and coworkers have suggested (*13*) the name *neo*-clerodane for $5\alpha,10\beta$-*trans*-fused clerodanes with Me-20 α-axial and Me-17 α-equatorial such as clerodin while the name *ent*-neoclerodanes was given to substances such as the ajugarins, clerodane-type diterpenes from *Ajuga remota* (*14a*), which were thought to have opposite chirality (*14b*). However, the absolute stereochemistry of ajugarin I was subsequently shown to be **5** (R=H) by an X-ray analysis of the 12-bromo-derivative (**5**, R=Br); hence the ajugarins also belong to the *neo*-clerodane family (*15*). The difference in the Cotton effect exhibited (*14a*) by the ajugarin derivative **6** ($\Delta\varepsilon_{298} - 3.41$) and the clerodin derivative **7** ($\Delta\varepsilon_{298} + 3.51$) which was responsible for the old assignment must therefore be attributed to the contribution of the C-11-C-16 functionalities of both products.

All clerodane diterpenes isolated so far from Verbenaceae and Labiatae species have the *neo*-clerodane absolute stereochemistry **8**. Exceptions with opposite configuration at the C-8 and C-10 chiral centres probably arise from epimerization produced by appropriate functions at C-1 and/or C-7. It should also be noted that the term *neo*-clerodanes for these diterpenes was not fully accepted until recently. Some workers have preferred the term *ent*-clerodanes for biogenetic reasons (*16*). There is also some confusion with regard to the numbering of the *neo*-clerodane

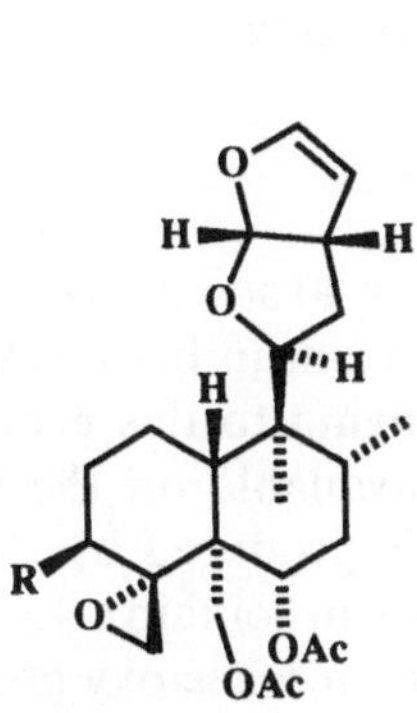

4

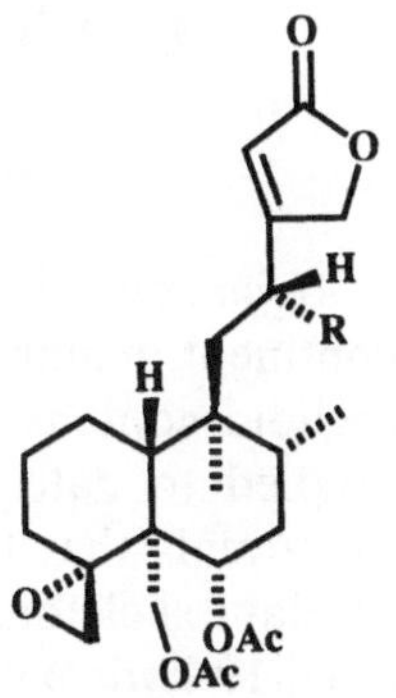

5

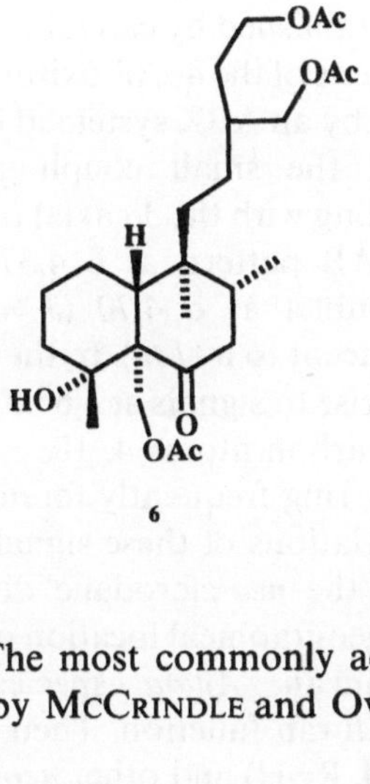

6

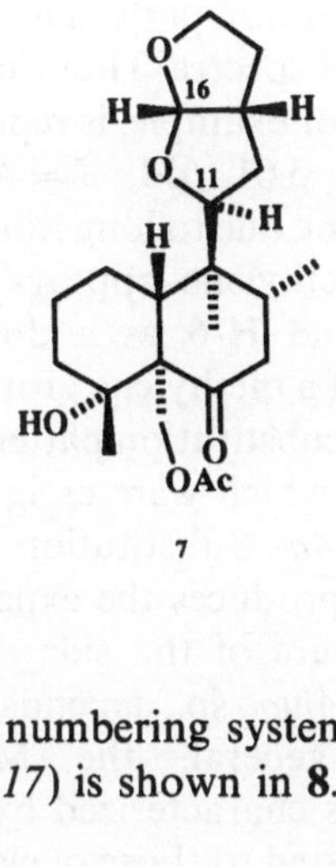

7

skeleton. The most commonly accepted numbering system, which was suggested by McCRINDLE and OVERTON (*17*) is shown in **8**.

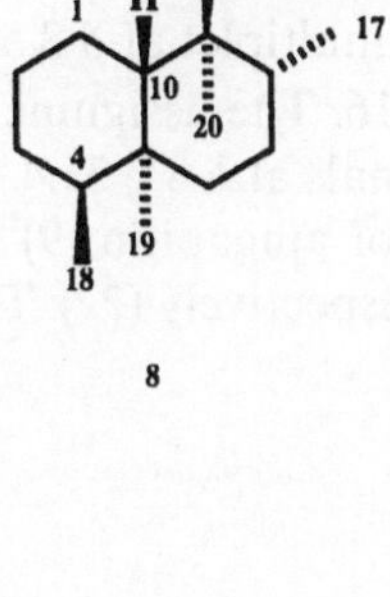

8

II. Clerodane Diterpenes

1. *Ajuga* Species

The genus *Ajuga* comprises approximately 40 species localized in the Euroasiatic continent, mainly in the Mediterranean basin (*18*).

The diterpenoid content of plants belonging to this genus has not been much studied to date. Information available on the few *Ajuga* species studied reveals that the diterpenoids produced by these plants have a *neo*-clerodane skeleton. Almost all of them contain a 4α,18-oxirane ring, an esterified 19-hydroxy methylene and an α-acetoxy group bound to C-6. Many of them contain one or two additional hydroxy groups in the A ring – free or esterified by acetic, 2-methyl-propanoic, 2-methyl-butyric, tiglic or 3-acetoxy-2-methyl-butyric acids. This clerodin type substitution pattern has been established by careful examination of the ^{1}H NMR spectra. Thus the methylene of the 4α,18-oxirane group of ajugarin I (**5**) for example, is represented by an ABX system at δ 2.23 (d, *J* = 4 Hz) and 3.01 (dd, *J* = 4, 2 Hz), the small coupling constant (*J* = 2 Hz) being due to long range coupling with the 3α axial proton. The 19-acetoxy methylene appears as an AB pattern at δ 4.37 and 4.85 (*J* = 12 Hz) and H-6 as a double doublet at δ 4.70 (*J* = 10, 6 Hz) indicating that a methylene group is adjacent to it (*14a*). In the ^{13}C NMR spectrum this substitution pattern gives rise to signals at δ 65.0, 48.5, 45.2, 61.8 and 72.2 which were assigned to carbon atoms 4, 18, 5, 19 and 6, respectively (*14a*). Substitution of the A ring frequently found in Ajuga diterpenoids, produces the expected variations of these signals.

The structure of the side chain of the *neo*-clerodane diterpenoids isolated from *Ajuga* sp., depends on the geographical location of the plant studied. In general, the Mediterranean *Ajuga* species contain *neo*-clerodanes characterized by a furo-furan function. Their structures are closely related to those of clerodin (**4**, R=H) and other *neo*-clerodane diterpenoids isolated from *Clerodendron* and *Caryopteris* species (Verbenaceae).

In the ^{1}H NMR spectrum of ajugapitin (**9**), for example the furo-furan moiety is responsible for a double doublet at δ 6.47 (*J* = 2.8, 2 Hz) and 4.81 (dd, *J* = 2.9, 2.8 Hz) assigned to the olefinic H-15 and H-14 respectively; H-13 is observed as a multiplet at δ 3.56 and a doublet (*J* = 6 Hz) at δ 6.05 is attributed to H-16. The assignments were proved by proton decoupling experiments. Signals at δ 84, 31.4, 46.0, 101.75, 147 and 107.7 in the ^{13}C NMR spectrum of ajugapitin (**9**) were assigned to the furo-furan carbon atoms 11–16 respectively (*22*). The mass spectrum was also of value. An important peak at m/z 111 was observed due to the furo-furan fragment.

In the ^{1}H NMR spectra of *neo*-clerodanes in which the furo-furan function is reduced, the H-16 signal appears at δ 5.67 (d, J = 6 Hz). In the ^{13}C NMR spectrum of 2-acetylivain I (13, R_1 =βOAc,H) for example, the 14,15-dihydrofuro-furan carbon atoms 11-16 were observed at δ 84.9, 32.4, 41.3, 33.2, 68.3 and 106.3 respectively. A peak at m/z 113, frequently the base peak of the mass spectrum, is due to the dihydrofuro-furan fragment ion.

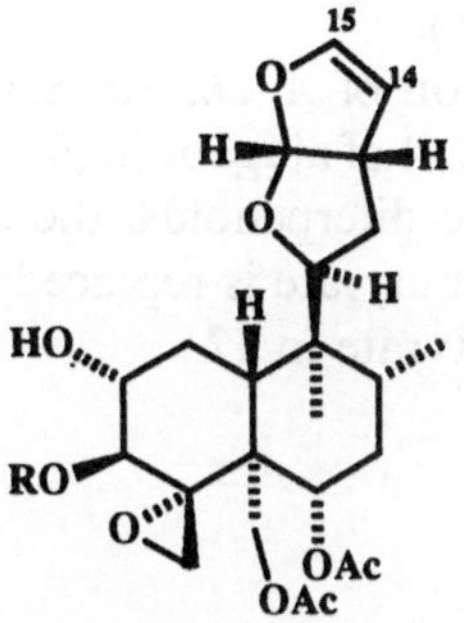

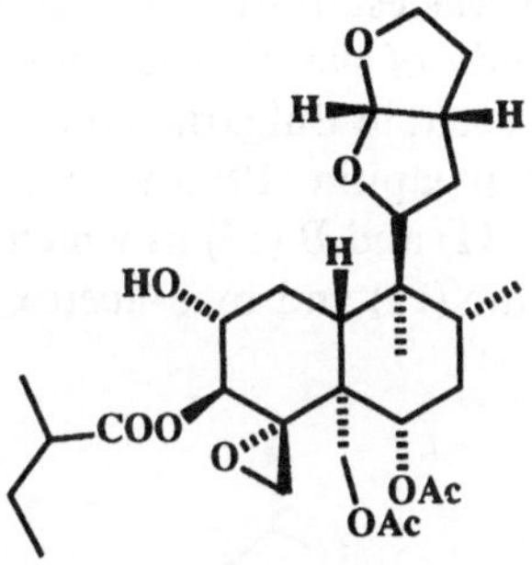

9: R = CO—⟨ (2S)

11: R = CO—⟨

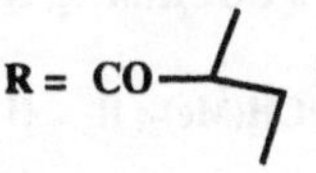

12: R = CO—⟨OAc

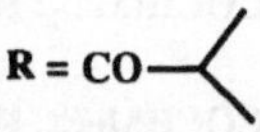

10

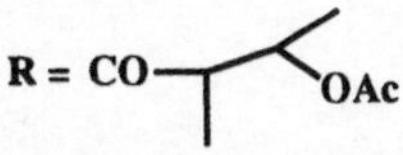

The three Japanese species studied (*A. nipponensis*, *A. ciliata* var. *villosior* and *A. decumbens*) which also occur in China and Korea contain *neo*-clerodanes with a terminal α,β-unsaturated-γ-lactone (β-butenolide) function. *Neo*-clerodanes of this type were first isolated from the African species *A. remota* collected in Nairobi (Kenya), which also contains clerodin (4 R=H) (*15*). *Neo*-clerodanes with a β-butenolide function have been also found as minor constituents of *A. reptans*, a species collected in Catalonia, (*19*). The terminal β-butenolide IR band was observed at 1780 cm^{-1}. A complex signal at δ 5.85 in the ^{1}H NMR spectrum of ajugarin I (5, R=H), was attributed to H-14 and a broad doublet at δ 4.75 to the C-16 methylene. When an hydroxy or an ester group is bound to

C-12, H-14 appears at δ 5.85 as a ddd (J = 1.8, 1.8, 1.2 Hz). The β-butenolide carbon atoms 13-16 of ajugarin I (5) were observed at δ 173.5, 115.7, 173.5 and 72.9 (*14a*).

A study of *Ajuga chamaepitys* a species which grows throughout Europe, collected in Santander, Spain, led to the isolation of two *neo*-clerodane diterpenoids ajugapitin, and the 14,15 dihydro derivative (20), whose structures 9 and 10 were determined on spectral evidence. The C-2 configuration (*R*) was established by Horeau's method (*21*). The config-uration at C-2 of the 2-methylbutyric ester was inferred as *S* identical with that of some other *Ajuga* diterpenoids containing this ester group, where it was established by X-ray analysis (*19*).

A study of the diterpenoids in a population of *A. chamaepitys* var. *chia* collected in Bulgaria led to the isolation (*22*) of ajugapitin (9), 14,15-dihydroajugapitin (10) and two *neo*-clerodane diterpenoids, the ajuga-chins A (11) and B (12) in which the 2-methylbutyrate is replaced by an isovalerate (11) and by 3-acetoxy-2-methylbutyrate in 12.

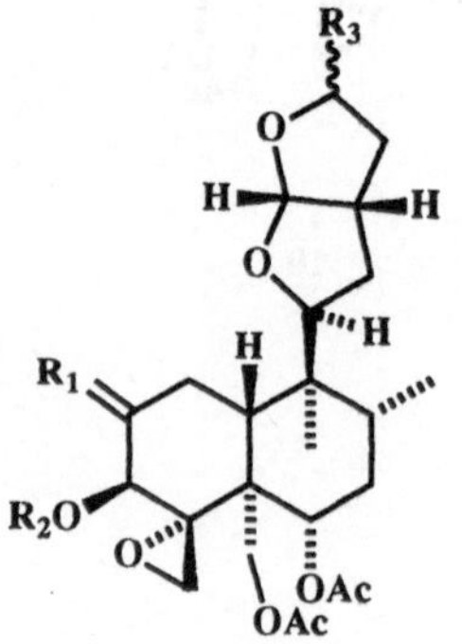

13: R_1 = β OH,H; R_2 = COCH(Me)$_2$; R_3 = H

14: R_1 = H$_2$; R_2 = COCH(Me)$_2$; R_3 = H

15: R_1 = β OH, H; R_2 = COCH(Me)$_2$; R_3 = OEt

16: R_1 = β OH,H; R_2 = COCHMeEt; R_3 = H

From a population of *A. chamaepitys* collected in Gerona, Spain, the 15-hydroxy and 15-ethoxy derivatives of ajugapitin (9) together with the 14,15-dihydro derivative (10) were isolated (*23, 24*) as well as the 15-hydroxy derivative of ajugachin (12). In both cases ether was used for the extraction (1 week at 35°); thus the 15-hydroxy and 15-ethoxy derivatives could be artifacts produced by hydration or addition of ethanol (present in the ether) to the 14,15-double bond of ajugapitin (9) and ajugachin B (12), products which have been isolated from other populations of *A. chamaepitys* (*20, 22*). The same mode of formation could be envisaged for ivain III, (15) from *Ajuga iva*, collected in Israel which also yielded the 14,15-dihydro derivatives, ivains I, II and IV (14–16) (*25*). In ivains I-IV the 2-hydroxyl is axially orientated and β as followed from the coupling constant ($J = 2$ Hz) of H-3 in all ivains. The structure and relative stereochemistry of ivain I were confirmed by X-ray analysis of the 2-oxo derivative (25). A population of *A. pseudoiva* collected in Murcia (Spain), gave 2-acetylivain I (13, $R_1 = \beta$ OAc,H) and 14,15-dihydroajugapitin (10) (26).

A study of a Bulgarian population of *Ajuga genevensis* led to isolation of the *neo*-clerodane diterpenoids ajugavensins A-C (27) for which structures 17–19 were deduced on spectral evidence. Originally it was thought that the ester group at C-1 was α-orientated on the assumption that ring A had a chair conformation. However X-ray analysis of ajugavensins A and B revealed (28) that both ester groups were β-orientated, with ring A in a twist conformation. The latter was attributed to strong steric hindrance between the bulky substituents on C-1β and C-9β if A ring were in the chair conformation. On the other hand ajugavensin C (19) possesses a free β-and equatorially orientated hydroxyl group at C-1, with ring A in the chair conformation, based on the coupling constants observed for H-1. The 1*R* configuration in 19 was confirmed by Horeau's method (21).

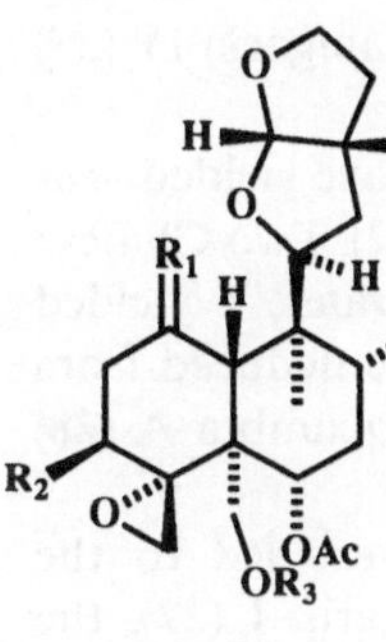

17: $R_1 = \beta$-OCO— , H; $R_2 = H$; $R_3 = Ac$

18: $R_1 = \beta$-OCO— , H; $R_2 = H$; $R_3 = Ac$

19: $R_1 = \beta$ OH, H; $R_2 = H$; $R_3 = CO$—

20: $R_1 = \beta$-OCO— , H; $R_2 = OH$; $R_3 = Ac$

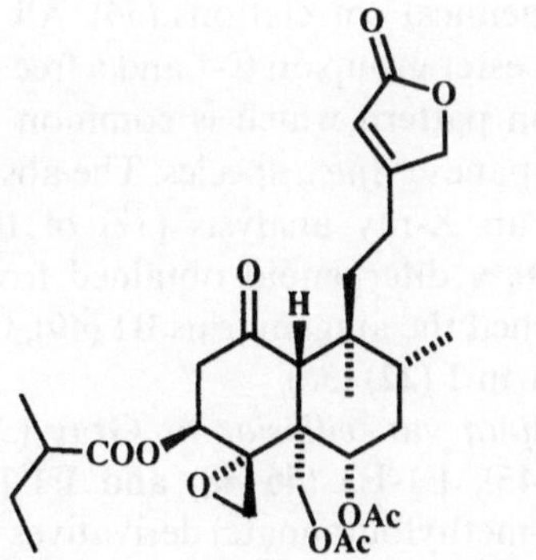

21a
21b (Δ^2 replaces C-3 substituent)

Finally from a population of *Ajuga reptans* (*29*) collected near Barcelona (Spain), the 3β-hydroxy derivative of ajugavensin A, ajugareptansin (**20**), was isolated, together with ajugareptansones A (**21a**) and B (**21b**) (*19*). The structure and absolute stereochemistry of ajugareptansin (**20**) were deduced on the basis of spectral data and an X-ray analysis of the *p*-bromo-benzoate ester, which also showed that ring A had a distorted boat conformation.

The isolation of ajugareptansones A (**21a**) and B (**21b**) from a Mediterranean *Ajuga* species is of interest from the chemotaxonomic point of view (*vide supra*). Such *neo*-clerodanes with a terminal β-butenolide group were first isolated (*14*) from *Ajuga remota*, a Kenyan species which also contains clerodin, and subsequently also from Far Eastern *Ajuga* species.

Ajugarin I (**22**) was the first *neo*-clerodane diterpenoid isolated from an *Ajuga* species. Its structure was deduced on spectral evidence which included an exhaustive analysis of nuclear Overhauser effects (*14a*) and X-ray analysis of the bromo derivative **23** (*15*). It was accompanied by ajugarins II (**24**), III (**25**), IV (**26**), and V (**27**) (*14a, 30, 31*). The structure of ajugarins IV (**26**) and V (**27**) is interesting as they represent *neo*-clerodanes in which the 19-methyl group is not functionalized. Recently stereospecific total syntheses of ajugarin I (*32, 33*) and of ajugarin IV (*33*) were described.

Japanese and Chinese *Ajuga* species studied to date have yielded *neo*-clerodanes with structures closely related to ajugarin I (**22**). Two Chinese populations of *A. decumbens*, Thunb. (Japanese "Kiranso") yielded ajugacumbins A-F (*34, 35*) whose structures **28–33** were deduced from their spectral data and from an X-ray analysis of ajugacumbin A (**28**) (*34*).

Study of a Japanese population of *Ajuga decumbens* led to the isolation of *neo*-clerodanes structurally related to ajugarin I (**22**), the ajugamarins A2, B2, G1, H1 and F4 whose structures **34–38** were deduced on spectral evidence and by chemical correlations (*36*). All of them contain a β, equatorially orientated ester group on C-1 and a free or esterified hydroxyl at C-12, an oxidation pattern which is common to most *neo*-clerodane diterpenoids from Japanese *Ajuga* species. The absolute stereochemistry at C-12 is *S* by an X-ray analysis (*37*) of the *p*-bromobenzoate of ajugamarin A1 (**39**), a diterpenoid obtained from *A. nipponensis* Makino, which also furnished the ajugamarins B1 (**40**), C1 (**41**), B2 (**38**), B3 (**42**), D1 (**43**), and ajugarin I (**22**) (*38*).

A Japanese population of *Ajuga ciliata* var. *villosior* A. Gray (*39*) yielded the ajugamarins B4 (**44**), B5 (**45**), E1-E3 (**46–48**) and F1-F3 (**49–51**), which could be considered 12-(2-methylbutanoate) derivatives of

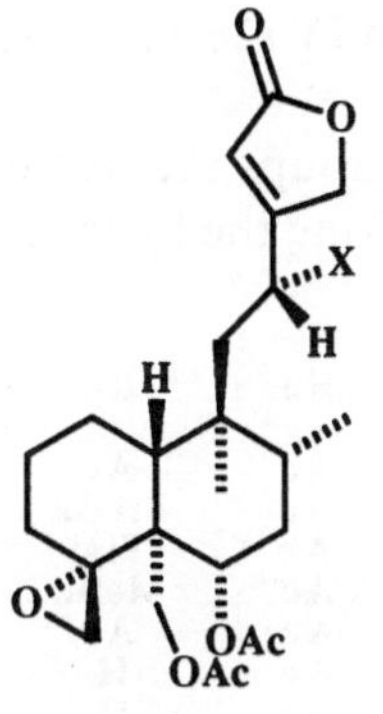

22 : X = H

23 : X = Br (12R)

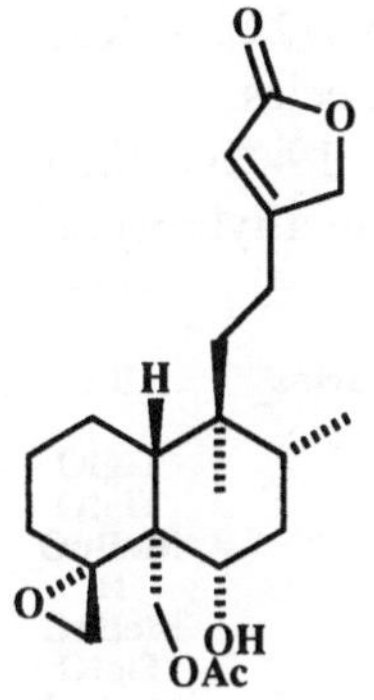

24

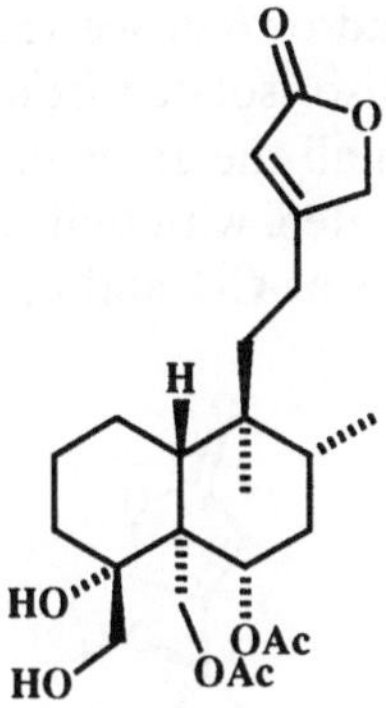

25

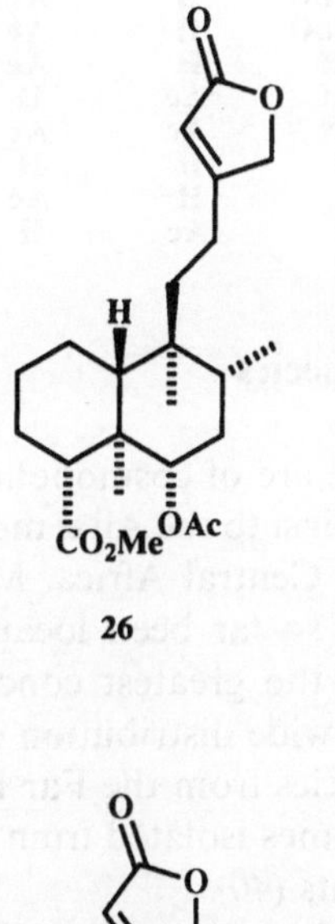

26

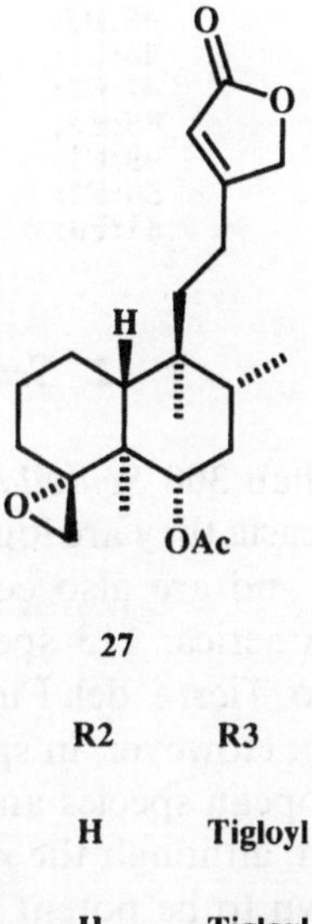

27

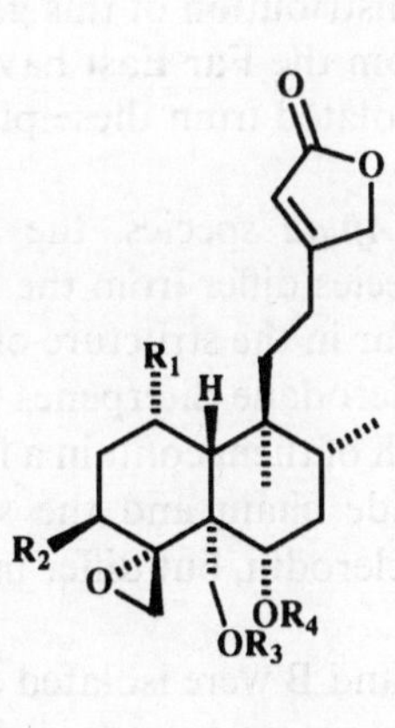

	R1	R2	R3	R4
28:	H	H	Tigloyl	Ac
29:	H	H	Tigloyl	H
30:	OAc	OAc	Tigloyl	Ac
31:	H	OH	Tigloyl	Ac
32:	OAc	OAc	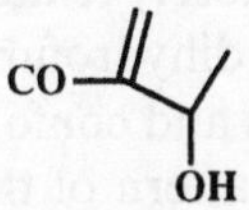 CO—〈	Ac
33:	H	H	Tigloyl	H
(4α,18 diOH)				

19- and/or 6-deacetylajugarin I (**22**). A deacetylajugarin IV (**26**, C-6 OH) was also isolated from this species.

In all the ajugamarins the 19- and the 6-hydroxy groups are free or acetylated, with tiglic and 2S-methylbutyric acid esterifying the hydroxyl groups at C-1 and/or C-12.

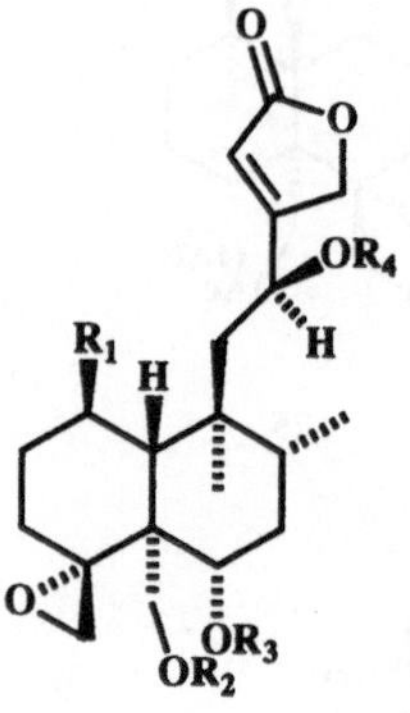

Ajugamarins	R1	R2	R3	R4
34:A2:	TiglO	Ac	Ac	Ac
35:G1:	TiglO	Ac	Ac	MeBu
36:H1:	MeBuO	Ac	Ac	Tigl
37:F4:	H	Ac	Ac	MeBu
38:B2:	MeBuO	Ac	Ac	Ac
39:A1:	TiglO	Ac	Ac	H
40:B1:	MeBuO	Ac	Ac	H
41:C1:	OH	Ac	Ac	H
42:B3:	MeBuO	Ac	H	H
43:D1:	MeBuO	Ac	H	H
(4αOH,18OAc)				
44:B4:	MeBuO	H	Ac	H
45:B5:	MeBuO	H	Ac	Ac
46:E1:	OH	H	Ac	MeBu
47:E2:	OH	Ac	H	MeBu
48:E3:	OH	Ac	Ac	MeBu
49:F1:	H	H	H	MeBu
50:F2:	H	H	Ac	MeBu
51:F3:	H	Ac	H	MeBu

2. *Scutellaria* Species

The more than 300 *Scutellaria* species are of cosmopolitan distribution (*18*). In Eurasia they are found from Siria to the Altai mountains and the Himalayas and are also common in Central Africa, Malaysia and Australia. In America, 113 species have so far been located from the Arctic Circle to Tierra del Fuego, with the greatest concentration in Central Mexico. However, in spite of the wide distribution of this genus only three European species and one species from the Far East have so far been studied, although the *neo*-clerodanes isolated from these plants have been shown to be potent antifeedants (*40*).

As in the case of the diterpenoids from *Ajuga* species, the *neo*-clerodanes of the three European *Scutellaria* species differ from the *neo*-clerodanes in the one Asiatic species studied so far in the structure of the side chain. The European species contain *neo*-clerodane diterpenes with structures closely related to clerodin (**4**, R=H). All of them contain a furo-furan or a dihydrofuro-furan moiety in the side chain and the same substituents and conformation of the B ring as clerodin, but differ in the oxidation pattern of the A ring.

From *Scutellaria woronowii* Juz jodrellins A and B were isolated (*40*). Their structures **52** and **53** were elucidated by spectral means which included 2D, COSY NMR spectra and nOe experiments. Jodrellin B (**53**) is the most potent clerodane antifeedant known to date (*40*). The presence of the esterified 19,2α-hemiacetal function in the jodrellins was deduced from the NMR data which showed the anomeric H-19 as a singlet at δ 6.74 and 6.70 respectively. A multiplet at δ 4.18 was assigned to H-2, while the 18-methylene protons were observed as doublets (*J* = 4.4 Hz) at δ 2.42 and 2.99.

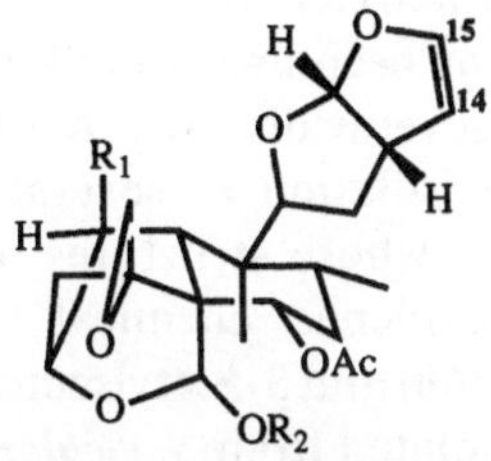

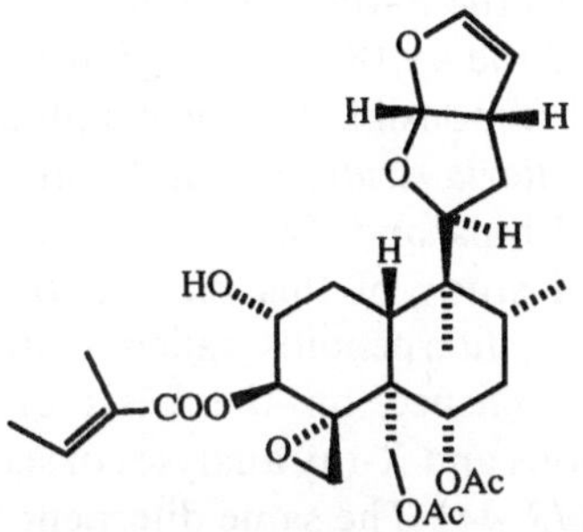

52: R₁ = H; R₂ = Ac

53: R₁ = H; R₂ = COⁱPr.

54: R₁ = Tigloyl-O; R₂ = Ac

55: R₁ = Tigloyl-O; R₂ = Ac; (14,15 dihydro 54)

57: R₁ = H; R₂ = CO—

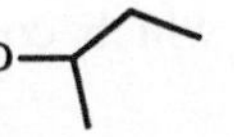

58: R₁ = H; R₂ = CO— 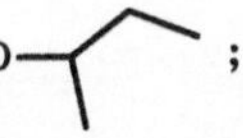   ; 14,15-dihydro

59: R₁ = H; R₂ = H ; 14, 15-dihydro

56

Scutellaria galericulata L. yielded (*41*) jodrellin B (**53**) and two other *neo*-clerodanes, jodrellin T (**54**) and its 14,15 dihydro derivative (**55**), which were shown to be 1β-tigloyl derivatives of jodrellin A (**52**). Galericulin (**56**) was obtained from the same source. Its structure is closely related to ajugapitin (**9**), a *neo*-clerodane isolated from *Ajuga chamaepitys* (*20*) in which the tiglate is replaced by a 2*S*-methylbutyrate.

The *neo*-clerodane diterpenoids scutecolumnins A, B, and C were isolated (*42*) from *Scutellaria columnae*. Their structures **57–59** were established on spectral evidence which included ^{13}C NMR spectra. Resonances of the hemiacetalic carbon atoms 2 and 19 were observed at δ 67.2 and 91.5. Comparison with the spectral data of jodrellin B established that scutecolumnins A and B contained a 2-methylbutyrate at C-19 in place of the iso-butyrate of jodrellin B.

The 19,2α-hemiacetal function present in most *neo*-clerodane diterpenoids from European *Scutellaria* species confers a high degree of tension to the A-ring which adopts a boat conformation. This functionality and the 4α,18-oxirane group have been proposed as the structural features responsible for their high antifeedant activity (*40*).

Scutellaria rivularis Wall ("Ban-zhilian") has been used in folk medicine in China and Taiwan for treatment of several diseases. A phytochemical study of this plant (*43*) led to the isolation of several *neo*-clerodane diterpenoids called scutellones A-F whose structures **60–65** were established on the basis of spectral evidence, chemical transformations and X-ray analyses of scutellone A (**60**) and 3-acetylscutellone D (**63**) (*43, 44*). The same diterpenoids were isolated from *S. rivularis* by Japanese workers (*47, 48*) who called them scuterivulactones. Several structural features of the scutellones deserve comment. In all of them the 18- and 19-methyl groups are unoxidized. This is very unusual in *neo*-clerodanes from *Labiatae*, especially from *Ajuga* and *Teucrium* species. The presence of a benzoate group at C-6 is also unusual; in fact scutellones are the only *neo*-clerodanes isolated so far which contain this group.

The isolation of *neo*-clerodanes with a conjugated diene in the side chain and a β-axial hydroxy group at C-8, such as scutellones D-F (**63–65**), and scutellones A-C (**60–62**) which contain, instead, a spiro-

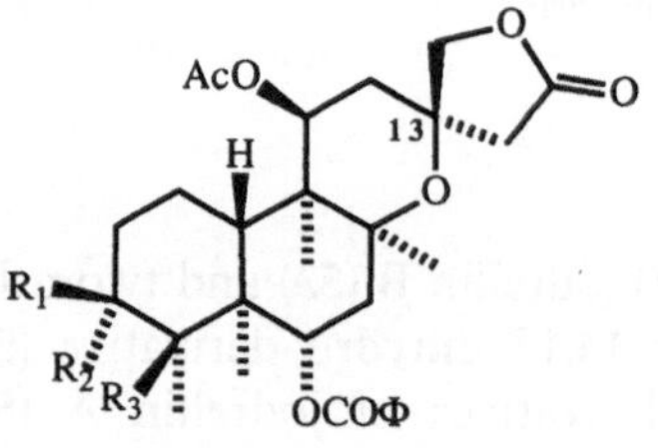

60: R_1 = H; R_2 = R_3 = OH

61: R_1R_2 = O; R_3 = H

62: R_1 = R_3 = OH; R_2 = H

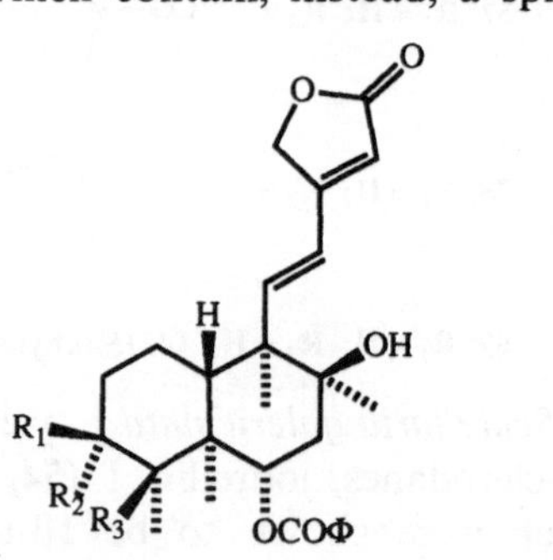

63: R_1 = H; R_2 = R_3 = OH

64: R_1R_2 = O; R_3 = H

65: 3α, 4α-epoxy

lactone moiety, suggest that the latter could be artifacts produced in the course of purification. The isolation of 13-epi-derivatives of scutellones A (**60**) and B (**61**) from the same source supports this possibility, although treatment of scutellone F **65** with *p*-toluensulfonic acid in acetone at room temperature yielded mainly the 3-ketoderivative, scutellone E **64** (*44, 46*).

3. *Teucrium* Species

The genus *Teucrium* comprises more than 300 species found mainly in regions of moderate climate. It is abundant in the Mediterranean Basin but scarcely represented in the Americas (*18*).

The *neo*-clerodane diterpenoids occurring in *Teucrium* species have been the subject of intensive study. Several review articles dealing with the literature through 1986 have appeared (*49–52*). The *neo*-clerodane diterpenoids isolated so far from *Teucrium* species, present several common structural features which have been firmly established (*53*).

1. All diterpenes show an oxidation pattern at carbon atoms 4, 6, 18 and 19, similar to that of clerodin (**4**) or chemically related to it.

2. Almost all *neo*-clerodane diterpenes bear an oxygen function at C-12. In most cases this is an hydroxy group involved in lactone or hemiacetal function with C-20, but has also been found as a free or acetylated hydroxy group. The diterpenes isolated from *T. fruticans* such as fruticolone (**66**) and deacetylajugarin II (**67**) from *T. massiliense*, are some exceptions.

3. All diterpenes from *Teucrium* contain a furan ring formed from C-13 to C-16. The only exception is deacetylajugarin II (**67**) in which C-13 to C-16 form a β-butenolide function.

4. Carbon atom 20 is usually oxidized. Some exceptions are represented by the constituents of *T. fruticans*, *T. massiliense*, *T. oliverianum* and *T. pyrenaicum* in which C-20 is a methyl group.

Most structural studies of the *neo*-clerodane diterpenoids from *Teucrium* species have been based on spectroscopic data which were confirmed in some cases by X-ray analysis. Chemical transformations were used in the early stages of structure elucidation and are of great value in understanding the chemical reactivity of some of the functionalities common to *neo*-clerodanes from *Teucria*.

The majority of the *neo*-clerodanes of *Teucrium* species have a terminal C-13, C-16 furan ring and a spiro-C-20, C-12 γ-lactone function. IR bands at 1505, 875 and at 1770 cm^{-1}, are assigned to the furan and the γ-lactone groups, respectively. In the ^{1}H NMR spectrum the furan protons are observed at δ 6.37 (dd, $J = 1.8$, 0.9 Hz, H-14) and 7.43 (m, H-15 and H-16). A triplet ($J = 8.6$ Hz) at δ 5.34 has been assigned to H-12. The ^{13}C NMR spectra resonances of the furan carbon atoms 13-16 were observed at δ 125.5, 108.1, 144.2, 139.5 respectively, while C-12 is usually found at δ 71.5 and C-20 at δ 176 (*51*). The mass spectrum has been also of value to establish the allylic relationship of the γ-lactone closure with respect to the furan ring. The m/z 81 and m/z 94 peaks are usually important and in some cases m/z 94 is the base peak.

m/z 81   m/z 94

Chemically the relationship of the furan and lactone ring was shown by catalytic hydrogenation, which produced a tetrahydrofuran-carboxylic acid derivative as a result of hydrogenolysis. For example, catalytic hydrogenation of picropolin **68**, a *neo*-clerodane isolated from *T. polium* (*54*), gave the hexahydro derivative **69**. This hydrogenolysis reaction was first observed and rationalized in the case of columbin (**70**), a clerodane diterpenoid isolated from Colombo root (*9, 55*).

Some diterpenoids from *Teucria* have a δ-lactone function involving C-19 and C-20. An IR band at 1700–1720 cm^{-1}, has been attributed to this functionality. Such is the case for teugnaphalodin **71** (R=H), teulepicephin **71** (R=OH), teusalvin E **72** (R=OH) and teubotrin **72** (R=H) from *T. gnaphalodes* (*56*), *T. lepicephalum* (*57*), *T. salviastrum* (*58*) and *T. botrys* (*59*) respectively. These *neo*-clerodanes have an hydroxyl on C-12, the signal of whose geminal proton appears in the ^{1}H NMR spectrum as a double doublet at δ 4.88–5.28 ($J = 10$, 2.5 Hz). In the 12-acetylated derivative H-12 is shifted to δ 6.05–6.35.

The structure proposed for teulamifin B from *T. lamiifolium* (*60*) was identical with that of teubotrin although the two differed in melting point; however direct spectroscopic comparison confirmed the identity of both compounds (*61*). Teulamifin B was shown to have the structure and absolute stereochemistry depicted in **72** (R=H), as sodium hydroxide treatment of 19-deacetyl-teuscorodol **73** (R=H) yielded a product identical with teulamifin B (*60*).

Teuchamaedrin C **74** from a Bulgarian collection of *Teucrium chamaedrys* var. *chamaedrys* also contains a C-19, C-20 δ-lactone group. An IR band at 1695 cm^{-1} was attributed to the hydrogen-bonded δ-lactone carbonyl, as it was shifted to 1740 cm^{-1} in the diacetyl derivative.

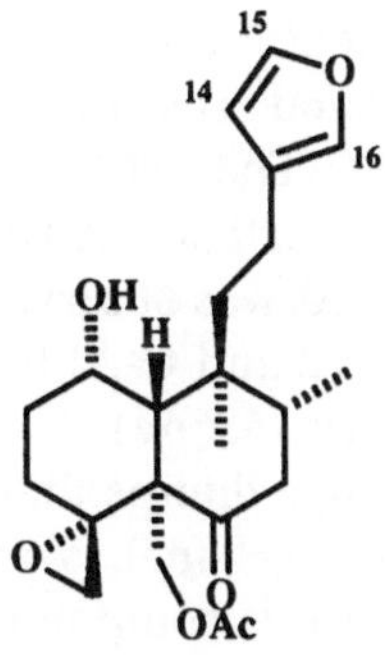

Teusalvin E **72** (R=OH) and teusalvin D **73** (R=OH) are equilibrated by base (*58*). This fact established the 12-*S* chirality for teusalvin E, as this configuration had been determined for **73** (R=OH) by nOe experiments. It is interesting that in all these cases the related *neo*-clerodane with a C-12, C-20 γ-lactone function and hydroxymethylene attached to C-5, was also isolated (*56–59*).

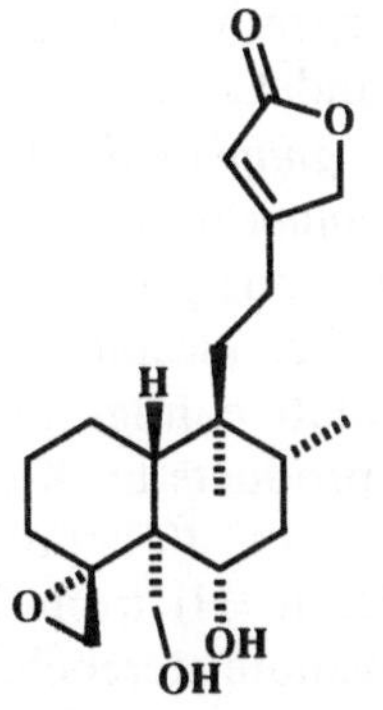

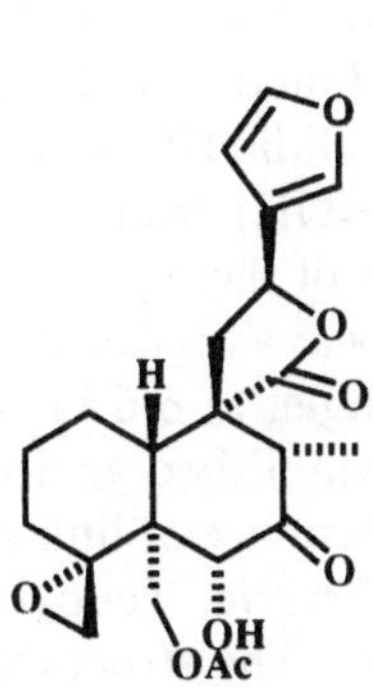

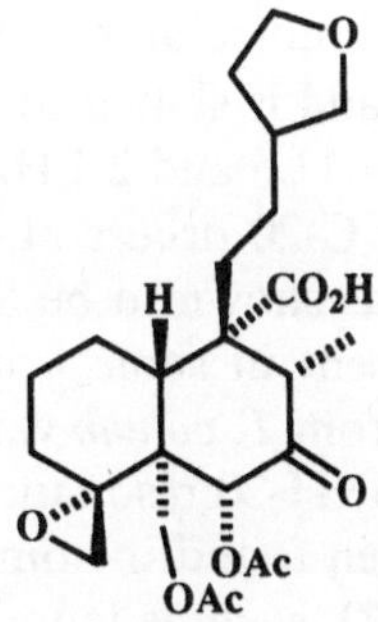

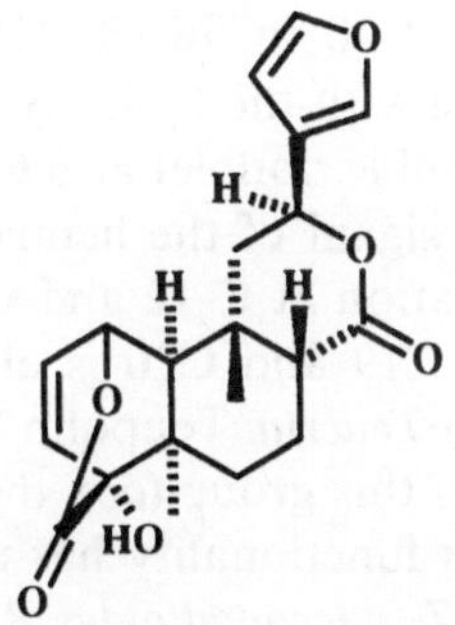

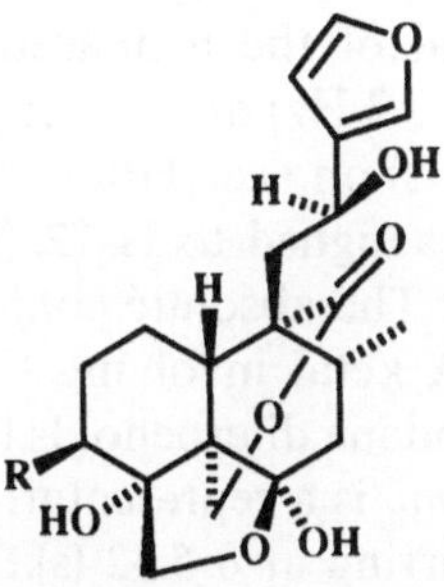

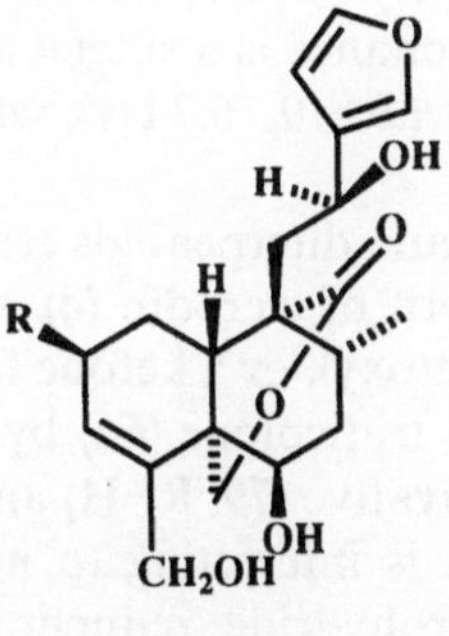

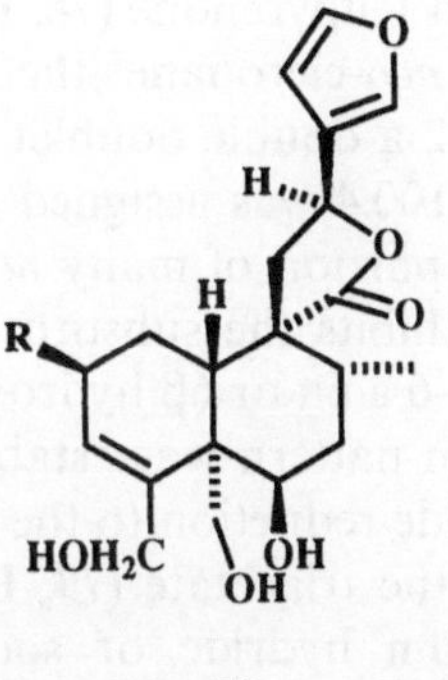

Several *neo*-clerodane diterpenoids from *Teucria* contain a hemi-acetal function linking C-20 and C-12 which is frequently acetylated. Gnaphalidin (**75**, R=H) from *T. gnaphalodes* L'Her (*62*) and eriocephalin (**75**, R=OH) from *T. eriocephalum* (*63*), are examples. The ^{1}H NMR spectra of these products show a one proton triplet at δ 5.20 ($J = 8$ Hz) which was assigned to H-12; the hemiacetalic proton, H-20, was observed as a singlet at δ 6.33. The absolute configuration at C-12 and C-20 has been established as *S* in both products by X-ray analysis (*63, 64*).

A hemiacetal linking C-20 and C-19 is present in some diterpenoids from *Teucria*. Teupyrin A (**76**, R=H) from *T. pyrenaicum* (*65*) is an example. The structure and absolute stereochemistry of this compound were established by spectral evidence which included nOe experiments and X-ray analysis of the 20-acetyl derivative (**76**, R=Ac). The NMR signal of the hemiacetalic H-20 in **76** (R=H) is a doublet at δ 5.01 ($J = 3.3$ Hz) due to coupling with the hydroxy group and is shifted to δ 5.99 (s) on acetylation. A double doublet at δ 6.02 ($J = 11.6$ and 2.4 Hz) was assigned to H-12. The signal of the hemiacetalic C-20 occurs at δ 94.5. The absolute configuration at C-12 and C-20 was shown to be *S*.

A ketal involving C12, C19 and C20 is also present in some *neo*-clerodane diterpenoids from *Teucria*. Teupolin V, **77**, from *T. polium* var. *polium*, is a representative of this group (*66*), the ketalic H-20 resonance occurring at δ 5.32 (s). This functionality has also been found in some diterpenoids isolated from *T. micropodioides* Rouy (*67*), such as teumicropin (**78**, R_1=OH, R_2=βOH, H), 3-acetyl teumicropin (**78**, R_1=OAc, R_2=βOH,H) and teupyrenone (**78**, R_1=OAc, R_2=O) from *T. pyrenaicum* L. (*68*). In these *neo*-clerodanes the H-20 resonance is a singlet at δ 5.10 and that of H-12 a double doublet at δ 5.11 ($J = 9$, 6.7 Hz), whereas a ^{13}C-signal at δ 100.4 was assigned to C-20.

The decalin portion of many *neo*-clerodane diterpenoids from *Teucrium* species exhibits the substitution pattern of clerodin (**4**), with the substituent at C-6 a 6α or 6β hydroxyl or acetoxy, or a ketone function. This substitution pattern was established in fruticolone (**66**) by lithium aluminum hydride reduction to the triol derivative (**79**, R=H) and by X-ray analysis of the triacetate (**79**, R=Ac). It is interesting to note that lithium aluminum hydride or sodium borohydride reductions of a 6-keto group in these *neo*-clerodanes, usually produces the 6α-equatorial hydroxy derivative (*64*). Isofruticolone (**80**, R_1=O, R_2=βOH, H; R_3=H) and 8β-hydroxyfruticolone (**80**, R_1=αOH, H; R_2=O; R_3=OH), were also isolated from *T. fruticans* (*69, 70*)

The decalin portion of the *neo*-clerodanes from *T. massiliense* (*71*), teumassilin (**81**, R_1=H; R_2=αOH,H; R_3=βOH,H) and 6,19-diacetyl teumassilin (**81**, R_1=Ac; R_2=αOAc,H; R_3=αOH,H) has a substitution

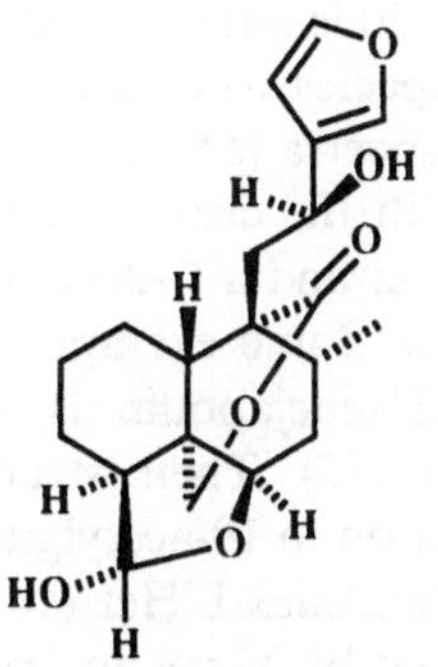

74

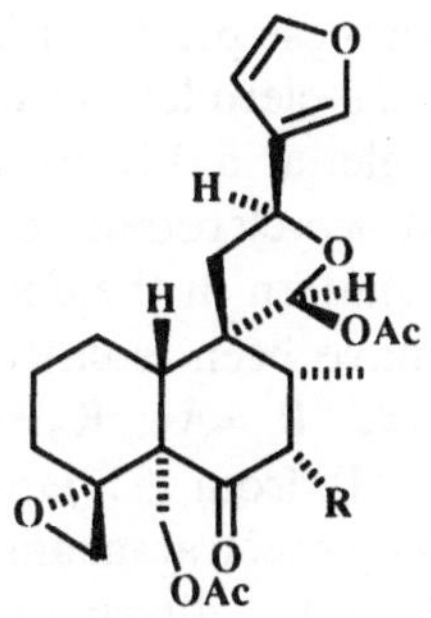

75

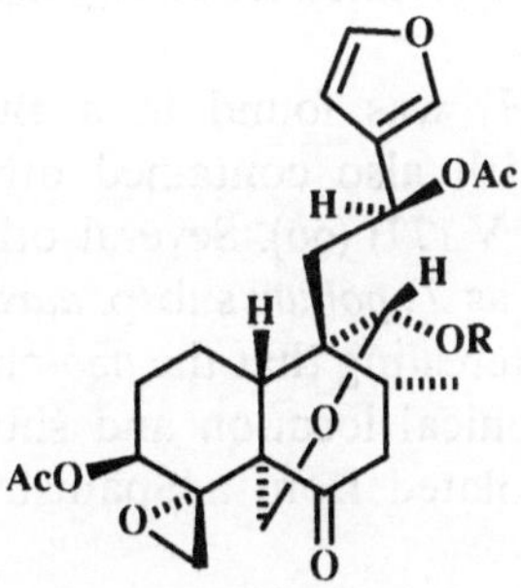

76

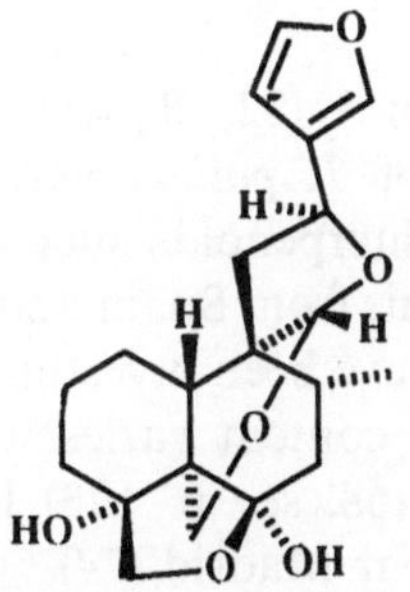

77

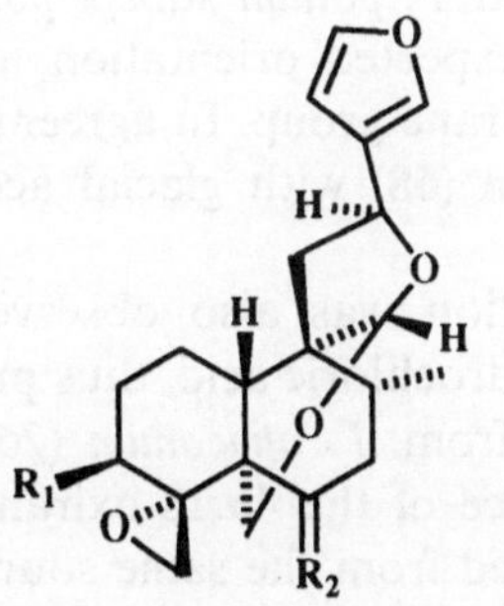

78

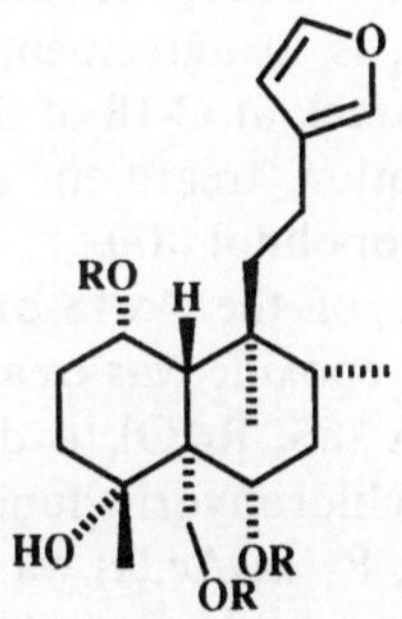

79

pattern similar to that of fruticolone with a hydroxy (or acetoxy) group at C-12 and no oxidation at C-1. Teupyrin B from *T. pyrenaicum* (65) (**81**, 3βOH, R_1=H; R_2=αOAc,H; R_3=OH,H) could be considered the 3β-hydroxy, 6-acetyl derivative of teumassilin.

In all of these products (**66, 79–81**) Me-20 is not oxidized, an unusual feature in *neo*-clerodanes from *Teucrium* species, as evidenced by a three proton singlet at δ 0.70 in the ^{1}H NMR spectra (*65, 69–71*).

Several *neo*-clerodane diterpenoids with the clerodin or fruticolone oxidation pattern in the decalin ring system and a C-20-C-12 γ-lactone function, have been isolated from *Teucria*. Some examples are teucjaponin A (**82**, R_1=Ac; R_2=βOH, H) and teucjaponin B (**82**, R_1=Ac; R_2=αOH, H) from *T. japonicum* Houtt (*72*). Their structures were elucidated by spectral means and by oxidation to 19-acetylgnaphalin (**82**, R_1=Ac; R_2=O), a diterpene from *T. gnaphalodes* L'Her (*64*) and other *Teucria* species, whose structure was proved by X-ray analysis (*62*). The 19-hydroxy derivative, gnaphalin (**82**, R_1=H, R_2=O), gnaphalidin (**75**, R=H) and other *neo*-clerodanes were also isolated from *T. gnaphalodes* (*64*).

Teupolin I (**82**, R_1=Ac, R_2=αOH,H) was found in a Bulgarian collection of *T. polium* var. *polium* which also contained other *neo*-clerodane diterpenoids such as teupolin V (**77**) (*66*). Several other subspecies of this from Spain and Italy, such as *T. polium* subsp. *aureum* and *capitatum* have been investigated. It is interesting that the *neo*-clerodane diterpenoid content varies with geographical location and subspecies. Picropolin (**68**, see p. 123) has been isolated from a Spanish sample collected near Madrid (*74*).

Some *neo*-clerodane diterpenoids from *Teucria* contain a 4,18-diol in place of a 4α,18-oxirane. The α configuration assigned to the hydroxyl at C-4 in picropolinol (**83**) and teupolin III (**84**), two products obtained from *T. polium* subsp. *capitatum* (*75*) and *T. polium* subsp. *polium* (*73*) respectively, is in agreement with the expected orientation for a nucleophilic attack at C-18 of the 4α,18-oxirane group. In agreement with this assumption, treatment of picropolin (**68**) with glacial acetic acid yielded picropolinol (**83**).

Opening of the 4α,18-oxirane function was also observed when tafricanin A epoxide was treated with hydrochloric acid, thus producing tafricanin A (**85**, R=O), a diterpenoid from *T. africanum* (*76*), which contains a chlorohydrin function in place of the 4α,18-oxirane. Tafricanin B (**85**, R=αOAc,H) was also isolated from the same source. Both diterpenoids were shown to be the true natural products and not artifacts (*76*).

Teuvincentin A (**86**) also contained a 18,4α-chlorohydrin. This product from *T. polium* subsp. *vincentinum* (*77*), was also shown to be a true natural product. The structure and absolute stereochemistry were established on the basis of spectral data and an X-ray analysis of the 7-acetate. With these data in mind the stereochemistry at C-4 proposed for montanin E (**87**), a product isolated from *T. montanum* subsp. *skorpillii* (*79*) should be revised. The chemical shift of the signal ascribed to H-10β (δ 2.80,dd) is in agreement with the deshielding expected as a result of its 1,3-diaxial interaction with the 6β-hydroxy group.

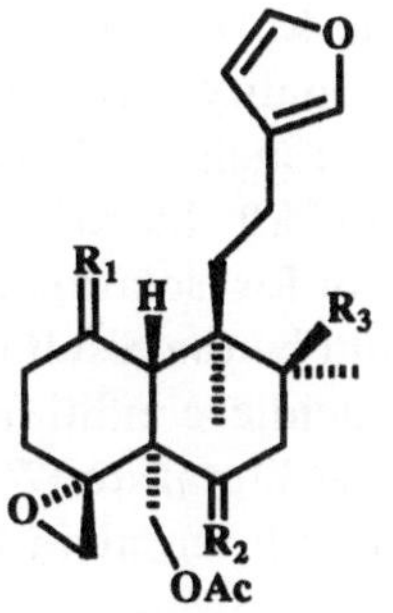

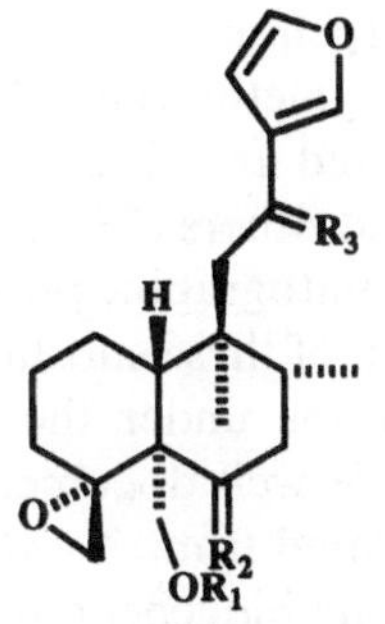

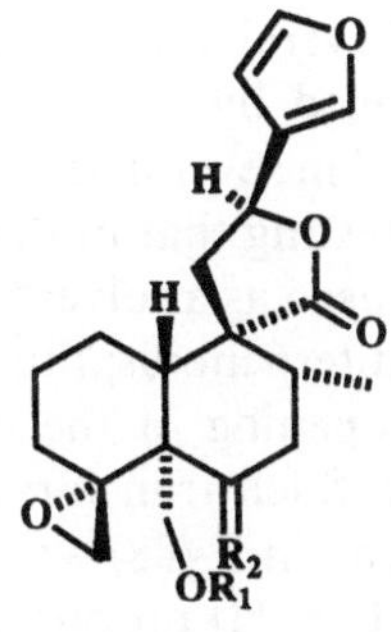

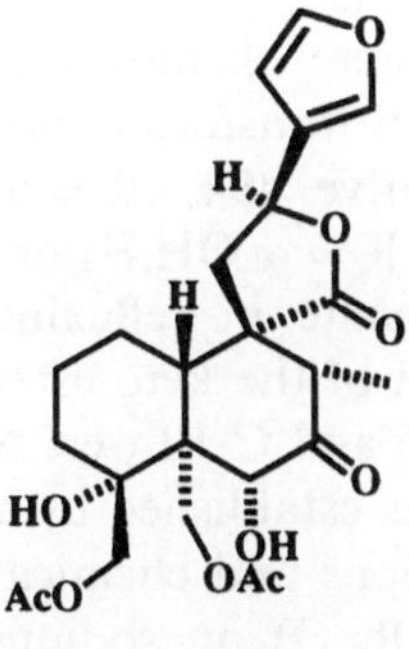

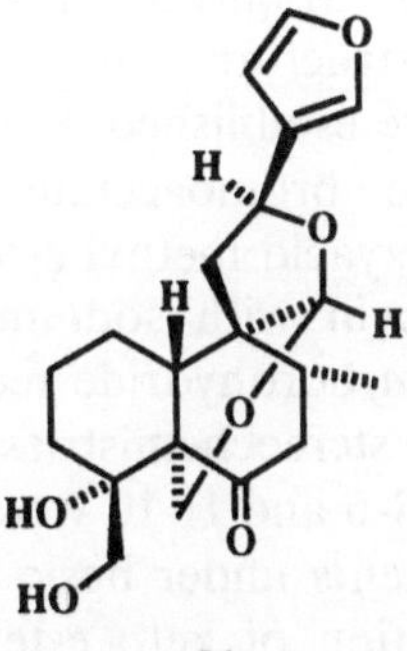

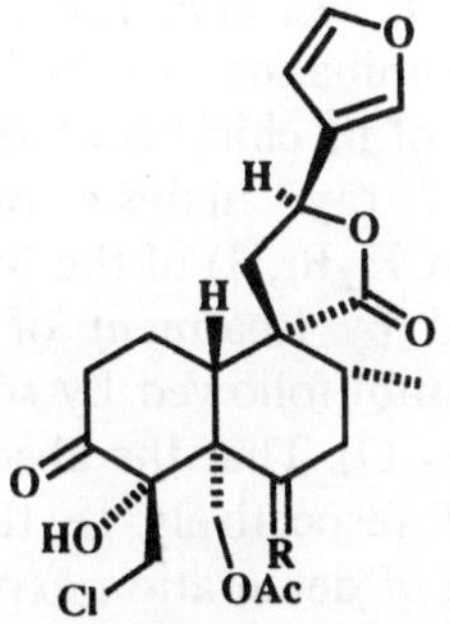

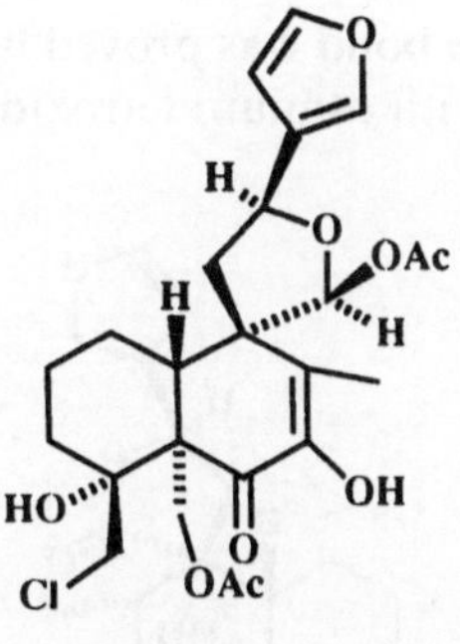

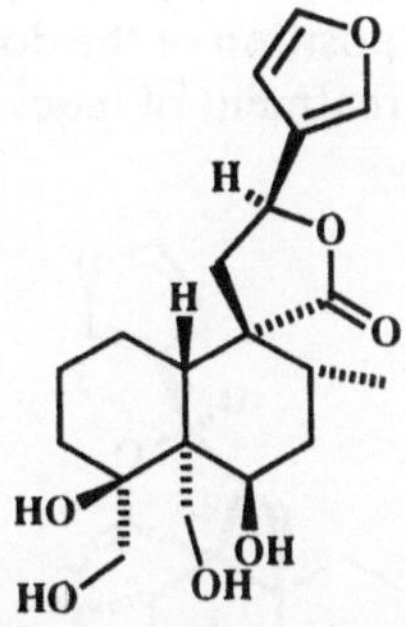

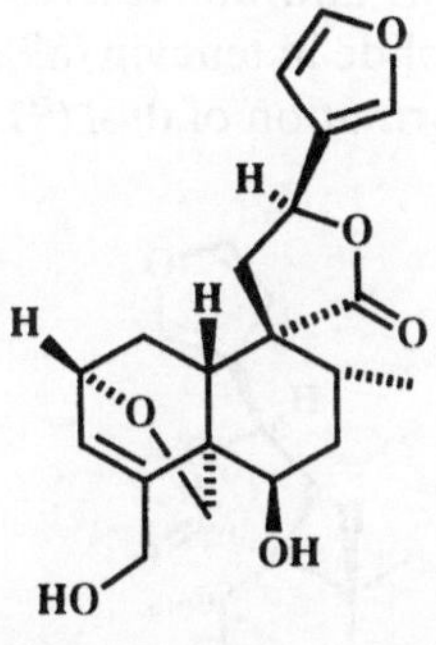

Several *neo*-clerodane diterpenoids exist in which the 4α,18-oxirane is replaced by a 3-ene-4-hydroxymethylene. This functionality has been found in several products isolated from *T. salviastrum* (*58*) such as **72**. It is interesting that in this case and others (**73**, for example) (*60*) chloroform was used as a solvent for chromatographic purification or for isolation of the diterpenoids; thus isolation of these substances could be the result of the opening of the 4α,18-oxirane under the slightly acidic conditions used. Such transformations are well documented in the literature (*78*). Teusalvin F (**88**) was also isolated from *T. salviastrum* with teusalvin D (**73**, R=OH) proposed (*58*) as its biogenetic precursor.

Some diterpenoids of *Teucrium* species have a 19-*nor*-clerodane skeleton. The first member of this group to be described was teucvin (**89**) from *T. viscidum* Blume var. *Miquelianum* (Maxim) (Japanese name "Tsurunibakusa") (*80*). The structure and absolute stereochemistry of most of its chiral centers were established by chemical transformations and X-ray analysis of the bromoacetate derivative (**90**, R = α-OCOCH$_2$Br,H) of the hydroxyacid methyl ester (**90**, R = α-OH,H) obtained by treatment of teucvin with sodium carbonate in refluxing methanol followed by sodium borohydride reduction of the keto-ester (**90**, R=O). That the absolute stereochemistries at C-6 and C-10 were *S* and *R* respectively, i.e. that H-6 and H-10 were β, was established by a series of deuteration experiments under basic conditions and chemical transformations (*80*). Formation of keto ester (**90**, R=O) on sodium carbonate treatment, and careful examination of the spectral data of teucvin and derivatives suggested the presence of the α,β-unsaturated 18,6-olide in teucvin (**89**). The position of the double bond was proved by the formation of diol (**91**) on treatment of teucvin with osmium tetroxide (*80*).

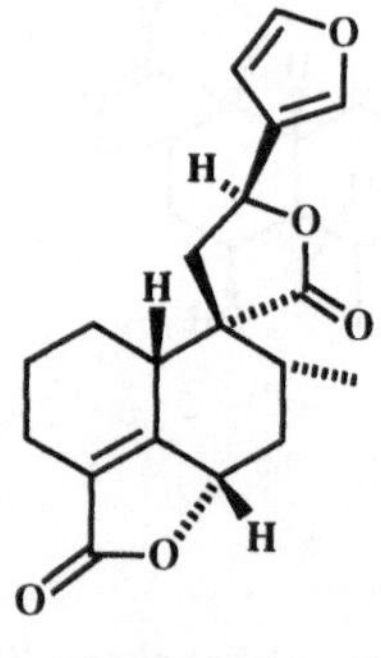

89 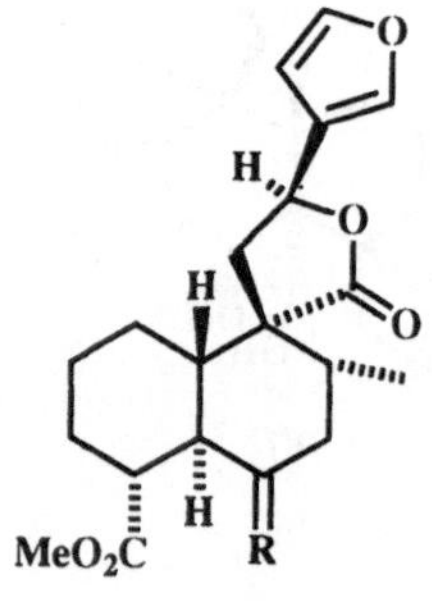   90 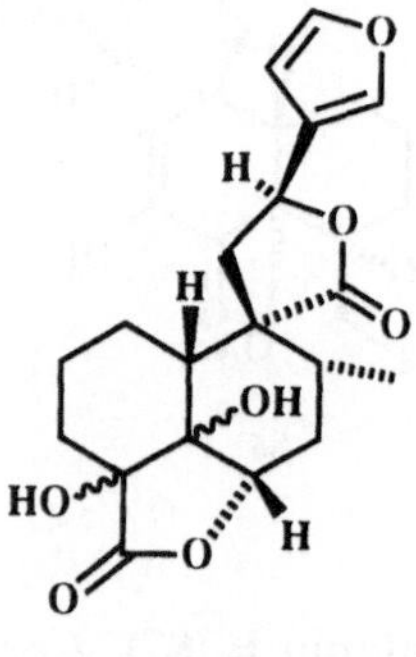   91

From the same source, teucvidin (**92**), was obtained. Its absolute stereochemistry was established by comparison of its CD spectrum with that of teucvin (**89**). The Cotton effects were of opposite signs and nearly the same intensities which suggested an antipodal relationship of the A, B and E rings system between teucvidin and teucvin assuming that they had the same conformation. This was confirmed by INDOR experiments. Therefore teucvidin (**92**) was the 6*R*,10*S* epimer of teucvin (**89**) (*81*), a conclusion proved by X-ray analysis (*82*). Several 19-*nor-neo*-clerodane diterpenoids with a 4-ene-18,6-lactone function, have been isolated from *Teucrium* species differing in the configuration of the C-6 and C-10 chiral centres.

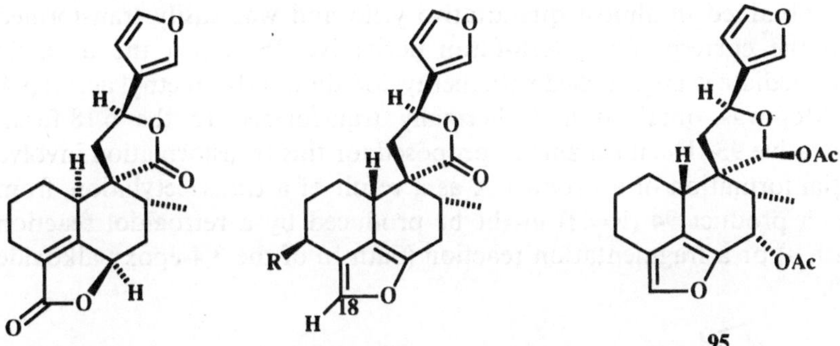

Structure and stereochemistry of another 19-*nor*-clerodane, montanin A (**93**) from a Bulgarian *T. montanum* collection (*83*) were deduced on the basis of spectral data and correlation with teucvin (**89**). The resonance of the furan proton, H-18, was a singlet at δ 6.98. When montanin A was kept for five days in chloroform, autoxidation occurred and teucvin (**89**) was obtained in 70% yield, showing the high stereoselectivity of this reaction (*83*). This type of oxidation of a trisubstituted furan derivative has been described earlier (*84*). A possible mechanism involves nucleophilic attack of hydrogen peroxide at C-18 as shown in Scheme 1.

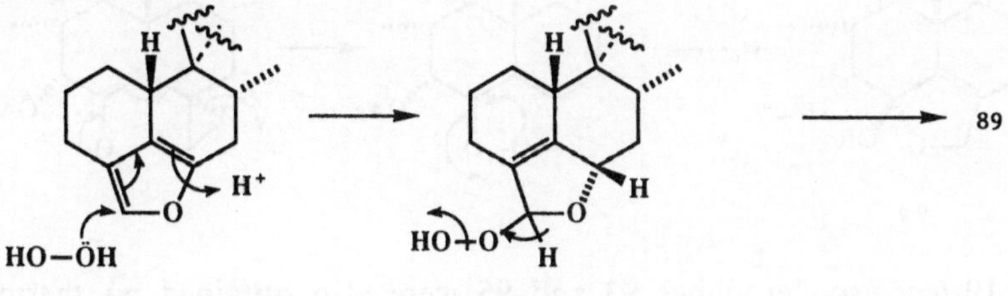

Scheme 1

Montanin A (**93**) could be an artifact as it was also obtained when gnaphalin (**82**, R_1=H; R_2=O; R_3=H) was kept in solution in contact with silica gel (*62*). This transformation was also produced by base treatment of 19-acetylgnaphalin (**82**, R_1=Ac, R_2=O, R_3=H) (*57*).

The mechanism of the transformation of the 19-hydroxy (or acetoxy)-4α,18-epoxy-6-oxo-*neo*-clerodane derivatives to the corresponding 19-*nor*-clerodanes containing a 6,18-furan ring such as montanin A (**93**, R=H) under basic conditions has been studied recently (*85*). To this purpose, eriocephalin (**75**, R=OH) was exposed to potassium tert-butoxide in dry THF under mild conditions. An unstable compound **94** (R=H) was obtained in almost quantitative yield and was easily transformed into the corresponding 6,18-furan derivative **95**. When the unstable intermediate was quenched with methyl iodide, a stable methyl acetal (**94**, R=Me) was obtained and thermally transformed to the 6,18-furan derivative **95**. The mechanisms proposed for this transformation involve initial formation of a product **A** as a result of a transacetylation, from which product **94** (R=H) might be produced by a retroaldol reaction (path a) or a fragmentation reaction (path b) of the 3,4-epoxy alkoxide (*86*).

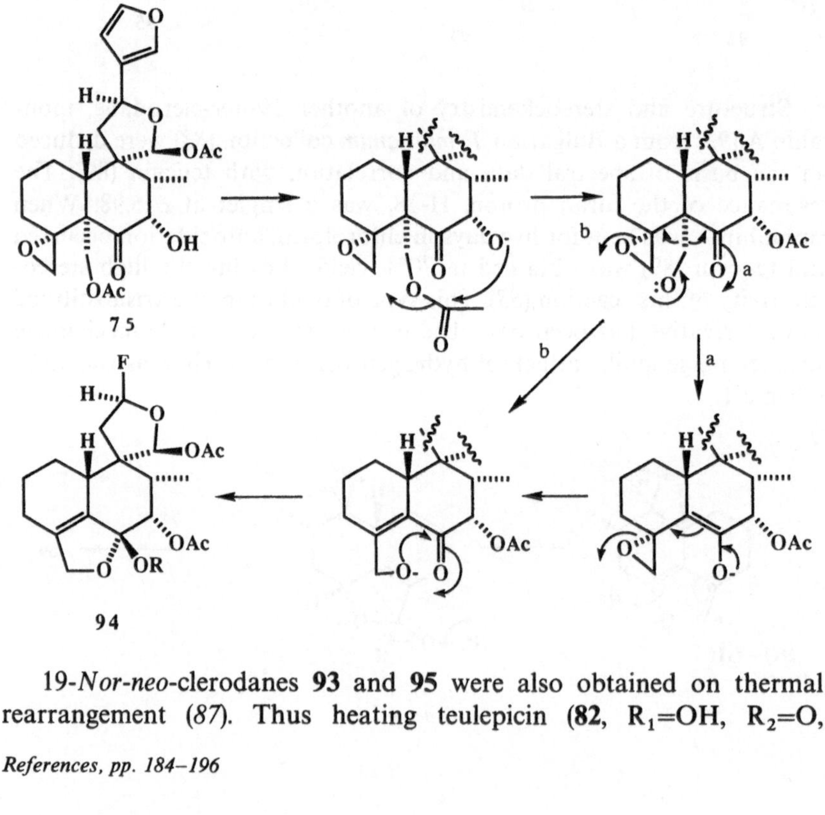

19-*Nor*-*neo*-clerodanes **93** and **95** were also obtained on thermal rearrangement (*87*). Thus heating teulepicin (**82**, R_1=OH, R_2=O,

R_3=OH) (*57*) at 185° for 5 min. yielded 3β-hydroxy-montanin A (**93**, R=OH) in almost quantitative yield. This transformation could be envisaged as the result of a retroaldol reaction and subsequent opening of the oxirane ring, nucleophilic attack at the C-6 ketonic group and dehydration. The necessary condition for its occurrence is the presence of a keto group at C-6 and a hydroxyl at C-19 (*87*).

Some *neo*-clerodane diterpenoids from *Teucrium* species have an α,β-unsaturated-γ-lactone function in ring A, i.e. a 3-ene-18,19-olide. This is so in the case of the 6α and 6β-hydroxyteuscordins, (**96**, R_1,R_2=α or β OH,H; R_3=H) and teuscordinon (**96**, $R_1 R_2$=O, R_3=H), diterpenoids of *T. scordium* (*88, 89*). This functionality was first described for olearin (**97**), a *neo*-clerodane of *Olearia heterocarpa* (Compositae) whose structure and stereochemistry were established by chemical transformations and spectroscopic data (*90*). The ^{1}H NMR spectrum of olearin contained a multiplet at δ 6.71 due to H-3 and an AB system at δ 4.10 and 3.92 (*J*=8 Hz) ascribed to the C-19 methylene. The upfield signal (δ 3.92) was long range coupled (*J* = 1.5 Hz) with axial H-6 which required that the A/B ring system be *trans* (θ = 180 − 205°). The H-20 signal (δ 0.65) also showed a W coupling (*J* = 1.2 Hz) which indicated that this methyl group was axial. The absolute stereochemistry of olearin shown in formula (**97**) was proved by correlation with a substance from a *Dodonea* species, whose absolute stereochemistry had been unequivocally established (*90*). Whenever a W coupling has been observed for the 19-pro-*S* proton in a *neo*-clerodane diterpenoid containing a 3-ene-18,19-olide function, it has been attributed (*91*) to coupling with the 6β axial proton in an A/B *trans*-fused system. However W coupling is also possible in an A/B *cis*-fused system if the dihedral angle is appropriate.

Recently (*92*) a partial synthesis of teuscordinon (**96**, $R_1 R_2$=O; R_3=H) has been accomplished starting from teubotrin (**72**, R=H). Thus treatment of the latter with sodium hydroxide in aqueous methanol produced

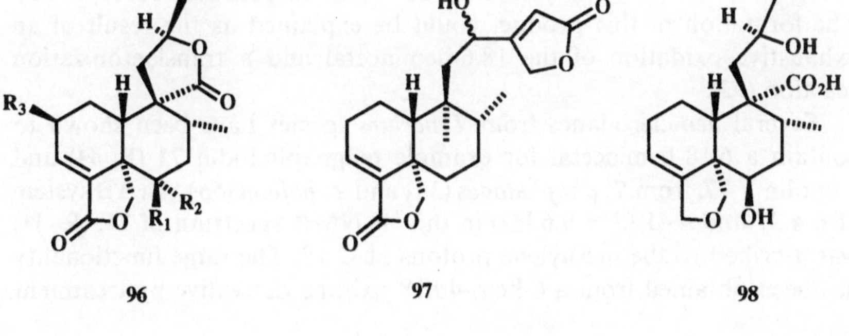

96   **97**   **98**

the 18,19-ether derivative (98) which could be transformed into teuscordinon (96) by lactonization with phosphorus pentoxide and oxidation with chromium trioxide in pyridine.

3,4-Dihydro-18,19-olide *neo*-clerodane diterpenoids have been also isolated from some *Teucrium* species. An example of this type is dihydroteugin (99, R_1=OH, R_2=βOH), a diterpenoid isolated from *T. chamaedrys* (93). Its structure and stereochemistry were based on spectral data and a correlation with teugin (96, R_1=OH=R_3,R_2=H). Reduction of the latter with sodium borohydride yielded dihydroteugin (99). Double doublets at δ 3.76 and 3.42 in the ^{1}H NMR spectrum of 99 (R_1=OH, R_2=βOH,H) were assigned to the 4β and 10β axial protons respectively, both strongly deshielded by 1,3-diaxial interactions with the secondary hydroxyl groups at C-2 and C-6.

A 18,19-hemiacetal functionality is also present in some *neo*-clerodane diterpenoids from *Teucria* for example in teuscorodin (100 R_1=H, R_2=OH) from *T. scorodonia* (94). Its structure was deduced from spectral data and chemical transformations. The hemiacetal proton H-18 was observed in its ^{1}H NMR spectrum as a broad singlet at δ 5.07. Oxidation of teuscorodin with chromium trioxide-pyridine yielded 6-keto-teuscordin (100, R_1R_2=O), a substance isolated from *T. scordium* and obtained by oxidation of teucrin E (99, R_1=H, R_2=αOH,H) and teuchamaedryn B (or teucrin H-2; 99 R_1=H; R_2=βOH,H) (94).

T. scorodonia also furnished the 3-ene-18,6-olide-19-hydroxy derivative teuscorodonin (101, R=H_2) (94). Two double doublets at δ 6.75 (J = 5.9, 2.7 Hz) and 5.39 (J = 10.7, 6.2 Hz) in its ^{1}H NMR spectrum were ascribed to H-3 and H-6α respectively. An analysis of the coupling constants for H-6, H_2-7 and H-8 led to the conclusion that ring B must have a boat conformation (94).

A partial synthesis of the 2-keto derivative of teuscorodonin (101 R=O) was achieved from teubotrin (72, R=H) as starting material (92). Thus manganese dioxide oxidation of teubotrin in acetone yielded the expected aldehyde 102 (R=H_2) its 2-keto derivative 102 (R=O) and a minor product which was proved to be 2-ketoteuscorodonin (101, R=O). The formation of this product could be explained as the result of an exhaustive oxidation of the 18,6-hemiacetal and a translactonization reaction (92).

Several *neo*-clerodanes from *Teucrium* species have been shown to contain a 6,18-hemiacetal for example teugnaphalodin 71 (R=H) and teupolin V 77, from *T. gnaphalodes* (56) and *T. polium* (66). An AB system at δ 4.27 and 4.41 (J = 9.6 Hz) in the ^{1}H NMR spectrum of 71, (R=H) was ascribed to the methylene protons at C-18. The same functionality has been obtained from a 6-keto-4α,18-oxirane derivative by treatment with phosphoric acid (66). The drastic conditions necessary for this transformation, exclude the possibility that compounds of this type are artifacts. A possible mechanism for this transformation (66) is shown below.

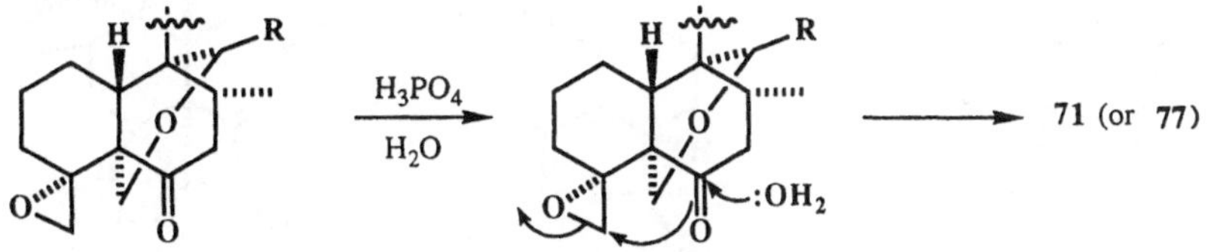

An orthoacetate is present in teulanigeridin (103, R_1=OH, R_2=H, R_3=αOH,H) from *T. lanigerium* (95). It was purified as the 3,20-diacetate. Presence of the ortho-acetate group was established by NMR spectrometry. The diacetate exhibited a singlet at δ 1.47 (3H), which was also present in the ^{1}H NMR spectrum of the mixture of diterpenoids prior to acetylation. A singlet at δ 106.1 and a quartet at 23.3 in the ^{13}C NMR spectrum were attributed to the orthoacetate group (95).

Thermal treatment (87) of 19-acetylgnaphalin (104, R_1=O; R_2=H_2) and 19-acetylteupolin (104, R_1=O, R_2=βOAc,H), produced the corresponding ortho-acetate derivatives 103 (R_1=H=R_2) and 103 (R_1=H, R_2=βOAc,H) in 49 and 34% yields respectively, together with the mixture of the corresponding C-4 epimeric C-18 aldehydes (105). Thermal cleavage of the oxirane ring followed by hydride migration has been proposed as a possible mechanism for this transformation (92), but thermal rearrangement of the 4α,18-oxirane group to enol A shown below followed by tautomerization to the most stable derivative (105, R_1=H, R_2=CHO, R_3=H_2 or βOAc,H), in which the aldehyde group is equatorial, cannot be discarded. The formation of the 6β,18-tetrahydrofuran-orthoester moiety could be rationalized as proceeding through intermediate B.

Pyrolysis of acetylteucjaponin (104, R_1=αOAc,H; R_2=H_2) yielded the C-4 epimeric mixture of aldehydes; while 7-deacetylcapitatin (104, R_1=O; R_2=αOH,H) and capitatin (104, R_1=O, R_2=αOAc,H) suffered an α-ketol rearrangement giving picropolin (104, R_1=αOH,H; R_2=O) and its acetate (104, R_1=αOAc,H; R_2=O) in almost 100 and 30% yields, respectively. This thermal isomerization is rationalized (87) as a consequence of the 1,3-diaxial steric interaction between the C-7α substituent and the C-20, C-12 lactone function. This interaction does not exist in 19-acetylgnaphalin (104, R_1=O, R_2=H_2).

Neo-clerodane diterpenoids containing an oxetane ring have been found in *Teucria*. Chamaedroxide (106, R=OH) was isolated from

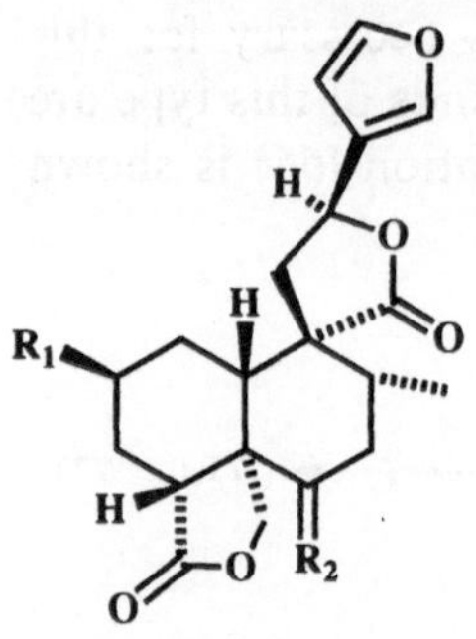

99

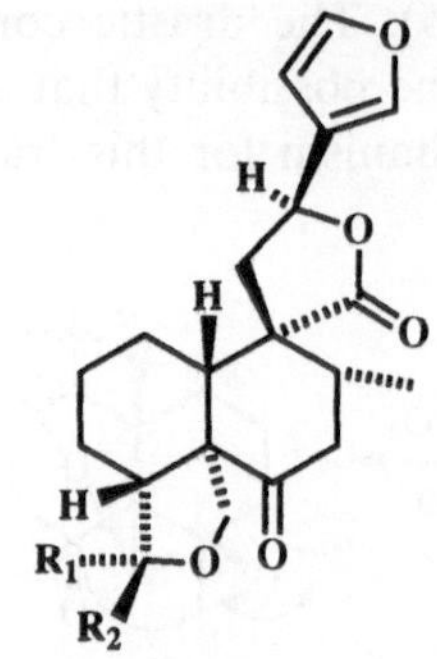

100

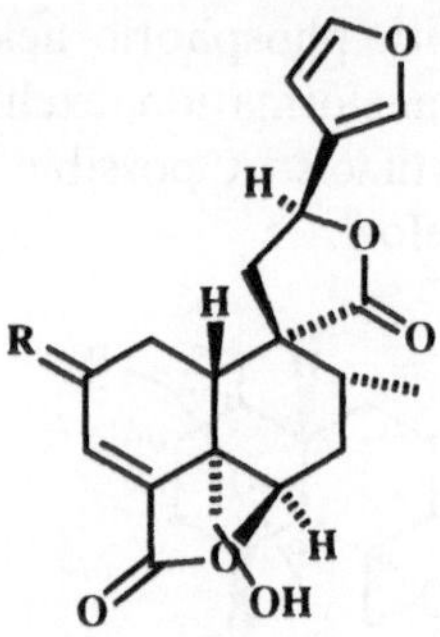

101

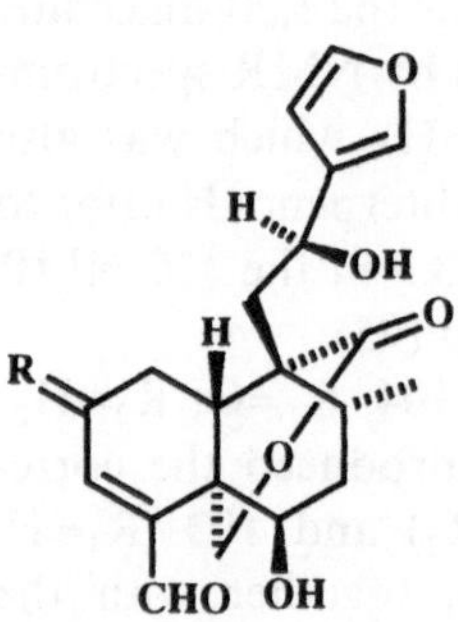

102

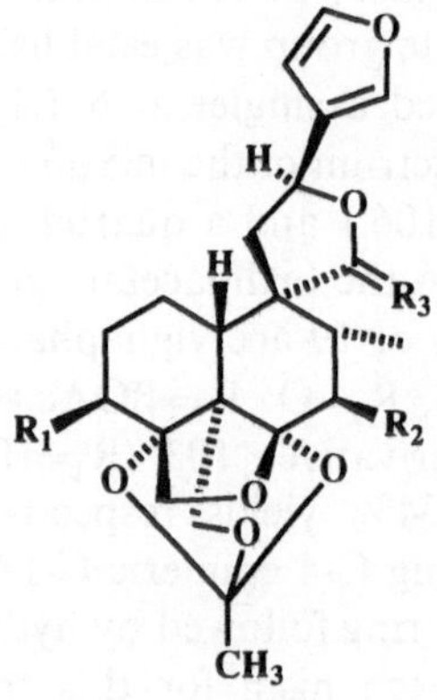

103

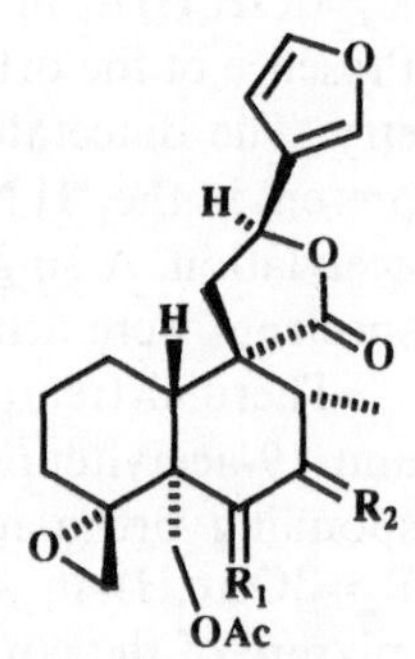

104

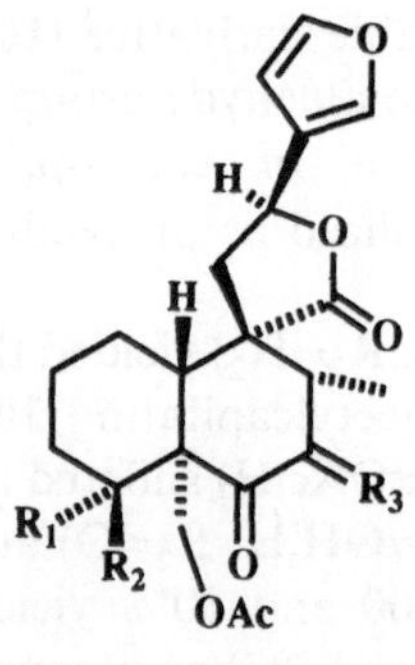

105

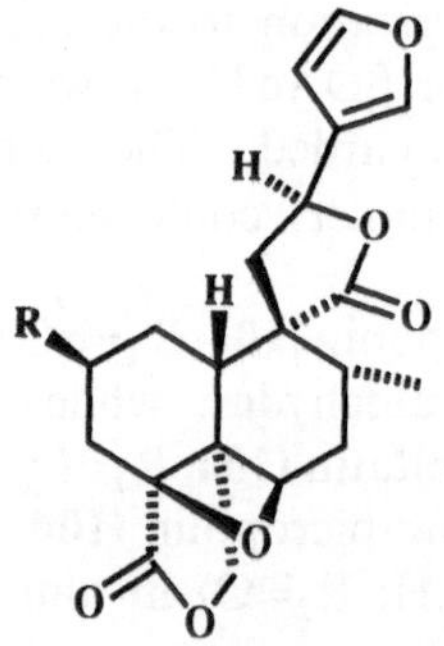

106

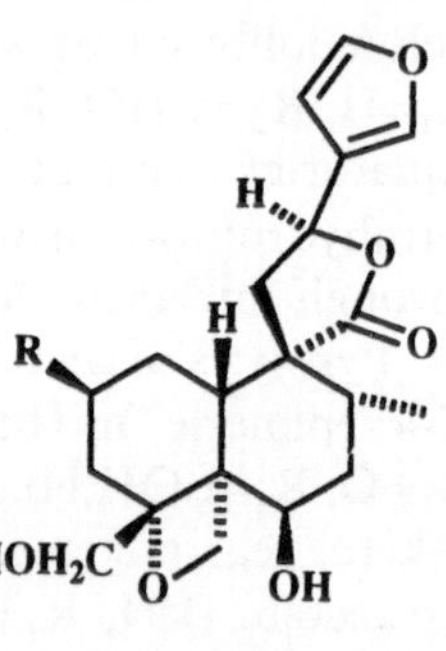

107

T. chamaedrys (*96*). Its structure and absolute stereochemistry were established by spectroscopic analysis. The presence of a β-orientated ether oxygen linking C-6 and a trisubstituted carbon atom C-4 (or C-10), was deduced from the ^{1}H NMR spectrum which showed the H-6α signal as a double doublet at δ 5.17 ($J = 7.5$, 6 Hz). A triplet at δ 3.25 ($J = 10$ Hz) was attributed to the C-10 (or C-4) β-axial proton. Structure **106** (R=OH) was finally established by X-ray analysis (*96*).

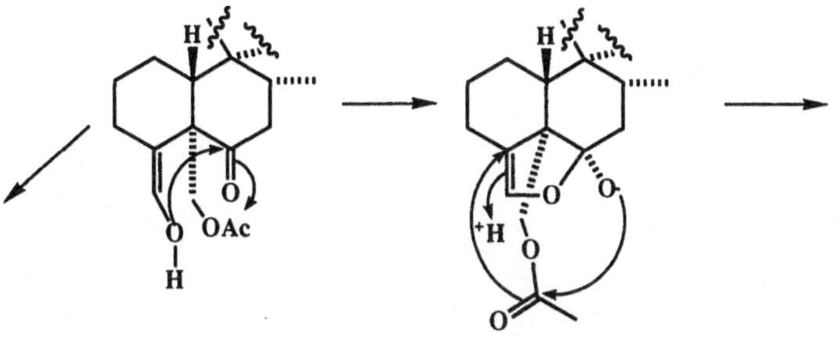

Teucroxide (**107**, R=OH) also from *T. chamaedrys* (*97*) contained an oxetane whose ethereal oxygen was α bound to C-4 and to the C-19 methylene group. This functionality had been previously found in montanin D (**107**, R=H), from *T. montanum* (*98*). In the ^{1}H NMR spectrum of **107**, the 4α,19-oxetane was responsible for an AB system, at δ 4.69 and 4.20 ($J = 8$ Hz), ascribed to the methylene protons at C-19. In the ^{13}C NMR spectrum a singlet at δ 88.7 and a triplet at 71.6 were assigned to C-4 and C-19 respectively. The 4β,6β-oxetane group has been also found in 2-deoxychamaedroxide (**106**, R=H) from *T. divaricatum* Sieber ex Boiss. subsp *canescens* (Celak) Holmboe (*99*). Montanin D (**107**, R=H) was also isolated from this *Teucrium* species.

The oxetane functionality found in these *neo*-clerodane diterpenoids is rare in natural products. It is a fundamental structural feature of the diterpenoid taxol, a known antitumor agent (*100*).

Recently several C-10 oxygenated *neo*-clerodane diterpenoids have been isolated from *Teucrium* species. The structure of teupestalin A (**108**) was established (*101*) on spectroscopic evidence and by an X-ray analysis. Two AB systems in the ^{1}H NMR spectrum, at δ 3.02, 3.86 ($J = 5.2$ Hz) and δ 3.68, 5.00 ($J = 9.5$ Hz), were ascribed to the methylenes at C-18 and C-19 respectively. The location of the tertiary hydroxyl at C-10 as β-axial was deduced by a careful analysis of the ^{1}H and ^{13}C NMR spectra with a singlet at δ 77.92 being attributed to C-10. The hemiacetalic nature of C-3 was also deduced from the ^{13}C NMR spectrum which contained a singlet at δ 101.54 attributed to C-3.

Teupestalin B (**109**) also from *T. pestalozzae* (*101*) also contained an hemiacetal involving C-3 and C-10, as deduced from the ^{13}C NMR spectrum, which contained two singlets at δ 105.17 and 87.16 attributed to C-3 and C-10 respectively. The β configuration of the C3-C10 oxygen bridge, was established by a careful analysis of the ^{1}H NMR data.

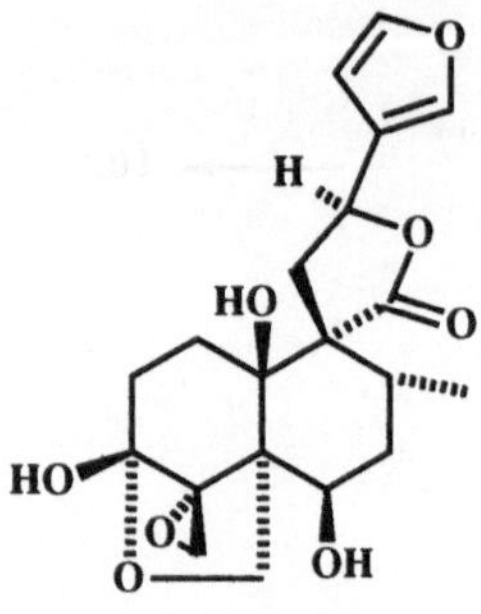

108 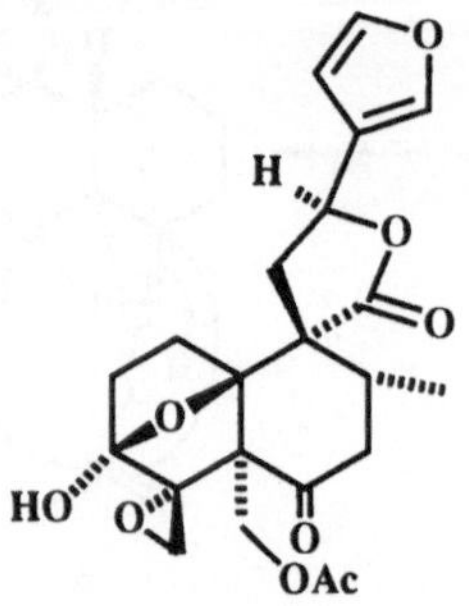   109 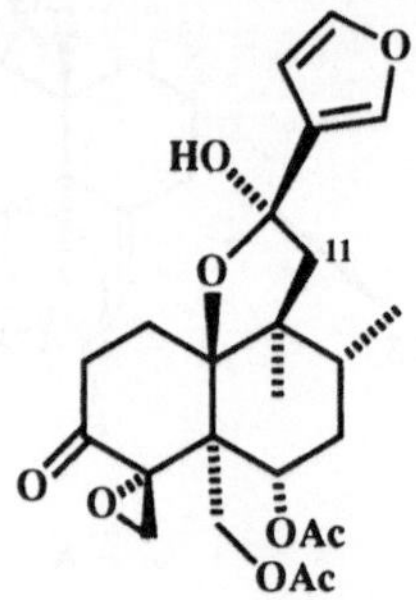   110

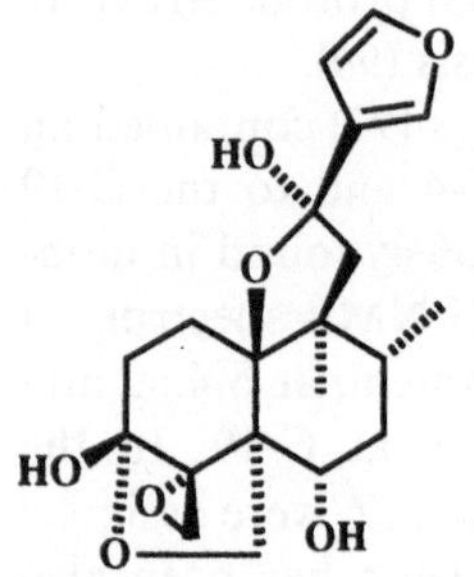

111 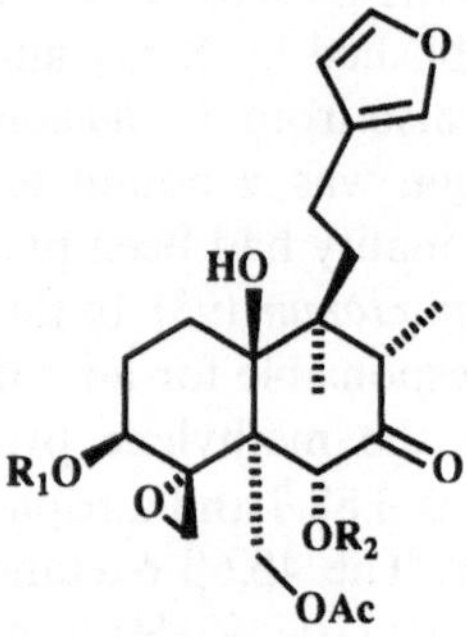   112 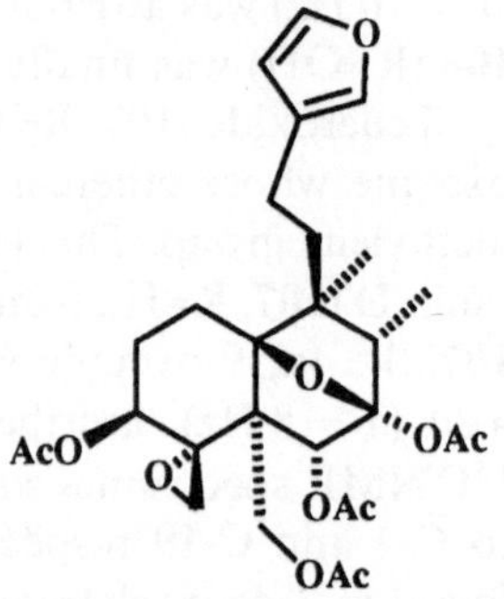   113

Teupestalins A (108) and B (109) are the first C-10 oxygenated *neo*-clerodane diterpenoids from *Teucria*. In both compounds the C-10 oxygen atom has been shown to be β-orientated.

Teucrium oliverianum, a species used in Saudi Arabia for treatment of diabetes, also contained several C-10 oxygenated *neo*-clerodane diterpenoids (*102*). The structure and relative stereochemistry of teucrolivin A (110) was established on the basis of chemical transformations, spectroscopic evidence, which included nOe experiments, and an X-ray analysis. The methyl group at C-9 in the diterpenoids from this species, is not oxygenated as in the *neo*-clerodanes from *T. fruticans* (*69, 70*), *T. massiliense* (*71*) and *T. pyrenaicum* (*65*). The ^{13}C NMR spectrum of teucrolivin A contained two singlets at δ90.09 and 100.77, which were ascribed to C-10 and C-12, involved in a hemiacetal. Pyrolysis of 110 produced the 11,12-ene derivative and alkaline treatment yielded 111, which contained a second hemiacetal.

Teucrolivins B (112, R$_1$=Ac, R$_2$=H) and C (112, R$_1$=R$_2$=H) from *T. oliverianum* also contained a free hydroxyl at C-10. Acetylation produced the hemiacetal acetate **113** together with the normal acetylation product **112** ($R_1 = R_2 = Ac$) (*102*).

Two *neo*-clerodane diterpenoids oxidized at C-8 and C-10 have been recently isolated from *Teucrium oliverianum* (*103*). Their structures (**114,** R=H and Ac) were established on spectroscopic evidence which included an X-ray analysis of teucrolivin G (**114,** R=Ac). The unusual 2,6-dioxabicyclo-[2·2·1]-heptane part structure present in both substances was evident from the ^1H and ^{13}C spectral data. Absence of a doublet due to the secondary C-17 methyl group common to *Teucria neo*-clerodanes suggested substitution at C-8. A singlet at δ 103.02 in the ^{13}C NMR spectrum of **114** (R =Ac) was assigned to the ketalic C-12. Two singlets at δ 86.39 and 82.59, were attributed to the oxygen carrying carbon atoms 8 and 10. The relative stereochemistry of teucrolivin G was deduced by nOe experiments. The positive Cotton effect found in the CD curve, was used as evidence for the *neo*-clerodane absolute configuration shown in **114.** The 2,6-dioxabicyclo-[2·2·1]-heptane functionality present in teucrolivins G and H may be the result of an interaction between a ketonic function at C-12 and two β-orientated hydroxyls at C-8 and C-10. Therefore **115** could be a biogenetic (or chemical) precursor of such substances.

A tetra*nor-neo*-clerodane diterpene, teucrolivin F, was also isolated from *T. oliverianum* (*104*). Structure and stereochemistry as in **116** were established on spectral evidence and comparison with the spectral data of teucrolivin A (**110**). The β-substituted furan attached to C-12 in teucrolivin A was absent in teucrolivin F which had spectral properties appropriate for a 12,10β-γ-lactone. Teucrolivin F is the first tetra*nor-neo*-clerodane diterpenoid isolated from *Teucrium* species.

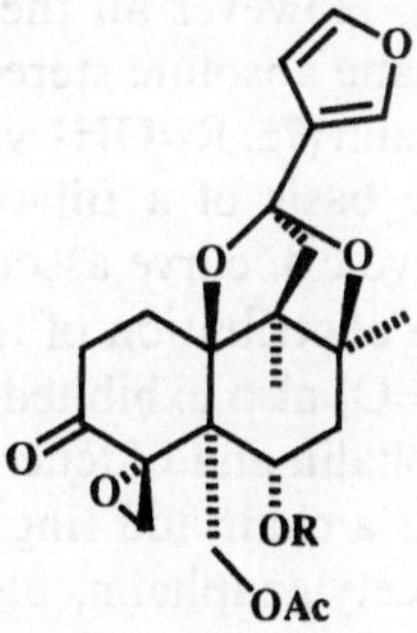

114 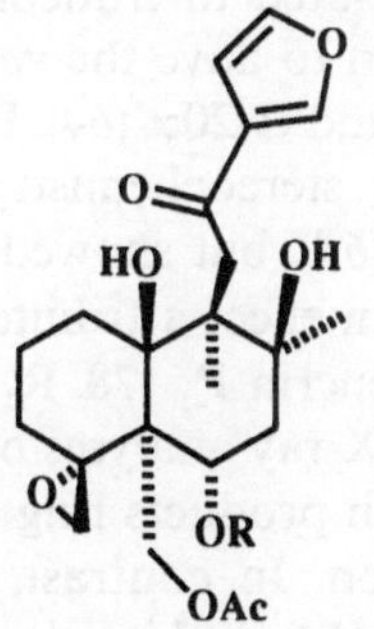   **115** 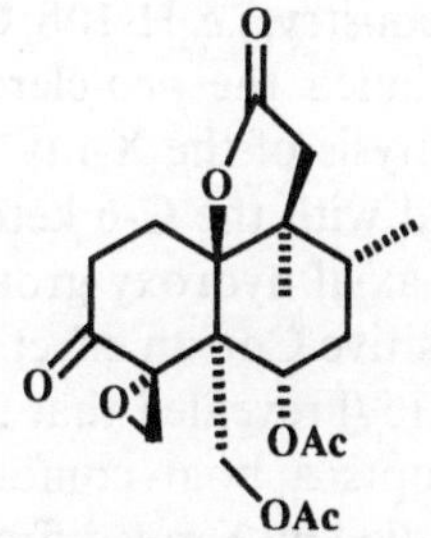   **116**

A *seco-neo*-clerodane, fruticolide, has been isolated recently from *T. fruticans* (*105*), the first such diterpene from *Teucrium* species. Its structure (**117**) was established by spectroscopic means, including an X-ray

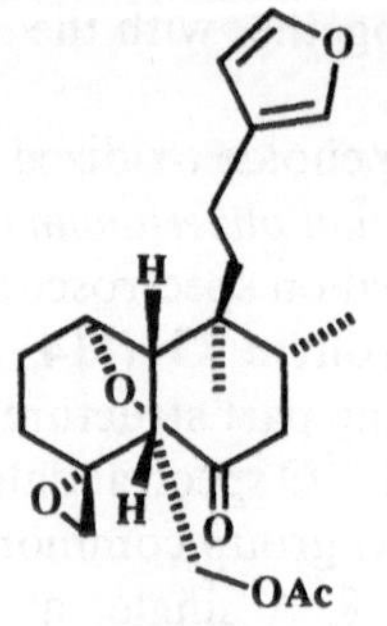

117

analysis. A multiplet at δ 4.82 in the ^{1}H NMR spectrum was ascribed to H-1β; C-1 was a doublet at δ 72.09 in the ^{13}C NMR spectrum.

Absolute Stereochemistry of Clerodane Diterpenoids from Teucrium Species

The absolute stereochemistry of the *neo*-clerodane diterpenoids from *Teucrium* species has been established by several procedures which include X-ray analyses, chiroptical data and chemical correlations.

Assignments of absolute stereochemistry based on the sign of the Cotton effect of 6-oxo clerodanes have proved to be misleading. For example gnaphalin (**83**, $R_1=R_3=H$, $R_2=O$) and related compounds show a negative c.d. curve related to the C-6 ketone as do 6-oxoajugarins and diterpenoids structurally related to fruticolone (**66**). However all these substances have been shown to have the *neo*-clerodane absolute stereochemistry, *i.e.* H-10β, C-19 and C-20α. (*64*). Eriocephalin (**75**, R=OH) was assigned the *neo*-clerodane stereochemistry on the basis of a Bijvoet analysis of the X-ray data (*63*), but showed a positive c.d. curve associated with the C-6 ketone, an effect attributed to the contribution of the 7α-axial hydroxy group. Teucrin P_1 (**78**, $R_1=H$, $R_2=O$) also exhibited a positive Cotton effect. The X-ray analyses of eriocephalin and of teucrin P_1 (*64*) revealed that in both products ring A adopts a chair and ring B adopts a boat conformation. In contrast, in 19-acetylgnaphalin, also studied by X-ray diffraction (*64*), and in related compounds which show a negative Cotton effect, rings A and B both adopt a chair conformation.

Thus, the sign of the Cotton effect of a 6-oxoderivative is not sufficient to permit deduction of the absolute stereochemistry of a given clerodane.

As stated earlier many *neo*-clerodane diterpenoids from *Teucria* have a 20,12-γ-lactone function. The configuration of these compounds at C-12 was usually established as 12 *S* by X-ray diffraction and by analysis of ^{1}H and ^{13}C NMR spectra. However recently (*106*) several 20,12-olide *neo*-clerodanes have been shown to have the 12 *R* configuration as, for example montanin C (**118**, R=Ac) from *T. montanum* (*51*) and *T. massiliense* (*71*), whose structure and absolute stereochemistry were rigorously established by X-ray analysis (*106*). Products with the same structure but differing in the C-12 configuration show differences in physical properties and ^{13}C NMR spectra but have very similar ^{1}H NMR spectra. The most significant influence of C-12 chirality was observed in the chemical shifts of C-8 and C-10. A change of C-12R to C-12S resulted in shielding of C-8 ($\Delta\delta$ = approx. + 2.5) and deshielding of C-10 ($\Delta\delta$ = approx. − 2.3). Differences in the ^{1}H NMR spectra are not sufficient to establish the C-12 configuration because they can be affected by the substitution pattern in the rest of the diterpene skeleton (*106*).

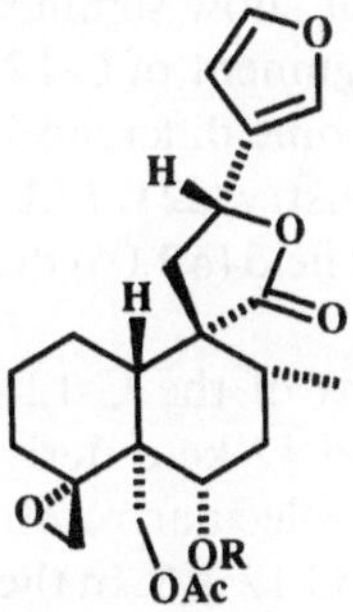

118 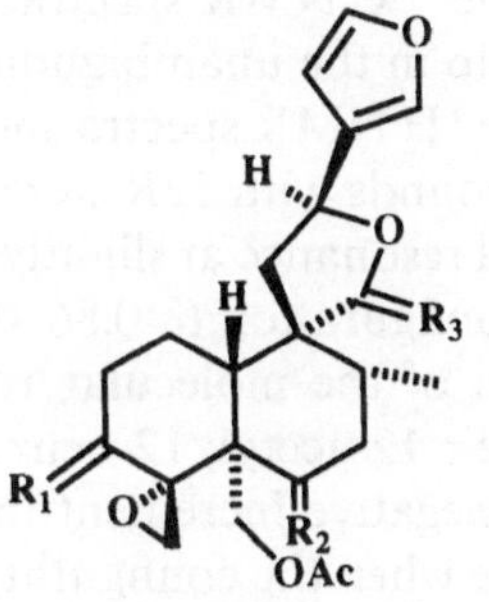   **119**

NOe difference spectrometry has also been proposed (*106*) as an adequate criterion for assigning the configuration at C-12. It has been observed that in compounds in which the C-12 methine proton and the C-17 methyl hydrogens are on the same side of the plane defined by the C-20, C-12-lactone ring (C-12 *R*), irradiation of the Me-17 protons produces a 8–12% nOe enhancement of the H-12 signal. This effect is not observed in compounds with the C-12 *S* configuration in which H-12 and the Me-17 are on opposite sides of the plane defined by the lactone ring. However a 2–6% nOe enhancement of the H-14 signal and, in some cases, a 1–5% nOe enhancement of the H-16 signal, were observed for compounds with the C-12 *S* configuration (*106*).

The C-20 configuration of *neo*-clerodanes incorporating a C-20, C-12-hemiacetal can be also established by nOe experiments (*107*). Irradiation of the H-17 signal produces a strong enhancement (10–15%)

of the H-20 signal when both are on the same side of the plane defined by the C-20-C-12 hemiacetal ring. This is so for teuflavin (119, R_1=O; R_2=βOH, H; R_3=αOH, H) whereas no enhancement of the H-20 signal is observed in compounds with opposite stereochemistry at C-20 such as eriocephalin (75, R=OH), (107).

In diterpenoids with a 12-hydroxyl or acetoxyl group, the C-12 configuration has been frequently established by X-ray diffraction or by Horeau's method (21), but there are some cases in which the configuration at C-12 could not be assigned. Recently (108) several methods have been investigated in order to find a useful technique for determining the absolute configuration at the C-12. To this purpose compounds 120 and 121 epimeric at C-12 were prepared from diterpenes of known stereochemistry. Acetylation gave the diacetyl derivatives 122 and 123. Treatment of 120 and 121 with benzoyl chloride in pyridine solution resulted in opening of the oxiranes to the chlorohydrins and selective benzoylation of the C-12 hydroxy group to furnish products 124 and 125 respectively. The ^{13}C NMR spectra did not show significant differences which could help in the unambiguous assignment of C-12 stereochemistry, whereas the ^{1}H NMR spectra showed some differences useful for this purpose. Compounds with 12R stereochemistry (121, 123, 125) exhibited the C-17 methyl resonance at slightly lower field (δ 1.05) than compounds with the 12S configuration (δ 0.86–0.97).

Comparison of the molecular rotations of the C-12 hydroxy-compounds and their 12-acetyl, 12-benzoyl and 12-keto derivatives showed that there is a negative increment in the molecular rotations of the 12-acetyl derivative when the configuration at C-12 is S. In the 12 R epimers, acetylation causes a strong positive increment (108).

Differences in the molecular rotations between the C-12 hydroxy-neo-clerodane and the related C-12 keto derivative have been also analyzed. In this case conversion of the 12S to the 12-oxo derivative results in a strongly positive increment of the molecular rotation, and only a weak positive increment in the case of the 12R epimer. Selective benzoylation at C-12 produced the expected differences in the molecular rotations of the carbinol and the benzoate derivatives. In the CD spectra of the furan alcohols 120 and 121, the π → π* absorption bands at 210 nm were of opposite sign. Also the CD exciton chirality method was applied to the 3-furylcarbinol benzoates 124 and 125 (108). The Cotton effects were found of opposite sign at 240 and 223 nm. Although of small amplitude these differences reflect the absolute configuration at C-12.

Because of the profusion of neo-clerodanes isolated so far from Teucrium species, an alphabetical list of sources and their constituents is provided in Table I at the end of this article.

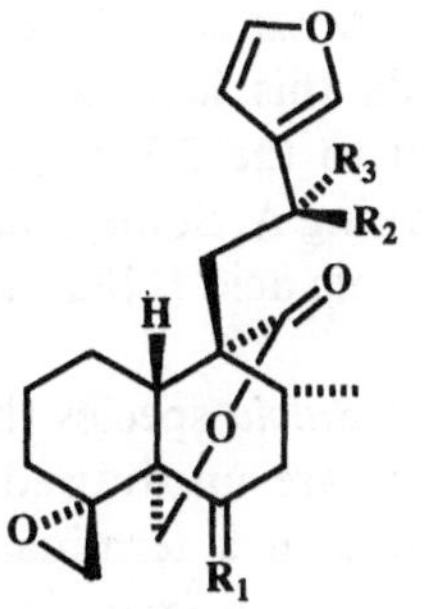

	R_1	R_2	R_3
120:	αOH,H	OH	H
121:	αOH,H	H	OH
122:	αOAc,H	OAc	H
123:	αOAc,H	H	OAc

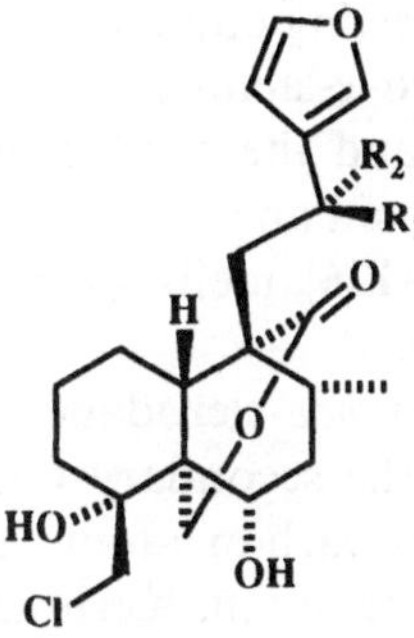

	R_1	R_2
124:	OBz	H
125:	H	OBz

4. *Salvia* Species

Salvia L. is the most diversified genus of the *Labiatae* family with over 900 species widespread all over the world. The genus has been divided (*154*) into the four subgenera *Salvia, Sclarea, Leonia* and *Calosphace* according to morphological characteristics and geographical distribution. European and Asiatic *Salvia* species belong to subgenera *Salvia* and *Sclarea* from which abietane diterpenes have been isolated (*155*). Neo-clerodane diterpenes have been found only in *Salvia* species of the American Continent which belong to subgenera *Leonia* and *Calosphace* (*154*). Members of subgenus *Leonia* are found mainly in North America, while the *Salvia* species of Mexico, Central and South America have been included in subgenus *Calosphace*, with over 500 species, which EPLING (*156*) has organized into over 105 sections.

The diterpenoid content of *Salvia*, subg. *Calosphace*, has been found to be related to the section to which it belongs. From some species such as those grouped in section *Erythrostachys* abietane diterpenes have been isolated (*157*). *Salvia* species included in section *Tomentellae* contain diterpenes with abietane or icetexane skeletons (*157*). However, more than 80% of the diterpenoids found in members of subgenus *Calosphace* studied so far have a *neo*-clerodane skeleton or a new skeleton biogenetically related to a clerodane precursor (*158*).

Despite the great structural diversity of the *neo*-clerodane diterpenoids isolated from subgenera *Calosphace* and *Leonia*, there can be found some interesting common structural features. Most contain a 18,19-α,β-unsaturated-γ-lactone. In some diterpenoids this lactone function is saturated and the double bond is displaced to the 2,3 position, while others contain an additional double bond in ring A. Some exception are lasianthin (**126**), melisodoric acid (**127a**), kerlinic acid (**128a**) and salvinorin (**129a**).

In many *neo*-clerodane diterpenoids from *Salvia* species the tertiary C-20 and the secondary C-17 methyl groups are unoxidized. In these compounds carbon atoms 13–16 are included in a terminal α- or β-butenolide function. Kerlinic acid (**128a**) is again an exception.

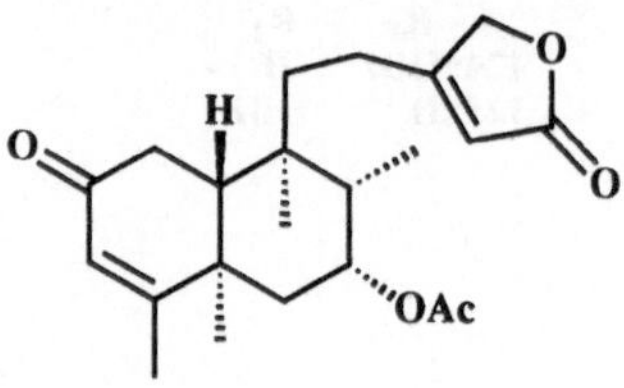

126

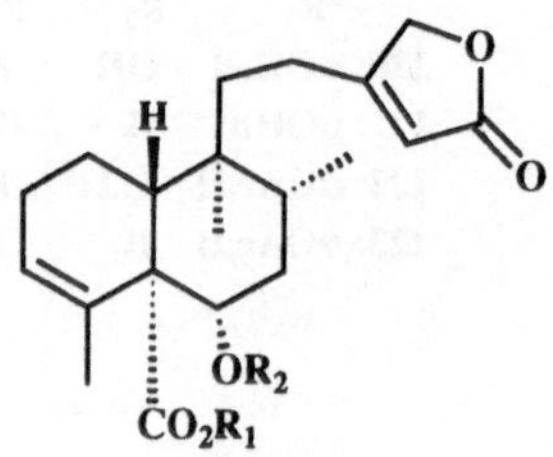

127 a: $R_1=R_2=H$
b: $R_1=Me, R_2=H$
c: $R_1=Me, R_2=Ac$

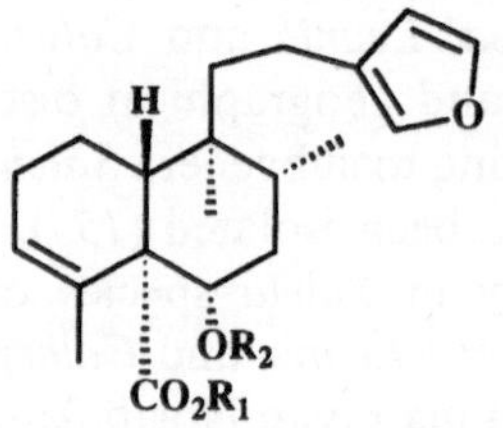

128 a: $R_1=R_2=H$
b: $R_1=Me, R_2=H$
c: $R_1=Me, R_2=Ac$

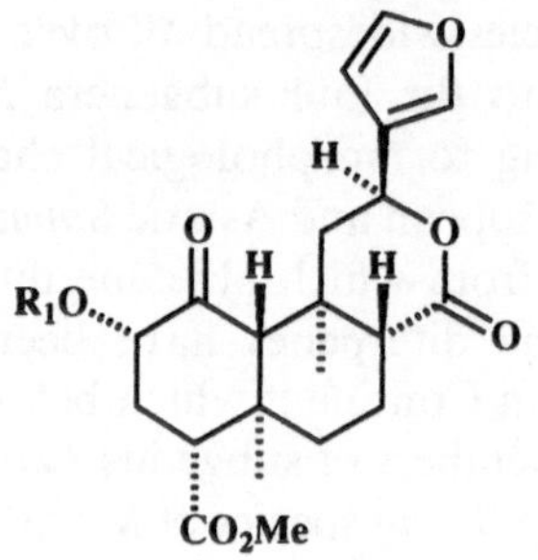

129 a: $R_1=Ac$
b: $R_1=H$

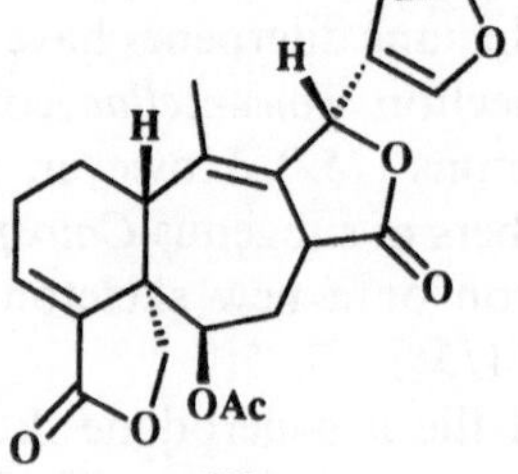

130

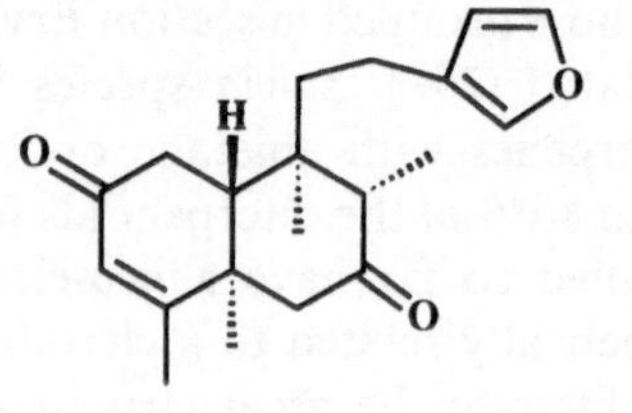

131

Frequently C-12 bears an oxygen atom. In most *Salvia* diterpenoids, it forms part of a δ-lactone function with C-17. In these *neo*-clerodanes carbon atoms 13–16 comprise β-substituted furan ring. Some exceptions are the 5,6-secoclerodanes (*159*) and the diterpenes isolated from *Salvia farinaceae* (*160*) in which C-20 forms part of a γ-lactone or a ketal function with C-12.

Some *neo*-clerodane diterpenoids from *Salvia* species contain an oxygen atom at C-7. In most cases this is an α-orientated axial hydroxy or acetoxy group. C-6 is not usually oxidized although an hydroxyl is present in this position in melisodoric (**127a**) and kerlinic (**128a**) acids. Salvigenolide (**130**) which has a rearranged *neo*-clerodane carbon skeleton has a β-orientated acetoxyl on C-6.

Structures and stereochemistries of the *Salvia neo*-clerodane diterpenoids have usually been determined by spectroscopic means. In some cases X-ray analysis has been used to corroborate the proposed structure or to clarify some structural problems. Thus, the structure and stereochemistry of lasianthin (**126**) from *S. lasiantha* (sect. *Mitratae*) (*161*) was deduced from spectral data and confirmed by X-ray analysis. Its IR spectrum showed the Fermi resonance (*162*) associated with the β-substituted α,β-unsaturated butenolide function and a strong band at $1658 \, cm^{-1}$ ascribed to an α,β-unsaturated ketone group. The ^1H NMR and ^{13}C NMR spectra revealed the presence of two tertiary and a secondary methyl groups. A doublet ($J = 1$ Hz) at δ 1.9 was attributed to the vinylic methyl C-18. Sodium borohydride reduction of lasianthin in the presence of cerium trichloride gave the 2α-hydroxy derivative. Structure **126** assigned to lasianthin is closely related to that of eremone, **131**, a *neo*-clerodane diterpene from *Eremocarpus setigerus* (Euphorbiaceae) (*163*) whose absolute configuration was based on comparison with hardwickiic acid isolated from the same source.

Melisodoric acid (**127a**) was isolated from a population of *S. melissodora*, Lag. (sect. *Scorodonia*) collected north of Mexico City (*164*). Its structure was established on spectral evidence and chemical transformations. It formed a methyl ester (**127b**) on treatment with ethereal diazomethane. Acetylation of **127b** gave the monoacetate derivative **127c**. Treatment of **127c** with *m*-chloroperbenzoic acid gave epoxide **132** which was submitted to ozonolysis followed by oxidative treatment of the ozonide and esterification to dimethyl ester **133**. In order to prove the relative disposition of the functionalities on carbon atoms 3-6, epoxide was converted to a diol with perchloric acid in methanol. Hydrolysis with potassium bicarbonate and diazomethane esterification yielded hydroxy-γ-lactone **134** which was oxidized with Jones' reagent to hexanone **135**. Dehydration with thionyl chloride in pyridine yielded **136** which on treatment with base followed by Jones oxidation and esterification gave the ene-dione **137**. Through this sequence of reactions the relative disposition of the decalin functionalities was established. The relative stereochemistry was deduced by correlating melisodoric acid (**127a**) with kerlinic acid (**128a**) and NMR data.

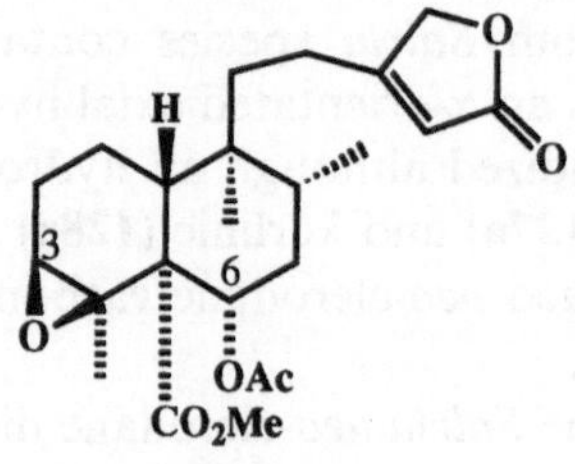

132

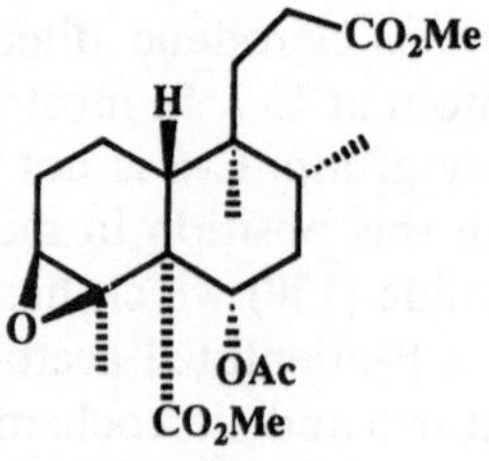

133

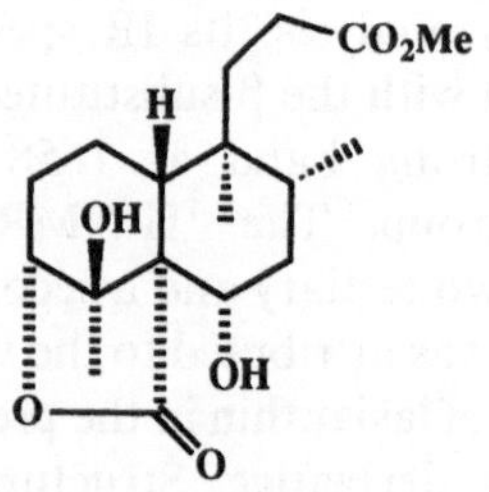

134

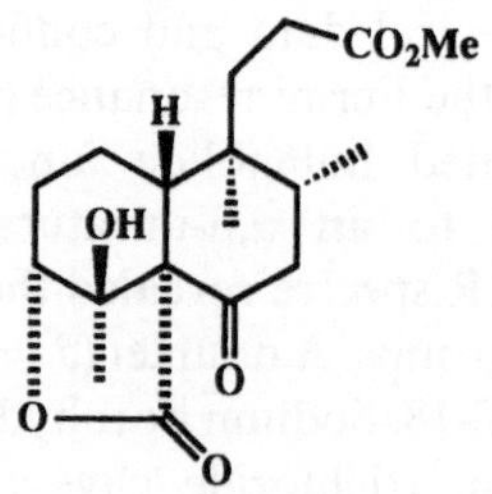

135

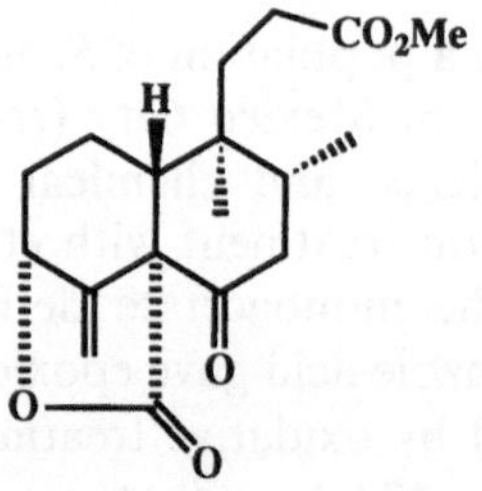

136

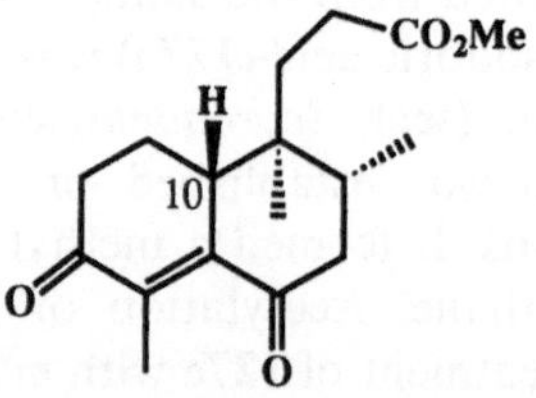

137

Kerlinic acid (**128a**) was isolated from a population of *S. keerlii*, Benth (sect. *Scorodonia*) from the State of Queretaro (Mexico) (*165*). The spectral data resembled those of melisodoric acid (**127a**) from which it differed in the presence of a terminal β-substituted furan ring in place of the β-butenolide function. Treatment of its acetate methyl ester (**128c**) with three equivalents of *m*-chloroperbenzoic acid gave a mixture of the 3,4-epoxy-hydroxy-lactones **138** and **139**, which when treated with sodium borohydride (*166*) yielded a mixture of the 3,4-epoxide of melisodoric acid (**132**) and the related α-butenolide (**140**). Ozonolysis of the mixture **138** and **139**, followed by oxidative treatment of the ozonide and esterification gave dimethyl ester **133** identical with the product obtained from melisodoric acid.

A phytochemical study of different populations of *S. melissodora* and *S. keerlii* led to the isolation of several *neo*-clerodane diterpenoids **141–151** which contained an α,β-unsaturated-18,19-olide function. Their structures were deduced on spectral evidence and confirmed, in some cases, by X-ray analysis.

The 3-ene-18,19-olide functionality was first described (*90*) in olearin **97** (*vide* p. 131). The C-19 methylene group is responsible for an AB system in the ^{1}H NMR spectrum. The *pro-S* diasterotopic proton of this group is found normally at δ 3.9 and W-coupled to the 6β proton which indicates α-axial orientation for the 19-methylene group, and A/B ring steroidal *trans* stereochemistry. The chemical shift of the 19-*pro-R* proton is deeply influenced by the substitution pattern at C-7. Thus, when an α-axial hydroxyl is attached to C-7, the *pro-R* H-19 is strongly deshielded, its signal appearing at δ 5.30. With an α-axial acetoxy group at C-7 its average chemical shift is δ 4.85 while in 7-oxo derivatives the 19-*pro-R* signal is observed at δ 4.0. In the absence of these substituents an average of δ 4.35 is expected for the 19-*pro-R* proton (*171*). In the ^{13}C NMR spectra of these compounds the sp^2 carbon atoms 3 and 4 are observed at δ 133–136 and 136–139 respectively, and C-5 at δ 45. Substitution at C-2 causes the expected shielding or deshielding depending on the substituent and its stereochemistry. The C-18 and C-19 signals were observed at δ 168–170 and 70–73 respectively.

In the terminal butenolide function of these *neo*-clerodane diterpenoids the carbonyl group can be on C-16 (α-substitution, **141–149**) (*171*) or C-15 (β-substitution, **150, 151**) (*168–170*). The β-butenolides can usually be detected in the IR spectrum because of its Fermi resonance (*162*). In the ^{1}H NMR spectra of β-substituted butenolides H-14 is observed as a split doublet at δ 5.9, whereas in α-substituted butenolide derivatives, the vinylic proton appears as a broad singlet at δ 7.1. The ^{13}C NMR spectra are also useful for establishing the nature of the terminal butenolide function. A singlet at δ 173–176 and a doublet at δ 112–114 are diagnostic for C-13 and C-14 of β-substituted butenolides, while in α-substituted butenolide derivatives the C-13 signal appears at δ 132–134 and C-14 at δ 143–147, the other signals being similar.

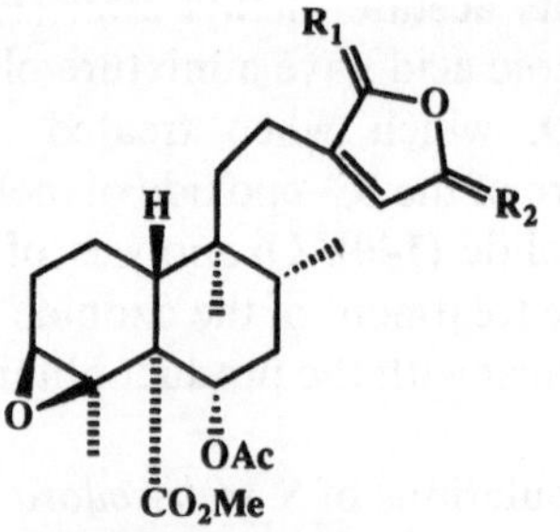

138: R₁=O; R₂=H, OH
139: R₁=H, OH; R₂=O
140: R₁=O; R₂=H₂

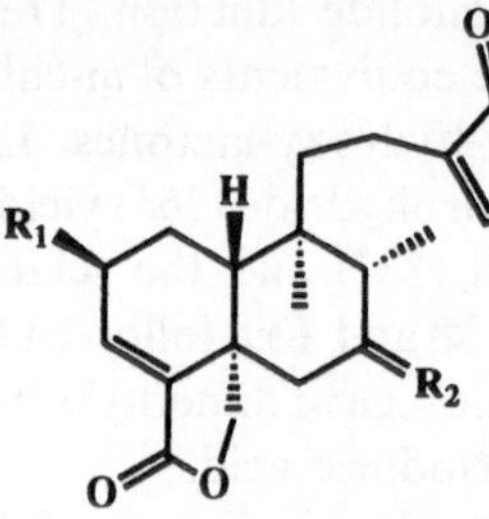

R₁	R₂
141:	OH αOAc, H
142:	H βOH, H
143:	H αOH, H
144:	H αOAc, H
145:	OH αOH, H
146:	OAc αOH, H
147:	H O
148:	OH H₂
149:	OH O

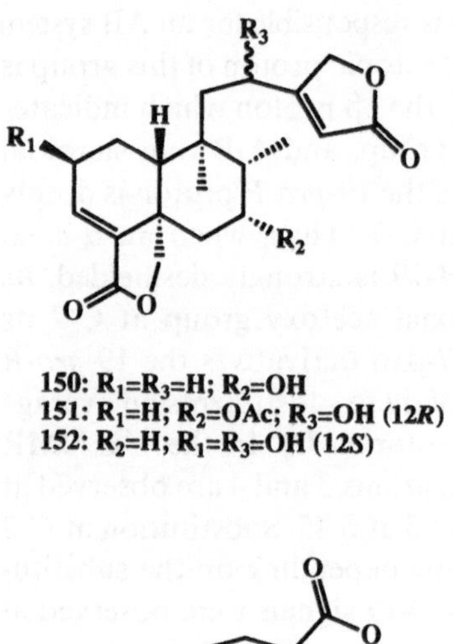

150: R₁=R₃=H; R₂=OH
151: R₁=H; R₂=OAc; R₃=OH (12R)
152: R₂=H; R₁=R₃=OH (12S)

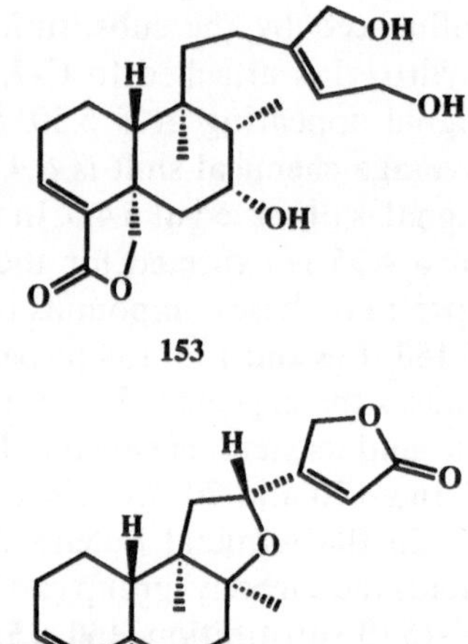

153

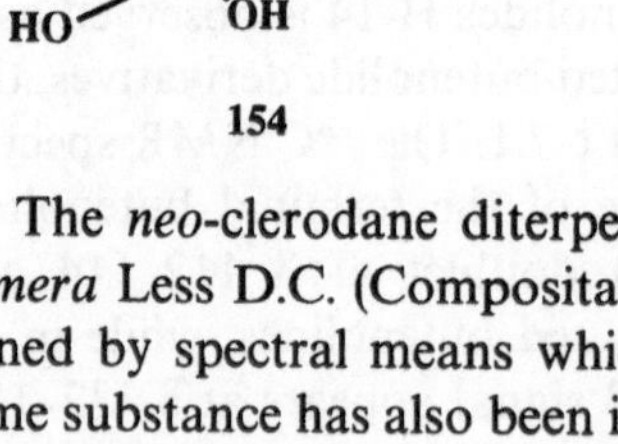

154 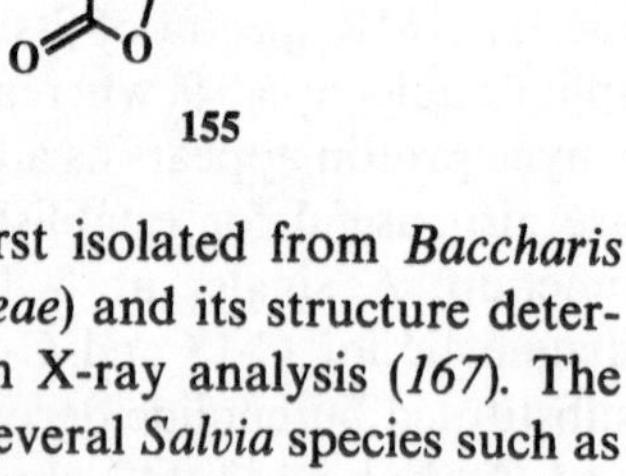   155

The *neo*-clerodane diterpene **150** was first isolated from *Baccharis trimera* Less D.C. (Compositae, Tribe *Astereae*) and its structure determined by spectral means which included an X-ray analysis (*167*). The same substance has also been isolated from several *Salvia* species such as *S. melissodora* (*168*), *S. microphylla* Kunt (sect. *Fulgentes*) (*169*) and

S. semiatrata Zucc. (sect. *Atratae*) (*170*). This population of *S. melissodora* contained also portulide C (**153**), a *neo*-clerodane diterpenoid previously found in *Portulaca* cv Jewel (*Portulaceae*). Portulide C may be a biogenetic precursor of the α and β-substituted butenolides.

A population of *S. melissodora* from San Luis Potosi (Mexico) yielded, together with the diterpenes **141–149**, the 18,19-dihydroxy derivative **154** (*171*) whose structure was proved by transformation into **142** on treatment with manganese dioxide.

From a population of *S. keerlii* collected in Oaxaca (Mexico) the *neo*-clerodane diterpenoids kerlinolide (**151**) and kerlin (**155**) were isolated (*172*). Their structures and relative stereochemistry were deduced on spectral evidence and an X-ray analysis of **155**. A *neo*-clerodane (**156**) structurally related to kerlin has been recently found in *S. rhyacophila* (sect. *Angulatae*) (*159*).

Brevifloralactone (**157**) has been found in a population of *S. melissodora* (*168*) and in *S. breviflora* Moc. and Sesse which also belongs to sect. *Scorodonia*. Its structure was established on spectral grounds and an X-ray analysis (*173*).

The *neo*-clerodane diterpene semiatrin, **152**, was isolated from *S. semiatrata* Zucc. together with **150** mentioned earlier (*170*). The *S* configuration at C-12 in semiatrin was established by X-ray analysis (*174*). In this connection it is interesting to note that the configuration at C-12 in the *neo*-clerodanes from *Salvia* species is most frequently *R*.

Salvinorin (**129a**) is also an exception to the 12*R* configuration. It was isolated (*175*) from *S. divinorum* (sect. *Dusenostachys*), together with its 2-deacetyl derivative **129b** (*176*). This plant has been used by the Mazatec Indians of Oaxaca (Mexico) in their divination rites and salvinorin has been claimed to be responsible for its hallucinogenic properties (*176*). Its structure was established on spectral evidence and an X-ray analysis (*175, 176*). As in all *neo*-clerodane diterpenes from *Salvia* sp. which contain a 17,12-δ-lactone group, carbon atoms 13–16 form part of a β-substituted furan ring. The absolute stereochemistry shown in **129** was deduced by applying the exciton chirality method to the dibenzoate **158** of the octahydrosalvinorin (*177*).

Neo-clerodane diterpenes **159–163**, which contain a 17,12-δ-lactone function, a furan ring at C-12 and a saturated 18,19-olide group, have also been isolated from *Salvia* species. All of them contain a 2,3-double bond which is responsible for a multiplet at δ 6.00 (H-2) and a doublet (*J* = 10 Hz) at δ 5.64 (H-3) in the ^{1}H NMR spectra. The β-axial proton at C-4 is usually found at δ 2.78. In the ^{13}C NMR spectra signals of the trigonal carbon atoms 2 and 3 appeared at δ 126–128 and 121 respectively while signals of C-4 occurred at δ 50–52 and those of C-5 at δ 41.4.

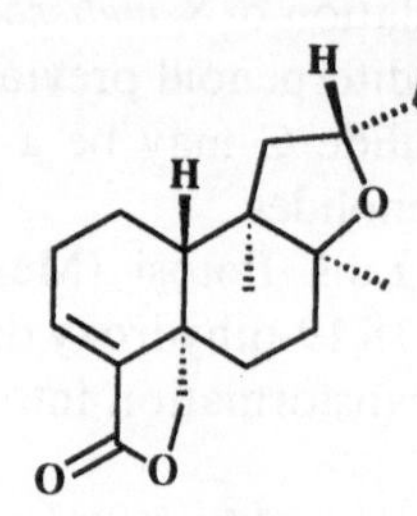

156

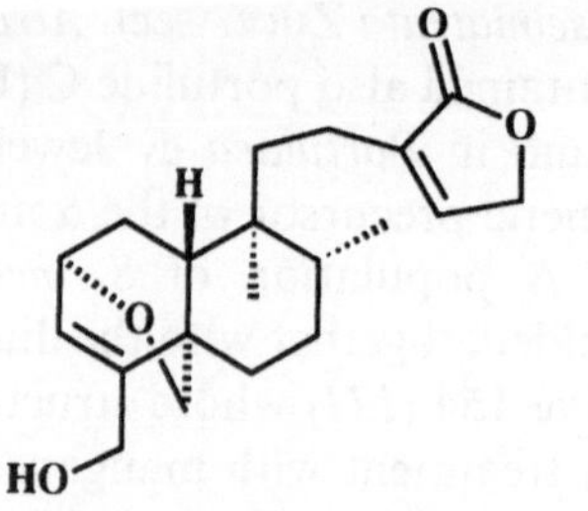

157

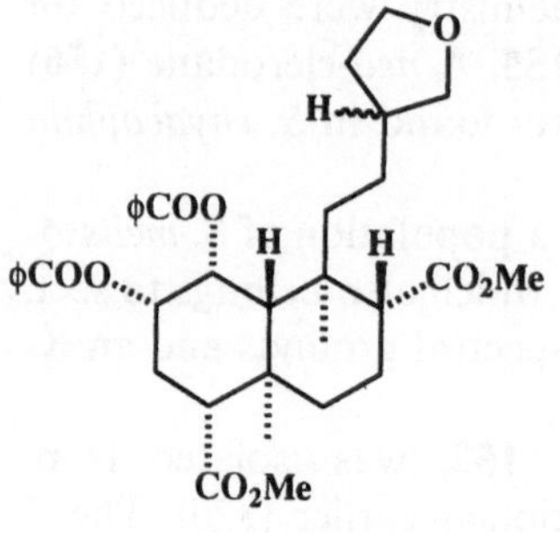

158

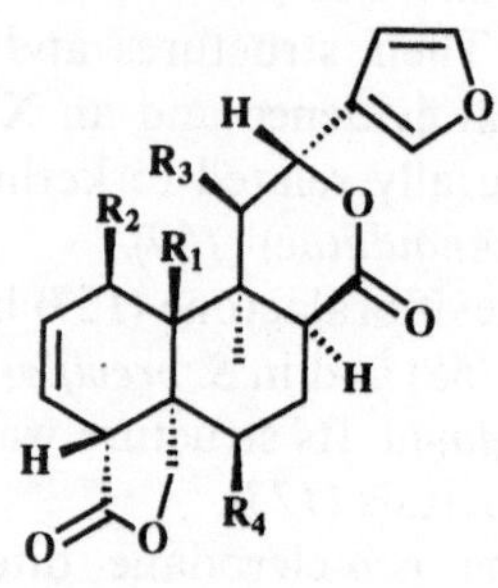

	R₁	R₂	R₃	R₄

159: H   H   H   H
160: H   OAc OAc H
161: H   H   H   OH
162: OAc H   H   H
163: H   H   H   OAc
164: OH   H   H   H (8epi)

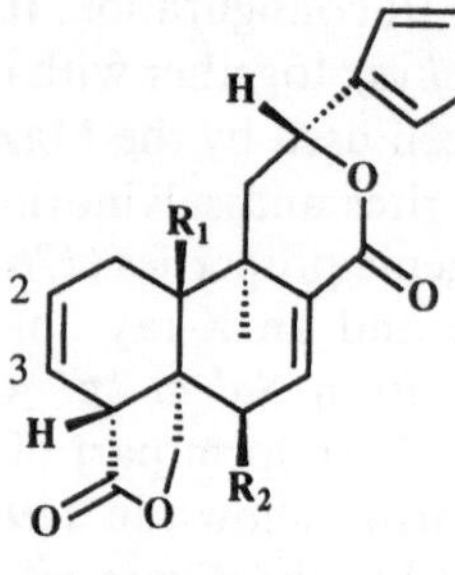

R₁ R₂
165: OH   H
166: H   OH
167: OH   H   2,3 β-epoxy

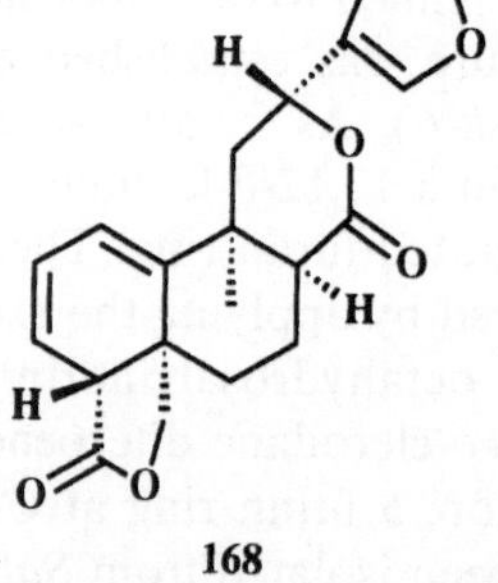

168

The C-12 signal usually appears at δ 71–72 irrespective of the configuration. The C-17 carbonyl is observed at δ 171–172 in diterpenoids with C-17 α or β-oriented.

The structure of salviarin (159) from *S. splendens* (sect. *Secundae*) was deduced on spectral evidence which included an X-ray analysis (*178*). The configuration at C-12 was established as R and at C-8 as S. Splendidin (160) was also isolated from the same source (*179*). The presence of a β-axial hydroxy or acetoxy groups at C-6 or C-10, produces a deshielding effect on the signal of H-4. This effect was observed in the *neo*-clerodane diterpenoids 6β-hydroxy-, 6β-acetoxy- and 10β-acetoxysalviarins (161–163) from *S. rhyacophila* (*159*).

Some *neo*-clerodane diterpenoids from *Salvia* species contain an additional 7,8-double bond. An example is salviacoccin (165) from *S. coccinea* Juss. (sect. *Subrotundae*) whose structure was established on spectral grounds and chemical transformations (*180*). Sodium borohydride reduction of salviacoccin yielded a 7,8-dihydro derivative (164) in which H-8 was found β-axial. The dihydro derivative has also been found as a natural product in *S. greggii* (sect. *Flocculosae*) (*181*). Alkaline treatment of salviarin (159) produces epimerization at C-8 (*159*).

A 2β,3β-epoxy derivative of salviacoccin (165) was isolated from *S. plebeia*, a *Salvia* species classified in subgenus *Leonia*. Structure 167 proposed for it was deduced on spectral evidence and confirmed by oxidation of salviacoccin, also present in this species, with *m*-chloroperbenzoic acid (*182*).

Neo-clerodane diterpenoids with a diene group in ring A of the decaline portion have been also isolated from *Salvia* species. 1(10)-Dehydrosalviarin (168), was found in a population of *S. lineata* (sect. *Fulgentes*) (*183*) and was identical with the dehydration product of a bitter principle 169 of *S. rubescens* Kunth (sect. *Rubescentes*), a plant endemic to Brazil. Structure 169 assigned to this *neo*-clerodane was established (*184*) on spectral evidence and a series of transformations which included formation of a dienone (170) on treatment with chromium trioxide in pyridine followed by dehydration. The relative stereochemistry shown in 169 was deduced (*183*) by comparison of the ¹H NMR spectral data with those of *neo*-clerodane diterpenoids structurally related to salviarin (*178*). Linearolactone 171 was also isolated from *S. lineata* (*183*). The X-ray analysis of 171 established (*185*) the A/B-*cis* fusion in this diterpenoid, although the *pro-S* H-19 was long-range coupled ($J = 1$ Hz), as found in all *neo*-clerodane diterpenoids containing an α,β-unsaturated 18,19-olide function with A/B *trans* ring fusion, α-axial 19-methylene and absence of 6β substituent.

Salvifaricin 174 and salvifarin 175, *neo*-clerodane diterpenoids (*186*) from *S. farinacea* Benth. (sect. *Farinaceae*) showed the same coupling of the *pro-S* H-19 although both compounds were shown to have an A/B-*cis* ring fusion. The structure and relative stereochemistry of salvifarin 175 was established on spectral grounds and by an X-ray analysis (*187*). Treatment of 175 with sodium iodide and *p*-toluensulfonic acid in

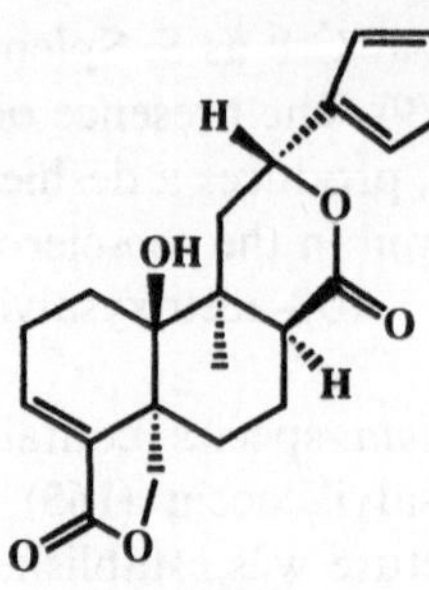

169

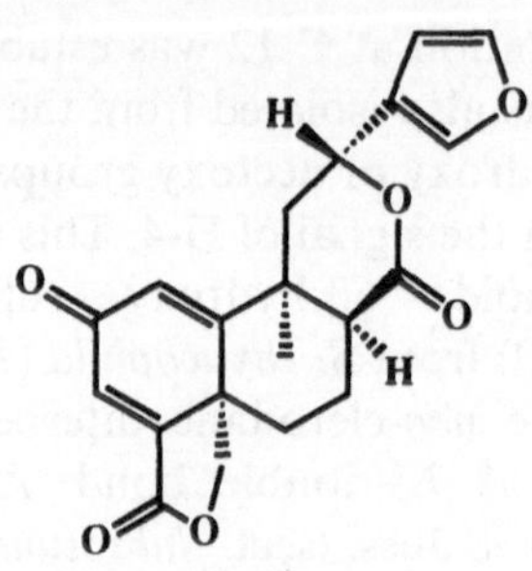

170

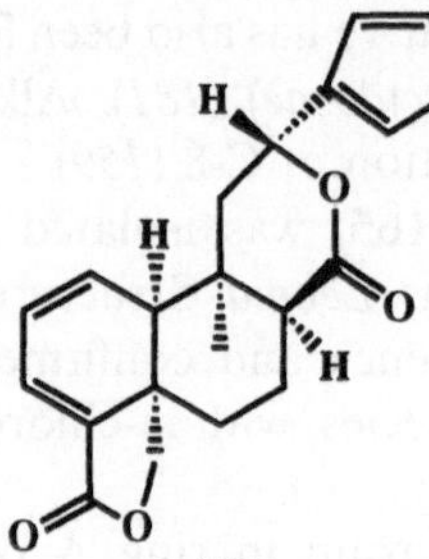

171

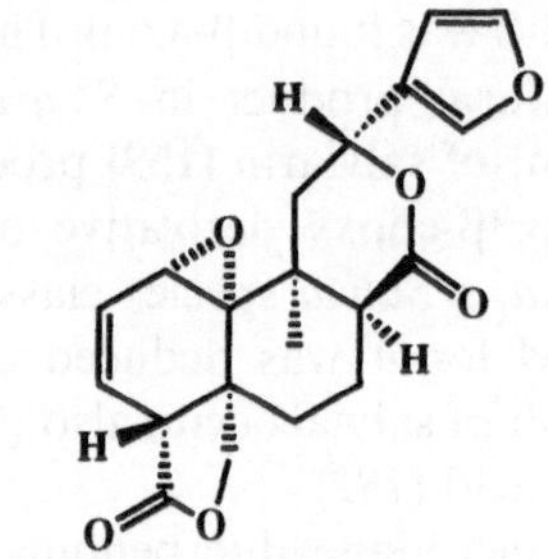

172

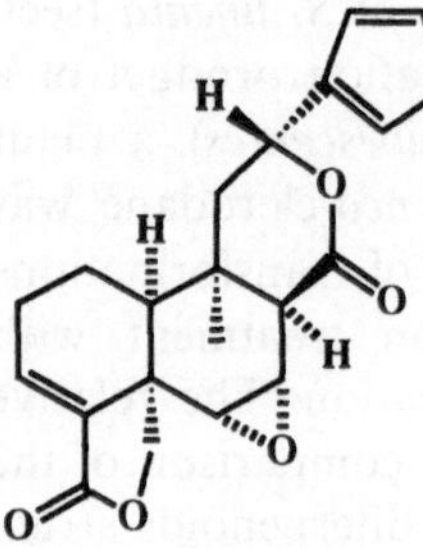

173

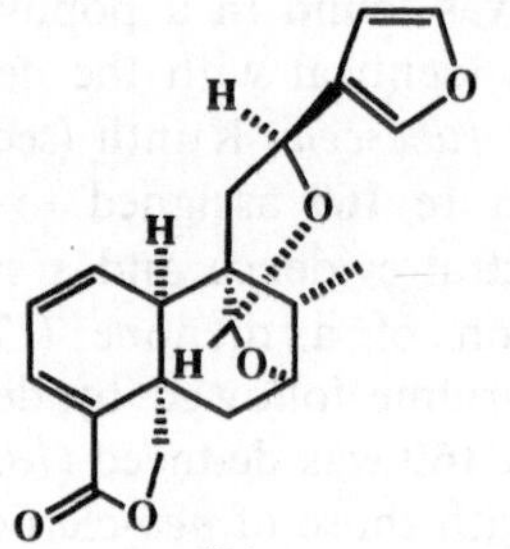

174

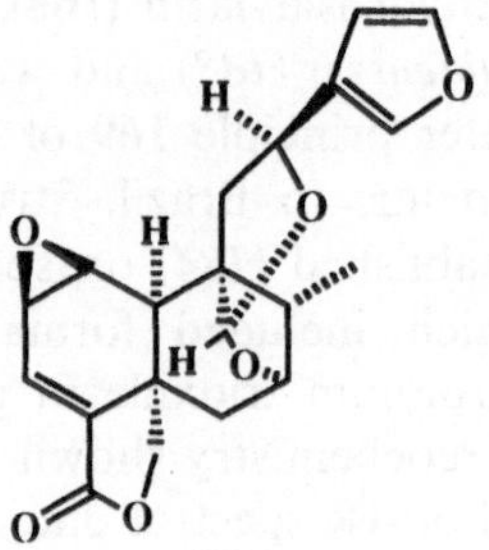

175

acetonitrile yielded salvifaricin **174**, thus proving its structure and relative stereochemistry (*160*). In these substances C-20 is oxidized and is involved in ketal formation with hydroxyls at C-7 and C-12 with a singlet at δ 5.2 in the NMR spectra being ascribed to the acetal proton. The configuration at C-12 was found to be *R*.

A clerodane diterpenoid **173** isolated from *S. sousae* Ramamoorthy (sect. *Polystachyae*) with an A/B-*cis* ring fusion also showed the W-coupling for H-19 *pro-S* (*188*). The 1α,10α-epoxysalviarin **172** was isolated from *S. lineata* (*183*).

A *neo*-clerodane **177** from *S. reptans* Jacq. (sect. *Farinaceae*) deserves special mention. The structure and relative stereochemistry shown in the

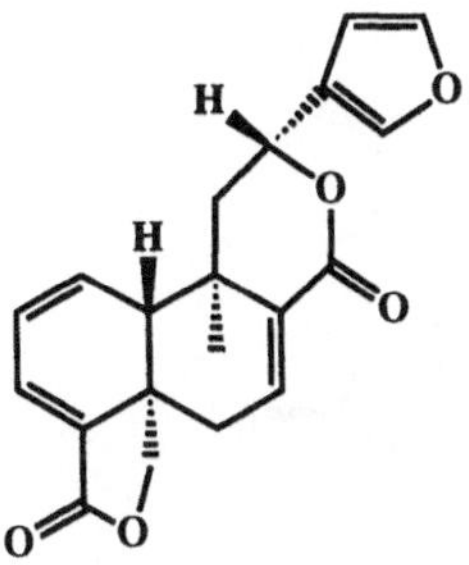

176 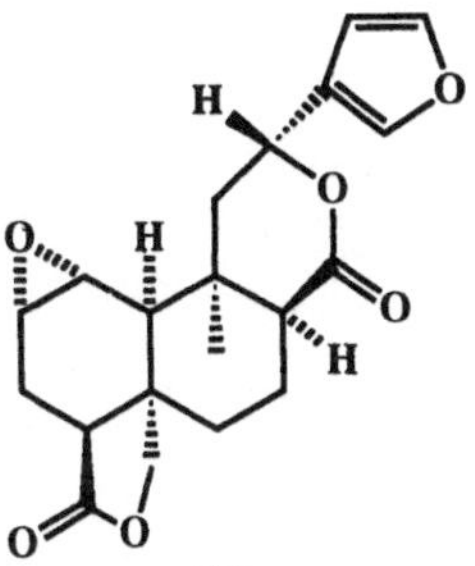   177

formula were deduced on spectral evidence and an X-ray analysis (*189*). This diterpene also had an A/B-*cis* ring fusion with the *pro-S* H-19 being W-coupled to H-6β. It is the only clerodane isolated so far in which the 18,19-olide function is *trans*-fused to ring A.

A *neo*-clerodane diterpene with an α,β-γ,δ-diunsaturated γ-lactone function in ring A and an additional 7,8 double bond, gesnerofolin B, **176**, was isolated from *S. gesneraeflora* (sect. *Nobiles*) (*190*). The relative stereochemistry shown was proposed on biogenetic grounds as salviarin (**159**) was isolated from the same source.

Some diterpenoids isolated from *Salvia* species within the subgenus *Calosphace* have been shown to possess new skeletons which can be biogenetically related to *neo*-clerodane precursors (*191*).

Several 5,6-*seco*-clerodane diterpenoids **178–180**, have been isolated from *Salvia* species. All of them contain a terminal β-substituted furan, an oxygen function at C-12 and in all of them C-20 is involved either in a γ-lactone function as in **178** or part of a ketal as in **179** and **180**. Ring A is aromatized and contains the 18,19-γ-lactone. The proton spectra of these diterpenoids show the signals of an ABC system arising from the 1, 2 and 3 protons of a trisubstituted aromatic ring and a phtalide methylene AB system. The rest of the spectra differed depending on the structure of the substituent at C-10.

The IR spectrum of salvireptanolide **178**, from *S. reptans* Jacq. (sect. *Farinaceae*) (*189*) had bands due to the γ-lactone and α,β-unsaturated ketone groups. The chemical shifts of the furan protons H-14 (δ 6.75) and H-16 (δ 8.20) indicated that the ketone function was at C-12. Two methyl doublets at δ 1.45 and 1.10 could be assigned to the vicinal secondary methyl groups C-6 and C-17. A double quartet (*J* = 10 and 7 Hz) at δ 4.55 could be ascribed to H-7. A related compound, rhyacophiline (**179**) from *S. rhyacophila* (sect. *Angulatae*) (*159*) contained a ketal function formed between C-20 on the one hand and C-7 and C-12 on the other, as indicated by a singlet at δ 5.97 due to H-20. In salvianduline C (**180**) from

178: R=

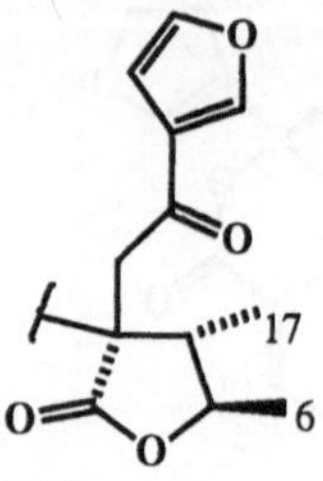

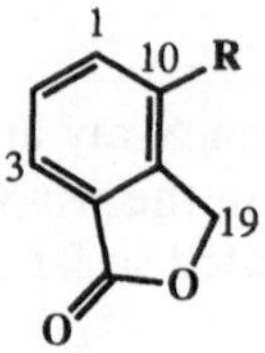

179: R=

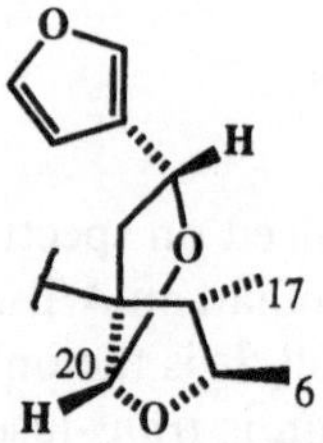

180: R=

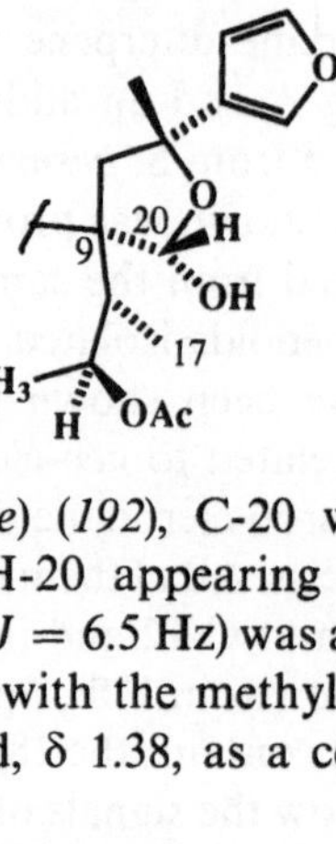

CH₃
6

S. lavanduloides L. (sect. *Lavanduloideae*) (*192*), C-20 was part of a hemiacetal also involving C-12, with H-20 appearing as a doublet ($J = 1.1$ Hz) at δ 5.77. A quartet at δ 5.28 ($J = 6.5$ Hz) was assigned to the proton under an acetoxy group on C-7, with the methyl singlet of the acetate appearing at unusually high field, δ 1.38, as a consequence of being shielded by the aromatic ring.

The structures and relative stereochemistries shown in **178–180** have been confirmed by X-ray diffraction. The common skeleton has been named (*159*) **rhyacophane** and can be derived from a 7-oxoclerodane precursor with a diene functionality in ring A and a *cis* A/B ring fusion (*191*). Salvifolin **181**, a *neo*-clerodane diterpenoid from *S. tiliaefolia*, shows this functionality (*193*).

The 9,10-*seco* clerodanes, salviandulines A and B, were also isolated from *S. lavanduloides* (*194*). Their structures **182** and **183**, were established on spectral evidence and an X-ray analysis of **183**. Salvianduline A, **182**, was shown to be the C-7 epimer of salvianduline B acetate since the same 7-keto derivative **184** was obtained from both substances. These 9,10-*seco*-clerodanes could be derived biogenetically by C-9,10 bond cleavage of a clerodane precursor such as 10β,8α-dihydroxysalviarin, in which the

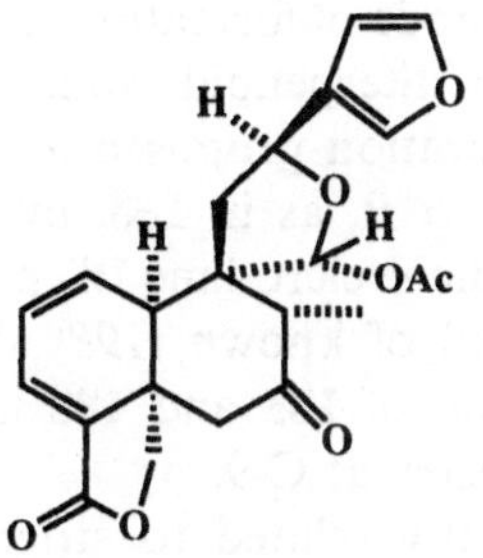

181

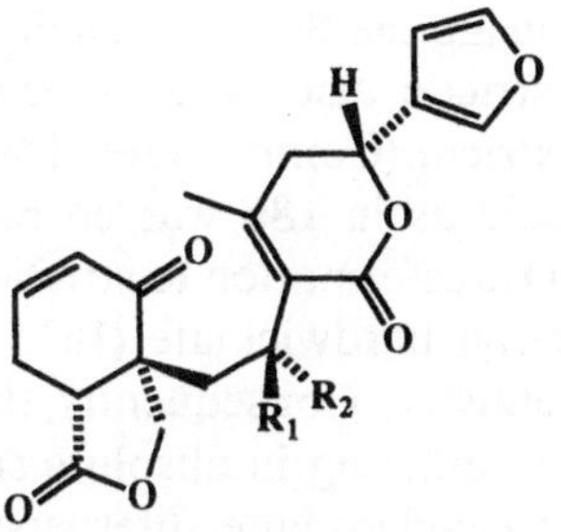

182: R₁=H, R₂=OAc
183: R₁=OH, R₂=H
184: R₁R₂=O

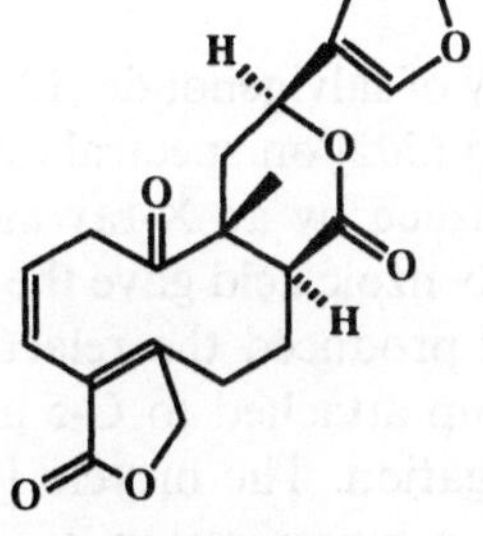

185

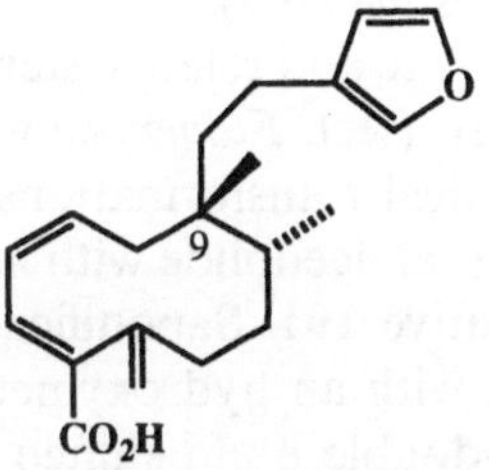

186

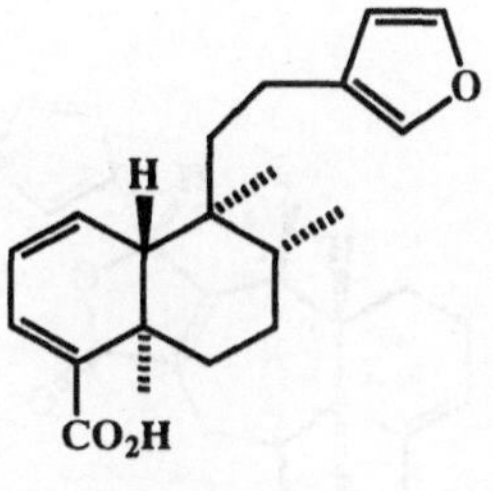

187

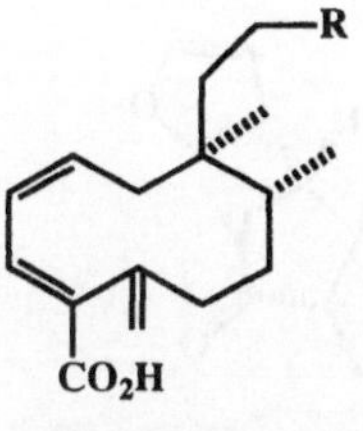

188: R=β-furyl
189: R=β-butenolide
190: R= α-butenolide

two hydroxy groups on C-8 and C-10 are antiperiplanar, followed by migration of the 2,3-double bond into conjugation with the 10-keto group so formed. 9,10-*Seco*-clerodanes have been also found in *Conyza* species (Compositae, Tribe *Astereae*) (*195*).

5,10-*Seco*-clerodanes have been isolated from several *Salvia* species. The absolute stereochemistry shown in **185** was assigned to cardiophyllidin from *S. cardiophylla* Benth. (sect. Rudes (*196*)) based on an X-ray analysis using the Bijvoet technique with a mode of formation similar to that of strictic acid, a 5,10-*seco*-clerodane diterpenoid isolated from *Conyza stricta* (Compositae). The *S* configuration proposed for C-9 of strictic acid as in **186** was corrected (*197*) to *R*, as in **188**, by photo-chemical transformation to strictic acid of a diene clerodane **187** prepared from methyl hardwickiate (**187**, 1,2-dihydro) of known (*198*) absolute stereochemistry. Consequently the formation of **185** and **188** involves precursors differing in absolute stereochemistry at C-9.

5,10-*Seco*-clerodane diterpenes structurally related to strictic acid with a terminal α or β-substituted butenolide function, **189** and **190**, have been isolated from *S. thymoides* (sect *Flocculosae*) (*199*) and *S. purpurea* (sect. *Purpureae*) (*200*). These products have also been found in members of the *Compositae* (*201*).

The structure and relative stereochemistry of salvigenolide (**130**) from *S. fulgens* Cav. (sect. *Fulgentes*) were deduced (*202*) on spectral evidence and by chemical transformations, and confirmed by an X-ray analysis. Oxidation of salvigenolide with *m*-chloroperbenzoic acid gave the 9,10β-epoxy derivative **191**. Saponification of **130** produced the relactonized product **192** with an hydroxymethylene group attached to C-5 and the former 9,11-double bond shifted into conjugation. The bicyclo [5·4·0] ring system of salvigenolide could have arisen by migration of the 8,9-bond of a normal clerodane to C-11 induced by an appropriate leaving group on C-11 (*191*); the corresponding carbon skeleton has been named **salvigenane**.

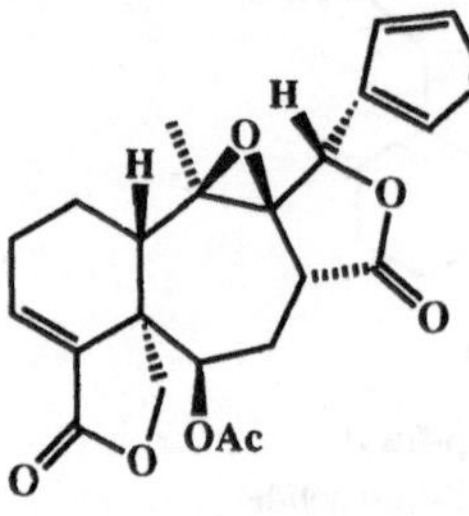

191 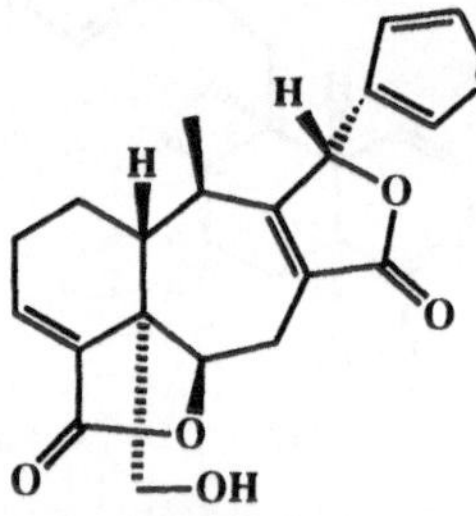   192

A phytochemical study of *S. puberula* Fern. yielded two diterpenoids with benzonorcaradiene and benzocycloheptatriene ring systems (*203*). *Salvia puberula* belongs to sect. *Holwaya* which is botanically related to sect. *Fulgentes*, a relationship which aided in structural elucidation of salvipuberulin (**193**) and isosalvipuberulin (**194**). The presence of a benzocycloheptatriene nucleus in isosalvipuberulin was deduced from the ^{1}H and ^{13}C NMR spectra (*203*) and the entire structure confirmed by X-ray diffraction. The presence of a benzo-*nor*-caradiene ring system in salvipuberulin **193**, was deduced from the ^{1}H NMR spectrum which showed signals due to a disubstituted cyclopropane ring (*203*). Salvipuberulin was slowly transformed into isosalvipuberulin in methylene chloride or chloroform solution. This transformation could occur by a disrotatory electrocyclic reaction followed by two suprafacial 1,5-sigmatropic hydrogen shifts or by acid catalyzed opening of the cyclopropane ring (*203*). The botanical relationship of sect. *Holwaya* in which *S. puberula* is included, and sect. *Fulgentes* of which *S. fulgens* is the type species (*156*), suggested that salvipuberulin and isosalvipuberulin could be biogenetically related to salvigenolide **130** (*191, 203*). The salvipuberulin carbon skeleton has been named **salvipuberulan**.

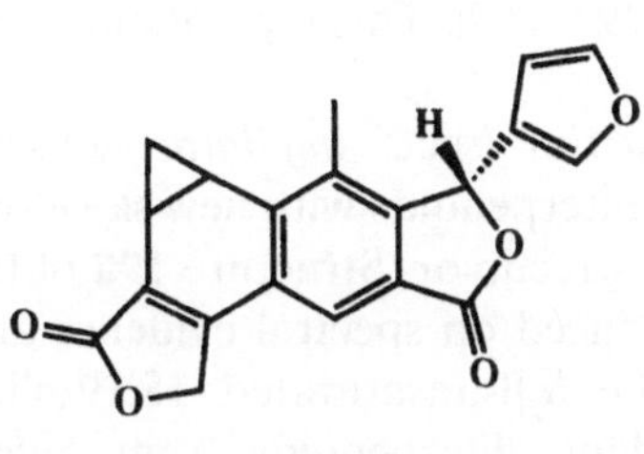

193 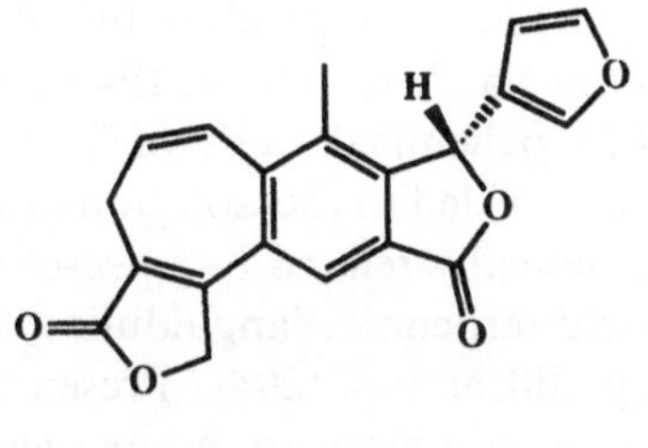   194

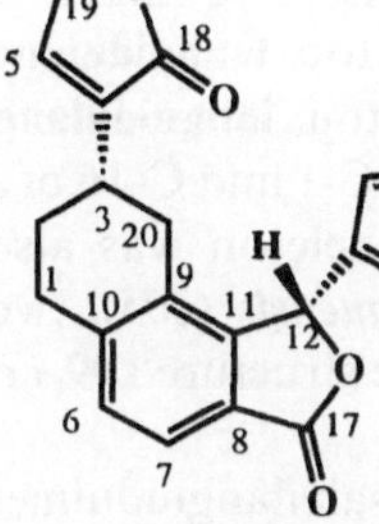

195 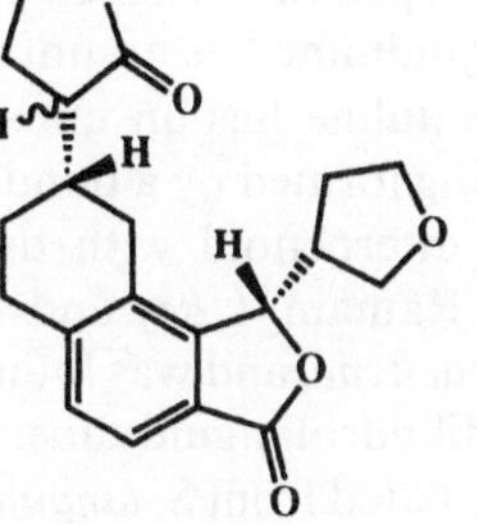   196 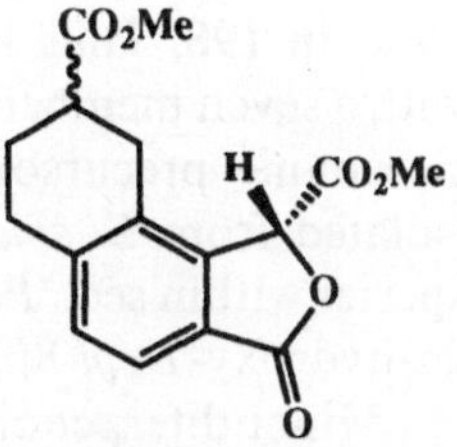   197

(Numbering following biogenesis)

Salvia tiliaefolia Vahl. (sect. *Angulatae*, subsect *Tiliaefoliae*) contained isosalvipuberulin **194** and a second diterpenoid, tilifodiolide, with a tetralin skeleton. Structure **195** for tilifodiolide is based on spectral data which included a thorough analysis of the ^{1}H and ^{13}C NMR data, chemical degradation and an X-ray analysis (*193*). The ^{1}H and ^{13}C NMR spectra indicated the presence of an α-substituted butenolide function, a β-substituted furan ring, the 17,12-γ-lactone group and a tetrasubstituted aromatic ring. Two doublets ($J = 8$ Hz) at δ 7.3 and 7.7 were ascribed to an aromatic AB system. Exhaustive catalytic hydrogenation of tilifodiolide gave a hexahydro derivative (196), whose ^{1}H NMR spectrum exhibited the signals of an aromatic AB system protons and the absence of vinylic and furan signals. Ozonolysis of tilifodiolide followed by oxidative treatment of the ozonide and esterification yielded a dimethyl ester (197), thus establishing presence of the tetralin skeleton. Tilifodiolide crystallizes with two independent molecules per asymmetric unit, which differ in the orientation of the α-butenolide and the furan groups with respect to the hydrobenzofuranone ring (193). The presence of isosalvipuberulin (194) and tilifodiolide (195) in S. tiliaefolia suggests a common biogenetic precursor such as salvigenolide (130). Presence of a good leaving group at C-20 in 130 could produce tilifodiolide (191, 193). This type of diterpene skeleton has been named **tilifolane**.

Phytochemical study of S. languidula Epl. (sect. Angulatae, subsect. Glumaceae) led to the isolation of several diterpenoids with new skeletons biogenetically related to a neo-clerodane precursor. Structure 198 of the major diterpenoid, languiduline, was deduced on spectral evidence and X-ray diffraction (204). Presence of the α,β-unsaturated 18,19-olide function common to many neo-clerodane diterpenoids from Salvia spectrum indicated the presence of an α,β-unsaturated cycloheptanone. Two doublets ($J = 2.1$ Hz) at δ 6.77 and 7.35 in the ^{1}H NMR spectrum were ascribed to protons 14 and 15 of a disubstituted furan ring. Analysis of the ^{1}H and ^{13}C NMR spectra indicated that C-1 must be linked to C-16 as in 198. Thus languiduline has an unusual skeleton, **languidulane**, 16 as in 198. Thus languiduline has an unusual skeleton, **languidulane**, with a seven membered ring formed by a bond between C-1 and C-16 of a clerodanic precursor. A diterpenoid with the same skeleton was also isolated from S. sousae, Ramam. (188) and S. zimapanensis (205), two species within sect. Polystachyae, and was found to have structure 199, i.e. 2α-hydroxy-7-epi-8β,17-dihydrolanguiduline.

Minor diterpenoids isolated from S. languidula, the salvilanguidulines A-D, were shown to have structures 200–203 by spectroscopic means which included an X-ray analysis of salvilanguiduline A (200) (206). The proton resonance spectra of these diterpenoids revealed that the disubstituted furan ring of languiduline (198) was replaced by a 14,15-epoxybutanolide with a fully substituted C-13, that is, an epoxyspirolactone (206). This was confirmed by the ^{13}C NMR spectrum of salvilanguiduline C (202) which contained a carbonyl signal at δ 170.1 ascribed to C-16, a singlet at δ 65.24 due to C-13 and two doublets at δ 68.16 and 105.27 assigned to C-14 and C-15, respectively, a singlet at δ 200.45 was attributed to the cyclohexanone carbonyl C-12. The location of the

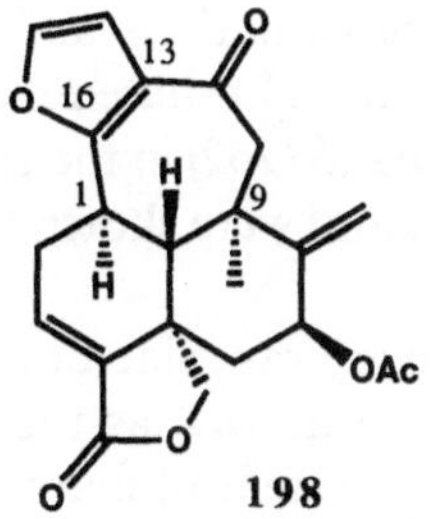

198 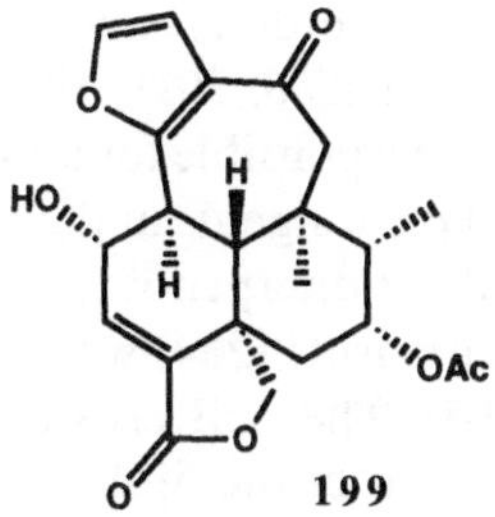   199

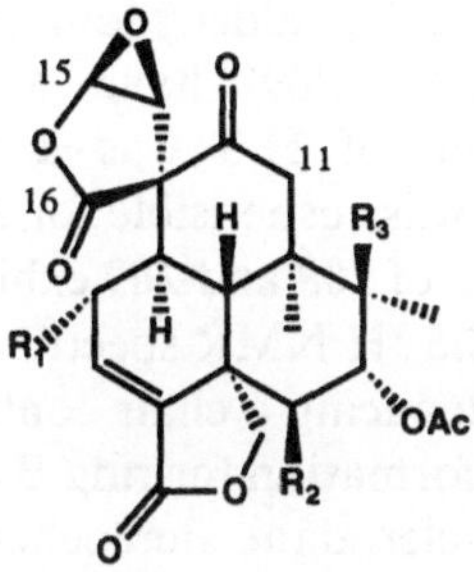

200 $R_1 = R_2 = R_3 = H$
201 $R_2 = R_3 = H$ $R_1 = OH$
202 $R_1 = R_3 = H$ $R_2 = OH$
203 $R_1 = R_2 = H$ $R_3 = OH$

hydroxy groups in salvilanguidulines B, C and D (**201–203**) was deduced from the ^{1}H NMR data (*206*).

The new carbon skeleton of the salvilanguidulines A-D, was named **salvilanguidulane**. These compounds as well as the languidulane diterpenoids **198** and **199**, could be related biogenetically to a clerodane precursor possessing a diene function in ring A or its 1,2-epoxy equivalent and a ketonic group at C-12. That furan ring is oxidized to the epoxylactone prior to formation of ring C has been suggested (*191*).

5. *Leonurus* and *Stachys* Species

Leonurus L. is a very small genus which comprises 10 species native to Europe and Asia. On the American Continent it is found mainly in the tropical zones of Central America (*207*). *Stachys* is a cosmopolitan genus of more than 200 species occurring in tropical and temperate climates (*207*). Only few species of these two genera have been studied up to now. From some of them clerodane diterpenoids have been isolated.

Some *Leonurus* species have been used in European folk medicine (*208*). A phytochemical study of the aerial parts of *Leonurus marrubiastrum* L. led to the isolation of several clerodane diterpenoids whose

structures were determined by chemical transformations and spectroscopic means (*208, 209*). All of them had a terminal β-butenolide group
which was responsible for a Fermi resonance band (*162*) in the IR spectra
and the typical signals in the ^{1}H NMR spectra and were shown to be A/B
cis-clerodane diterpenoids.

Marrubiaside (**204**) was the mono-βD-glucopiranoside of marrubiagenin (**205**). The A/B *cis*-steroidal arrangement as well as the absolute
configuration of the 5β,10β-*cis*-clerodane ring system of marrubiagenin
was established by structural correlation (*210*) with (−)-hardwickiic acid
of known (*198*) absolute stereochemistry.

Marrubialactone (**206**), marrubiastrol (**207**), aldehydomarrubia-
lactone (**208**), and desmethylmarrubiaketone (**209**) have an α,β-
unsaturated 18,19-olide function. The signal of H-3 was a triplet
($J = 7.0$ Hz) of δ 6.80 while the 19-methylene was responsible for an AB
system, with the *pro-S* 19 proton in the spectra of **208** and **209** exhibiting
W-coupling with the *axial* H-6 absent from the ^{1}H NMR spectra of **206**
and **207**. This observation was the basis for deducing a chair conformation for ring B of **208** and **209** and a boat conformation for ring B of **206**
and **207**. That **206–208** possessed 5β,10β-*cis*-clerodane stereochemistry
was also based on ^{1}H and ^{13}C NMR spectral analysis as well as on mass

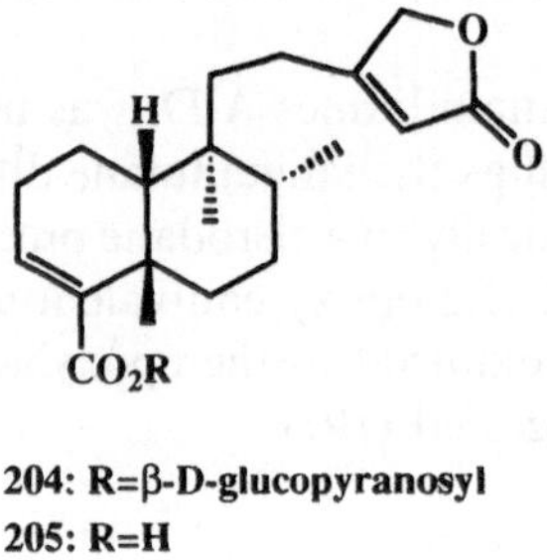

204: R=β-D-glucopyranosyl
205: R=H

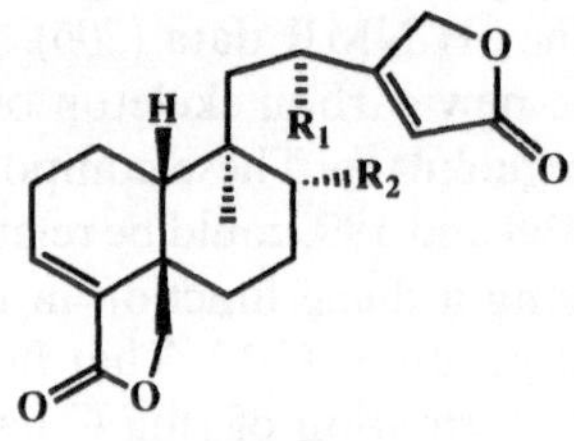

	R₁	R₂
206:	H,	CH₂OH
207:	OH,	CH₃
208:	H,	CHO

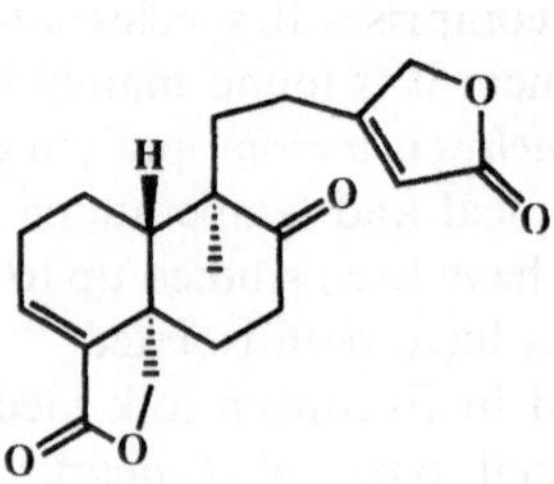

209

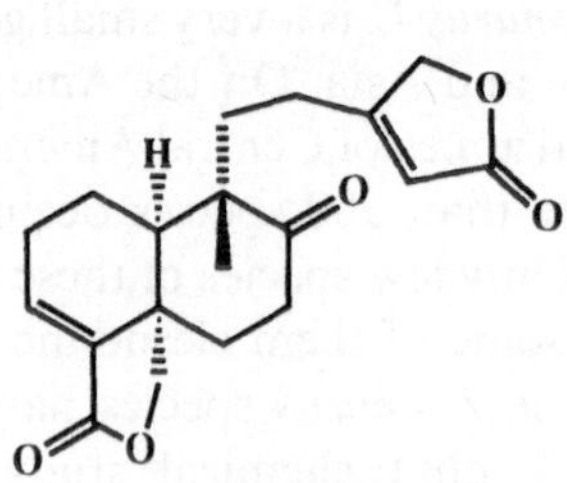

210

spectrometry. The relatively high abundance of the mass peaks, produced by loss of formaldehyde from the 18,19-γ-lactone group, was used (*208, 209*) as an argument in support of the proposed 5β-configuration in these diterpenoids. The CD spectra of the 1,2-dehydro-13,14-dihydro derivatives of marrubiagenin and of marrubialactone were in agreement with this assignment. Marrubiastrol (**207**) was shown to have the same basic structure as olearin (**97**) but differed from olearin in the 5β orientation of the 18,19-olide ring. The configuration at C-12 in **207** was established as *R* (*209*).

The mass spectrum of desmethylmarrubiaketone did not exhibit an intense peak due to the loss of formaldehyde from the 18,19-lactone group. This observation suggested (*209*) that the configuration at C-5 was the inverse of the C-5 configuration of **204–208**, *i.e.* that the C-5 oxymethylene was α and the A/B ring fusion *trans*. Structure presumed **209** was confirmed by X-ray diffraction. However, analysis of the perspective view shown in ref. *209*, p. 2136, indicates that desmethylmarrubiaketone is better represented by structure **210** with an A/B cis ring fusion the C-5 oxymethylene and H-10 α. In this structure C-20 is β-equatorial, while the side chain becomes α-*axial*. Several A/B cis 5α,10α- and 5β,10β-clerodane diterpenoids have been isolated from *Solidago* species (Compositae) (*210–212*).

A bitter principle from *Leonurus cardiaca* L. was a mixture of two unstable epimers $C_{22}H_{30}O_8$ (*213*). Structures **211** (without stereochemistry) were based on spectroscopic data of the acetylation products **212**, with one of the isomers having an A/B *cis* and the other an A/B *trans* fusion of the decalin ring (*213*). Analysis of the ¹H NMR spectral data given in (*213*) and comparison with the data of structurally related clerodane diterpenoids suggests that both epimers possess *trans*-A/B ring fusion with a *cis*-18α,19α-γ-lactone group and H-4 β-axially oriented as found in salviarin (**159**) (*178*) and its derivatives. The signal of H-3 under

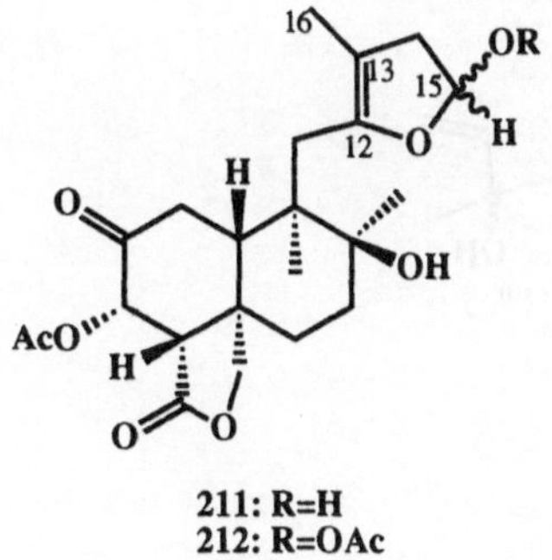

211: R=H
212: R=OAc

the acetoxy group (δ 5.00) showed a coupling constant of 6 Hz with H-4, in agreement with equatorial orientation for H-3, and the chemical shift of H-20 (δ 0.74) is similar to shifts (δ 0.6–0.9) for the H-20 in several *neo*-clerodane diterpenoids (*168, 171, 214*). The epimeric center at C-15 could be responsible for the presence of two epimers and their instability.

Neo-clerodane diterpenoids have also been isolated from three *Stachys* species *S. annua* (*215–218*), *S. recta* L. (*219*) and *S. rosea* (*220*). Structures were deduced on spectroscopic evidence and some structural correlations. Most of the *neo*-clerodane diterpenoids found in the three *Stachys* species have structures **213–217** related to kolavelool, whose absolute stereochemistry at all centres except C-13 has been rigorously established (*221*).

Stachysolone (**213**) and its 3,4-dihydro derivative as well as the 3,4,14,15-tetrahydro derivatives, stachylone and stachone, were isolated from *S. annua* (*215, 216, 218*). The stereochemistry at C-8 and C-13 in these diterpenoids was not ascertained. The clerodane diterpene annuanone (**218**), was also found in this species (*217*).

The 7- and 13-monoacetates **214** and **215**, as well as the 7,13-diacetate **216** of stachysolone (**213**), were found in *S. recta* (*219*). The configuration

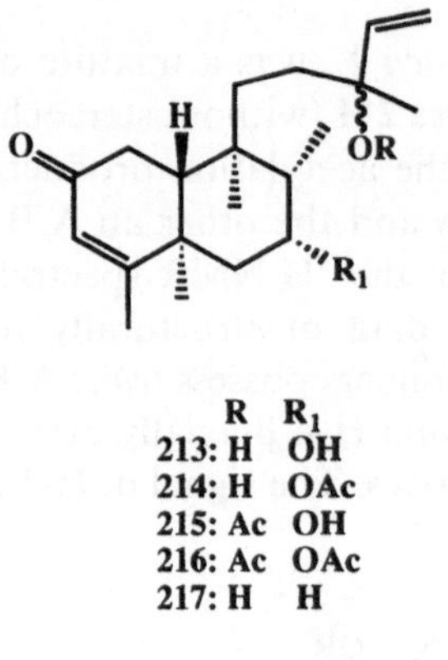

	R	R₁
213:	H	OH
214:	H	OAc
215:	Ac	OH
216:	Ac	OAc
217:	H	H

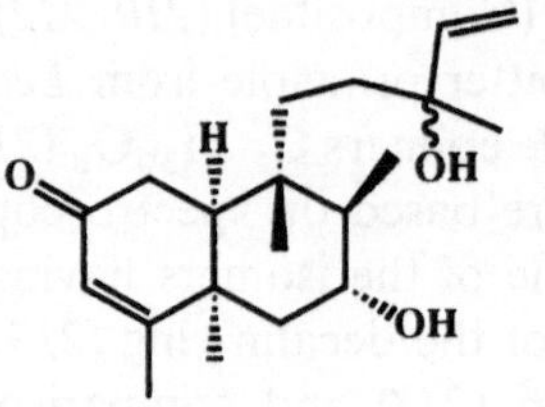

218

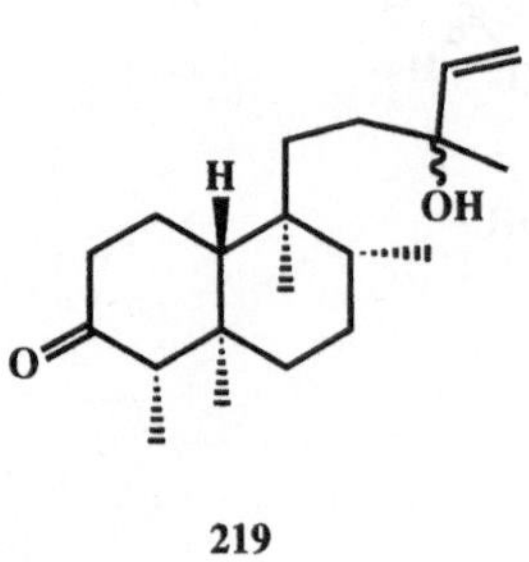

219

at C-8 in stachysolone diacetate shown in **216** was established by nOe experiments (*219*).

The 7-dehydroxystachysolone, roseostachenone (**217**), as well as the 3-keto derivative roseostachone (**219**) were isolated from *S. rosea* (*220*). The relative stereochemistry of the secondary methyl group at C-4 of **219** was deduced by nOe experiments (*220*).

III. Biological Activity

Neo-clerodane diterpenoids have attracted attention because some of them possess antifeedant activity which appears to be specific to certain insects.

The ajugarins I (**22**), II (**24**) and III (**25**) from *Ajuga remota* are moderately strong antifeedants against *Spodoptera exempta* (African army worm) and *S. littoralis* (Egyptian cotton leaf worm) (*14a*). Ajugarin I was also active against *Heliothis armigera* (Hubner) (*222*). Small structural changes cause variation in the biological activity. Ajugarin IV (**26**) (*30*) did not show antifeedant activity against *S. exempta* but was insecticidal against *Bombyx mori*, while ajugarin V (**27**) was not active as an antifeedant or as insecticide (*31*). The structurally related ajugacumbins A, B and C (**28–30**) isolated from *A. decumbens* (*34*) in which the acetyl group at O-19 is replaced by a tiglate, displayed insect antifeedant activity against larvae of *Pareba vesta* Fabricius. This plant is used in folk medicine of China and Japan for inflammation and infectious diseases.

Neo-clerodanes from *Clerodendron* and *Caryopteris* Verbenaceae, such as clerodin (**4**) also found in *Ajuga remota* (*15*), and clerodendrin (**3a**) are active against *Spodoptera exempta*. Both diterpenoids contain a furo-furan ring attached to C-9 in place of the terminal β-butenolide group found in the ajugarins I-V and ajugacumbins. The furo-furan ring has been claimed as essential for the insect antifeedant activity of these *neo*-clerodanes, but 14,15-dihydro derivatives such as the ivains (**13, 14, 16**) from *Ajuga iva* (*25*) and *A. pseudoiva* (*26*) as well as the ajugapitins (**10**) from *A. chamaepitys* (*23*) showed good activity against *Spodoptera littoralis* (*223*). Furthermore, the 15-methoxy derivative of clerodendrin had higher antifeedant activity than the parent compound (**3a**) against the tobacco cut worm *Spodoptera litura* F. (*224*).

Jodrellins A and B (**52, 53**) from *Scutellaria woronowii* (*40*) showed very potent antifeedant activity against *Spodoptera littoralis*; in fact jodrellin B has been claimed the most potent clerodane antifeedant described so far (*40*). 14,15-Dihydrojodrellin T (**55**) from *S. galericulata*, had less activity than jodrellins A and B and than clerodin (*41*).

Several *neo*-clerodane diterpenoids from *Teucrium* species have been studied as antifeedant agents against *Spodoptera littoralis* and *Heliothis armigera* (*222*). The most active compound against *H. armigera* was 6,19-diacetylateumassilin (**81**, R_1=Ac, R_2=αOAc, H,R_3=βOH,H) from *T. massiliense* (*71*) which differs from ajugarin I in the presence of a β-substituted furan group in place of the terminal β-butenolide function and an hydroxyl on C-12. Ajugarin I (**22**) was the only diterpenoid to show significant antifeedant activity against both insect species. 12-*Epi*-teucvin (**89**, 12-*epi*) from *T. flavum* (*106*) showed moderate activity against *S. littoralis*, while 7α-hydroxy-teucvin (teucrin, **89**, 7αOH) was inactive against both insect species. From these studies it was concluded (*222*) that small changes in the decalin and/or in the side chain of *neo*-clerodanes influence the antifeedant properties of the product and no substitution pattern can be described as essential for the antifeedant activity. Tafricanin B (**85**, R=αOAc,H) was as active as clerodin hemi-acetal as an antifeedant against *Locusta migratoria*, while the related 6-keto derivative tafricanin A (**85**, R=O) was not active (*76*).

Recently the antifeedant activity of some *neo*-clerodanes isolated from *Salvia* species has been studied (*225*). Compounds **141** and **146** from *S. melissodora* (*171*) as well as semiatrin (**152**) from *S. semiatrata* (*170*) showed important activity against *Spodoptera littoralis*. 1(10)-Dehydro-salviarin **168**, from *S. lineata* (*183*) was also active while linearolactone **171** in which the diene of the A-ring is conjugated to the 18,19-lactone group, from the same source did not show significant activity against *S. littoralis* or *S. exempta*.

Antifungal activities of clerodin (**4**) and the jodrellins A and B (**52, 53**), from *Scutellaria woronowii* (*40*) were studied against the plant pathogenic fungi *Fusarium oxysporum* f.sp. *lycopersici* and *Verticillium tricorpus* Issac (*226*). Clerodin was the most active agent against both fungi while jodrellins B and A were less active in this assay.

Salvia divinorum is used in divinatory rites by the Mazatec Indians of Oaxaca, Mexico. The hallucinogenic effect produced by this plant has been attributed (*176*) to the *neo*-clerodane diterpenoids salvinorins A and B (**129a,b**).

Acknowledgement

The authors wish to express their gratitude to Drs. Benjamin Rodriguez and Maria C. de la Torre of the Instituto de Quimica Organica, CSIC, Madrid, Spain, for their invaluable comments and suggestions and for information provided prior to publication.

Table 1. *Distribution of Clerodane Diterpenoids in Teucrium sp.*

Plant	Product	References
T. abutiloides L'Herit	montanin C (150) 12-epiteupolin II (153) teubutilin A (113) teubutilin B (51)	109
T. africanum	tafricanin A (121) tafricanin B (122)	76
T. asiaticum	teuflin (69) auropolin (119)	24
T. bicolor	montanin C (150) 12 epi-teucvin (156) teucvin (65) teucrin H-2 (77) teupolin I (18) 12-epiteupolin I (151) 12-epiteuscorodonin (159)	110
T. bidentatum Hemsl.	6-ketoteuscordin (79) teuscordinon (63) teucrin H-2 (77) teuflin (69) bidentatin (80)	111
T. botrys L.	6β-OH-teuscordin (61) teucvidin (67) montanin D (135) teuchamaedrin C (104) 19-deacetylteuscorodol (94) teubotrin* (106)	59
T. buxifolium Schreber	19-acetylgnaphalin (21) 19-acetylteulepicin (33)	57
T. canadense L.	teucvidin (67) teuflin (69) teucvin (65) 12-epi-teupolin II (153) teuscorodal (97) (12R)-teupolin I (151) 18-acetyl-montanin D (135b) isoteuflin (129) acetylisoteucrin H-4 (130)	115
T. capitatum L.	capitatin (26) teucapitatin (28) 19-acetylgnaphalin (21) lolin (96)	112, 113
T. carolipani C. Vicioso *ex* Pau	19-acetylgnaphalin (21)	116
T. chamaedrys	teucrin A (73) teucrin E (78) teuchamaedrin A (66)	119 118

Table 1 (*continued*)

Plant	Product	References
	teuchamaedrin B (**77**)	*120*
	teucrin B (**76**)	*96*
	teucroxide (**136**)	*97*
	6α-hydroxyteuscordin (**60**)	*121*
	teuchamaedryn C (**104**)	
	isoteuflidin (**72**)	*122*
	teucvin (**65**)	
	teucvidin (**67**)	
	teuflidin (**71**)	
	teucrin F (**88**)	
	teucrin G (**89**)	
	chamaedroxide (**137**)	
	teugin (**59**)	
	teuflin (**69**)	
	6-epiteucrin A (**74**)	
T. chartaginense	19-acetylgnaphalin (**21**)	*53*
Lange var.	eriocephalin (**53**)	
homotricum		
T. cossonii	teucossin A (**14**)	*141*
	teucossin B (**120**)	
	montanin H (**110**, 7-keto)	
T. creticum L.	19-acetylgnaphalin (**21**)	*114*
	teucjaponin B (**18**)	
	6,19-diacetylteumassilin (**9**)	
	teucretol (**13**)	
T. cubense	teucvin (**65**)	*117*
T. divaricatum	teuflin (**69**)	*99*
var. canescens	teucrin H-2 (**77**)	
	teuflidin (**71**)	
	teucrin A (**73**)	
	teucrin F (**88**)	
	teucrin G (**89**)	
	montanin D (**135**)	
	dihydroteugin (**76**)	
	2-deoxychamaedroxide (**138**)	
T. eriocephallum	eriocephalin (**53**)	*63*
T. flavum L.	teuflin (**69**)	*125*
subsp. glaucum	teuflavin (**50**)	*123*
Jordan and Fourr	teuflavoside (**141**)	
	teuflidin (**71**)	*124*
	12-epi-teucvin (**156**)	*106*
T. fragile	teugin (**59**)	*91*
T. fruticans	fruticolone (**5**)	*69*
	isofruticolone (**7**)	*70*
	8-β-hydroxyfruticolone (**6**)	
	fruticolide (**148**)	*105*

Table 1 (*continued*)

Plant	Product	References
T. gnaphalodes	gnaphalin (**20**)	*56*
L'Her	19-acetylgnaphalin (**21**)	
	gnaphalidin (**47**)	
	teucrin P$_1$ (**111**)	
	teugnaphalodin (**101**)	*62*
T. gracile	teugracilin A (**36**)	*138*
	3-O-deacetylteugracilin A (**37**)	
	teugracilin B (**38**)	
	teugracilin C (**12**)	
	teumicropodin (**39**)	
	teugracilin D (**45**)	*105*
	teugracilin E (**15**)	
	19-acetylteulepicin (**33**)	
T. grisebachii	6-acetyl-teucjaponin B (**19**)	*126*
	6,12,19-triacetylteumassilin (**11**)	
T. hyrcanicum	teucrin H1* (**71**)	*127*
	teucrin H4 (**70**)	
	teucrin H2 (**77**)	
	teucrin H3* (**21**)	
T. intricatum	teucvin (**65**)	*53*
Lange		
T. japonicum	teucjaponin B (**18**)	*72*
Houtt	teucjaponin A (**17**)	
	teucvin (**65**)	
	teuponin (**160**)	*128*
T. kotschyanum	teucrin H4 (**70**)	*129, 130*
Poech	isoteucrin H4 (**131**)	
	teucvidin (**67**)	
	teuscorodin (**83**)	
	teucrin H2 (**77**)	
	teuscorodonin (**98**)	
	montanin D (**135**)	
	12-epiteucvidin (**157**)	
	12-epiteuflin (**158**)	
	teukotschyn (**86**)	
	teuflin (**69**)	
T. lamiifolium	teuscordinon (**63**)	*131*
D'urv	teuflin (**69**)	
	montanin C (**150**)	
	19-acetylgnaphalin (**21**)	
	12-epiteupolin II (**153**)	
	teulamifin B (**106**)	*60*
T. lanigerum	eriocephalin (**53**)	*132*
Lag	20-deacetyleriocephalin (**54**)	
	iso-eriocephalin (**56**)	
	7,8-dehydroeriocephalin (**55**)	*95*

Table 1 (*continued*)

Plant	Product	References
	teulanigeral (**139**)	
	teulanigin (**48**)	
	20-epiteulanigin (**49**)	
	teulanigerin (**100**)	
	teulanigeridin (**105**)	
T. lepicephalum	19-acetylteulepicin (**33**)	*57*
Pau	teulepicin (**32**)	
	teulepicephin (**102**)	
T. lucidum L.	teucvidin (**67**)	*53*
	teuflin (**69**)	
	teucrin F (**88**)	
	teucrin G (**89**)	
	6α-hydroxyteuscordin (**60**)	
T. marum	teumarin (**23**)	*133*
T. massiliense	montanin C (**150**)	*71*
	teucjaponin A (**17**)	
	6,19-diacetylteumassilin (**9**)	
	deacetylajugarin II (**149**)	
	teumassilin (**8**)	
T. microphyllum	teucrin A (**73**)	*135*
Desf.	teucrin G (**89**)	
	teugin (**59**)	
	dihydroteugin (**76**)	
T. micropodioides	3-acetylteumicropin (**114**)	*134*
Rouy	teumicropin (**115**)	
	deacetylteupyrenone (**116**)	
	teumicropodin (**39**)	
	3-deacetyl-20-epi-teulanigin (**52**)	
T. montanum L.	montanin A (**134**)	*83*
var. skorpilii	montanin B (**140**)	
	teucvin (**65**)	
	montanin E (**123**)	*79*
	6-epiteupolin I* (**17**)	
	montanin C (**150**)	*136*
	montanin D (**135**)	
	montanin G (**154**)	*137*
T. montbretii var.	teucrin H2 (**77**)	*141*
heliotropiifolium	6β-hydroxyteuscordin (**61**)	
	montanin D (**135**)	
	teugin (**59**)	
T. odontites Boiss	teucrin H2 (**77**)	*135*
and Bal.	montanin D (**135**)	
	teuscorodol (**93**)	
T. oliverianum	teucrolivin A (**144**)	*102*
Ging ex Benth R. Br.	teucrolivin B (**1**)	
	teucrolivin C (**2**)	

References, pp. 184–196

Table 1 (*continued*)

Plant	Product	References
	teucrolivin D (**3**)	*103*
	teucrolivin E (**4**)	
	teucrolivin F (**145**)	*104*
	teucrolivin G (**147**)	
	teucrolivin H (**146**)	
T. oxylepis	gnaphalin (**20**)	*139*
Font. Quez. *var.*	19-acetylgnaphalin (**21**)	
marianum	teuscorodonin (**98**)	
	montanin D (**135**)	
	teucrin H4 (**70**)	
	19-deacetylteuscorodol (**88**)	
	teubotrin (**106**)	
	teucroxide (**136**)	
	teukotchyn (**86**)	
	teucroxylepin (**109**)	
	isoteucrin H4 (**131**)	
	12-O-acetylteugnaphalodin (**103**)	
T. pernyi Franch	teupernin A (**81**)	*140*
	teupernin B (**80**)	
	teupernin C (**82**)	
T. pestalozzae	tafricanin A (**121**)	*135*
Boiss.	4α,18 epoxy-tafricanin A (**35**)	
	20-oxo-teuflavin (**34**)	
	teupestalin A (**142**)	*101*
	teupestalin B (**143**)	
T. polium	picropolin (**29**)	*54*
var. album	montanin C (**150**)	*142*
var. aureum	gnaphalidin (**20**)	*144*
	teucrin P1 (**111**)	
	19-acetylgnaphalin (**21**)	
	auropolin (**119**)	
	picropolinone (**31**)	*74*
	picropolin (**29**)	
var. capitatum	picropolin (**29**)	*75*
	picropolinone (**31**)	
	19-acetylgnaphalin (**21**)	
	teucjaponin B (**18**)	
	7-deacetylcapitatin (**27**)	
	picropolinol (**124**)	
	20-epi-isoeriocephalin (**57**)	
	teucrin A (**73**)	
var. expansum	picropolinone (**31**)	*141*
	19-acetylteulepicin (**33**)	
	3-O-deacetyl-teugracilin A (**37**)	
var. pilosum	19-acetylteupolin IV (**25**)	*145*
var. polium	teupolin III (**125**)	*60, 73*

Table 1 (*continued*)

Plant	Product	References
	montanin B (140)	
	teupolin IV (24)	66
	teupolin V (99)	143
	teucrin P1 (111)	
	teulamifin B (106)	
var. vicentinum	19-acetylgnaphalin (21)	77
	isoeriocephalin (56)	
	3-deacetyl-20-epiteulanigin (52)	
	eriocephalin (53)	
	teuvincentin A (126)	
	teuvincentin B (118)	
	teuvincentin C (58)	
T. pumilum var. cariolipani	19-acetylgnaphalin (21)	153
T. pyrenaicum L.	teupyrenone (112)	68, 65
	teupyreinin (152)	107
	teupyreinidin (41)	
	teupyrin A (108)	
	teupyrin B (10)	
T. salviastrum Schreber	teucvidin (67)	58
	teucroxide (136)	
	teusalvin A (84)	
	teusalvin B (85)	
	teusalvin C (155)	
	teusalvin D (95)	
	teusalvin E (107)	
	teusalvin F (127)	
T. scordium L.	teuscordinon (63)	88
	montanin E (123)	
	teucrin E (78)	89
	teucrin H4 (70)	
	6α-hydroxyteuscordin (60)	
	6β-hydroxyteuscordin (61)	
	2β,6β-dihydroxyteuscordin (59)	
	2-keto-19-hydroxyteuscordin (92)	
	6-ketoteuscordin (79)	147–149
	20-epi-deacetylteupyreinidin (44)	
	20-epi-6,20-deacetylteupyreinidin (46)	
	2,3-dehydroteucrin E (87)	
	teugin (59)	
	dihydroteugin (76)	
	teucjaponin B acetate (19)	
	teucroxide (136)	
T. scorodonia L.	teupolin I (18)	146
	teuscorolide (132)	

References, pp. 184–196

Table 1 (*continued*)

Plant	Product	References
	teuscorodal (**97**)	
	teuscorodol (**93**)	
	2α-hydroxyteuscorolide (**133**)	*94*
	teuscorodin (**83**)	
	teuscorodonin (**98**)	
var. *euganeum* (Vis)	teuflin (**69**)	*53*
var. *scorodonia*	teuflin (**69**)	*53*
	teuscorolide (**132**)	
	teuscorodin (**83**)	
T. spinosum	teuspinin (**91**)	*150*
	19-acetyl-teuspinin (**90**)	
	19-acetyl-gnaphalin (**21**)	
T. subspinosum	teucvin (**65**)	*65*
	6α-hydroxyteuscordin (**60**)	
	teuflin (**69**)	
	teucrin H2 (**77**)	
T. turredanum	19-acetylgnaphalin (**21**)	*153*
	eriocephalin (**53**)	
	iso-eriocephalin (**56**)	
T. viscidum var.	teucvin (**65**)	*80*
miquelianum	teucvidin (**67**)	*81, 82*
	teuflin (**69**)	*151*
T. webbianum	teuflidin (**71**)	*152*
Boiss.	teucrin A (**73**)	
	2β-hydroxyteucvidin (**68**)	

* Teucrin H2 = Teuchamaedryn B (**77**)
Teubotrin = Teulamifin B (**106**)
Teucrin H1 = Teuflidin (**71**)
Teucrin H3 = 19-Acetylgnaphalin (**21**)
Teucrin B = Dihydroteugin (**76**)
Teugin = 2β,6β-Dihydroxyteuscordin (**59**)
6-epi-Teupolin I = Teucjaponin A (**17**)

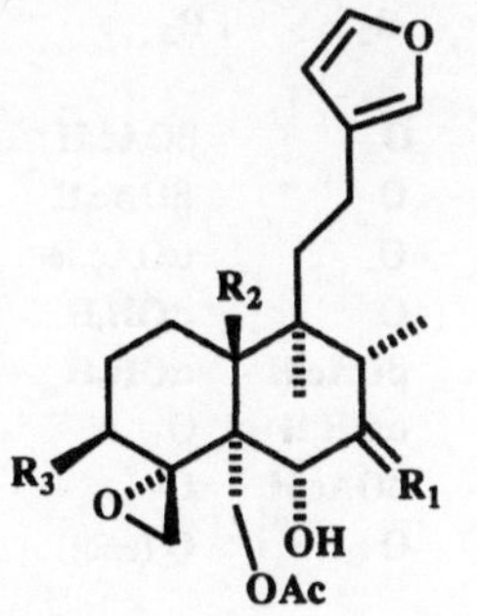

	R_1	R_2	R_3
1:	O	OH	OAc
2:	O	OH	OH
3:	βOAc,H	H	OAc
4:	O	H	OAc

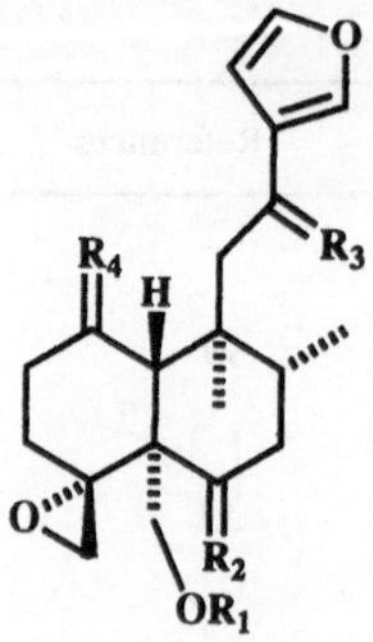

	R₁	R₂	R₃	R₄
5:	Ac	O	H₂	αOH,H
6:	Ac	O	H₂	αOH,H (8β-OH)
7:	Ac	βOH,H	H₂	O
8:	H	αOH,H	βOH,H	H₂
9:	Ac	αOAc,H	βOH,H	H₂
10:	H	αOAc,H	OH,H	H₂ (3β-OH)
11:	Ac	αOAc,H	βOAc,H	H₂

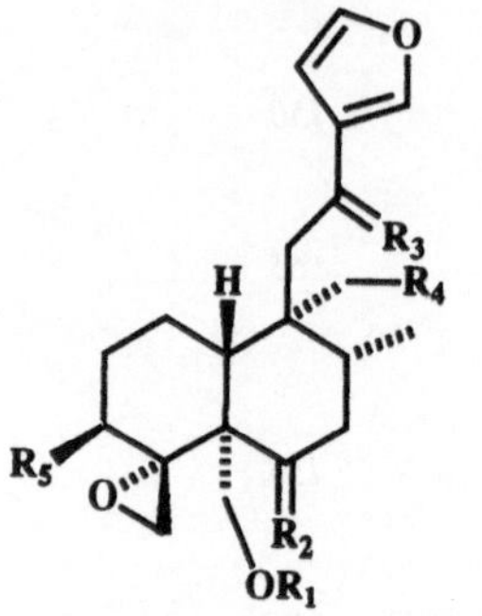

	R₁	R₂	R₃	R₄	R₅
12:	Ac	αOH,H	OAc,H	H	OH
13:	Ac	αOAc,H	OH,H	OH	H
14:	Ac	αOH,H	OAc,H	OAc	H
15:	Ac	αOAc,H	OAc,H	OAc	OAc
16:	Ac	αOH,H	OH,H	OH	H

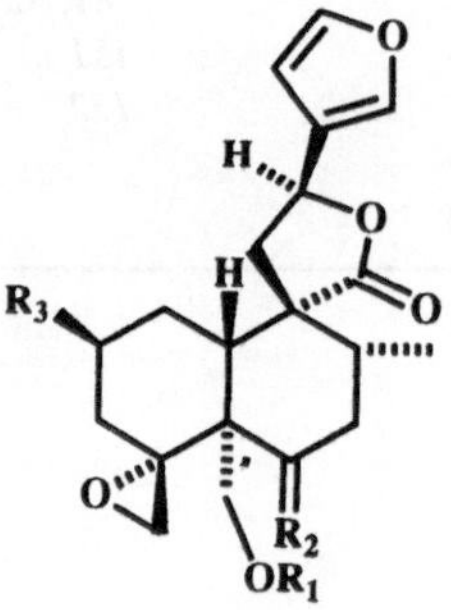

	R₁	R₂	R₃
17:	Ac	βOH,H	H
18:	Ac	αOH,H	H
19:	Ac	αOAc,H	H
20:	H	O	H
21:	Ac	O	H
22:	H	αOAc,H	H
23:	Ac	βOH,H	OH

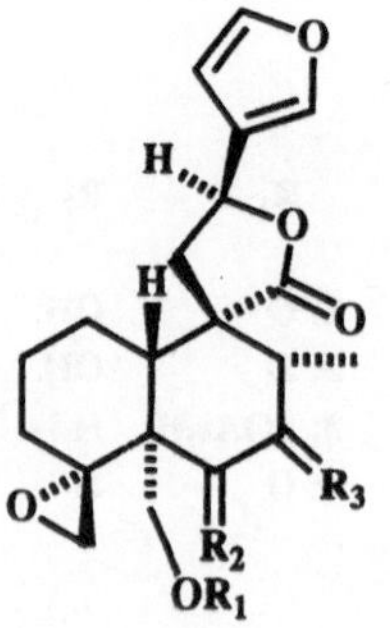

	R₁	R₂	R₃
24:	H	O	βOAc,H
25:	Ac	O	βOAc,H
26:	Ac	O	αOAc,H
27:	Ac	O	αOH,H
28:	Ac	βOAc,H	αOH,H
29:	Ac	αOH,H	O
30:	Ac	βOAc,H	O
31:	Ac	O	O (enol)

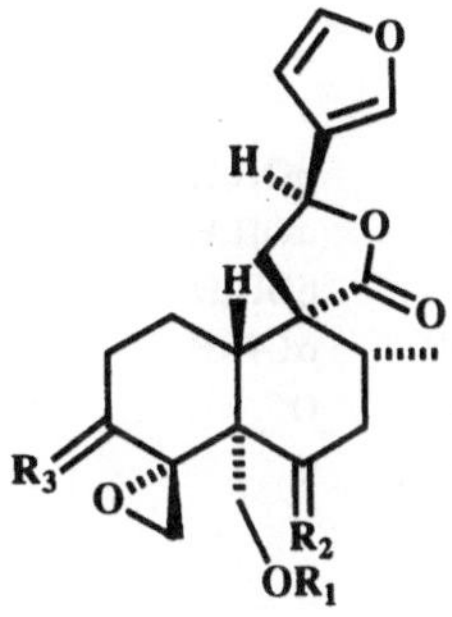

	R₁	R₂	R₃
32:	H	O	βOH,H
33:	Ac	O	βOH,H
34:	Ac	βOH,H	O
35:	Ac	O	O
36:	Ac	βOH,H	βOAc,H
37:	Ac	βOH,H	βOH,H
38:	Ac	αOH,H	βOH,H
39:	Ac	αOH,H	βOAc,H
40:	Ac	αOAc,H	βOAc,H

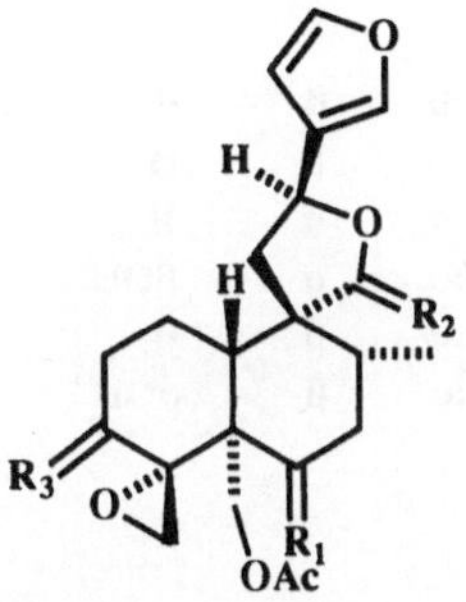

	R₁	R₂	R₃
41:	αOAc,H	βOAc,H	βOAc,H
42:	αOH,H	αOAc,H	βOAc,H
43:	αOH,H	αOH,H	βOAc,H
44:	αOH,H	βOAc,H	βOAc,H
45:	αOH,H	αOAc,H	βOH,H
46:	αOH,H	βOH,H	βOAc,H
47:	O	βOAc,H	H₂
48:	O	αOH,H	βOH,H
49:	O	βOAc,H	βOAc,H
50:	βOH,H	αOH,H	O
51:	αOAc,H	βOAc,H	H₂
52:	O	βOAc,H	βOH,H

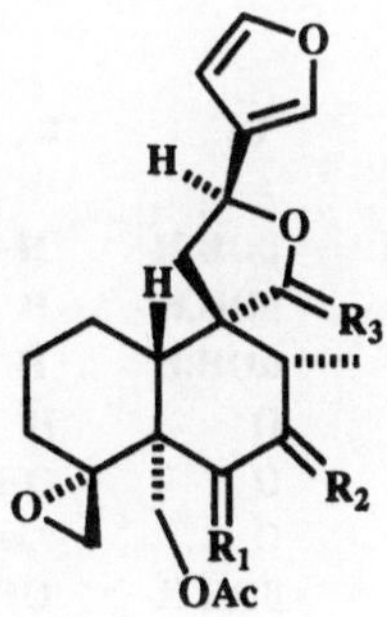

	R₁	R₂	R₃
53:	O	αOH,H	βOAc,H
54:	O	αOH,H	βOH,H
55:	O	O (enol)	βOAc,H
56:	αOH,H	O	βOAc,H
57:	αOH,H	O	αOAc,H
58:	αOAc,H	O(8 epi)	βOAc,H

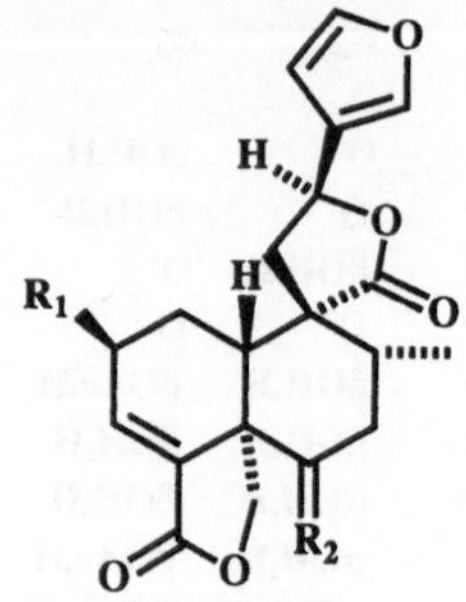

	R$_1$	R$_2$
59:	OH	βOH,H
60:	H	αOH,H
61:	H	βOH,H
62:	OH	αOH,H
63:	H	O
64:	OH	O

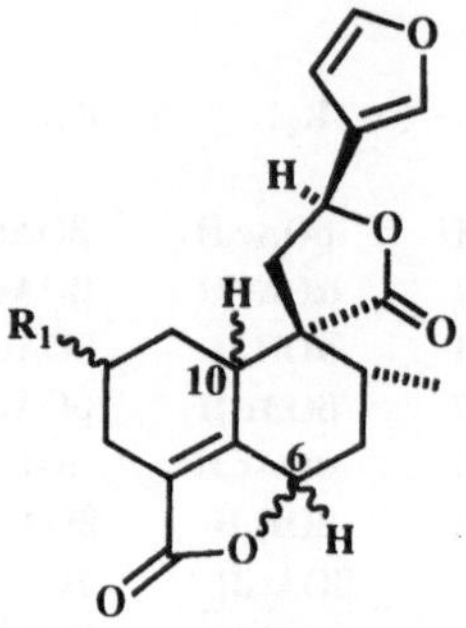

	H-6	H-10	R1
65:	β	β	H
66:	α	β	H
67:	α	α	H
68:	α	α	βOH
69:	α	β	H
70:	α	β	αOH

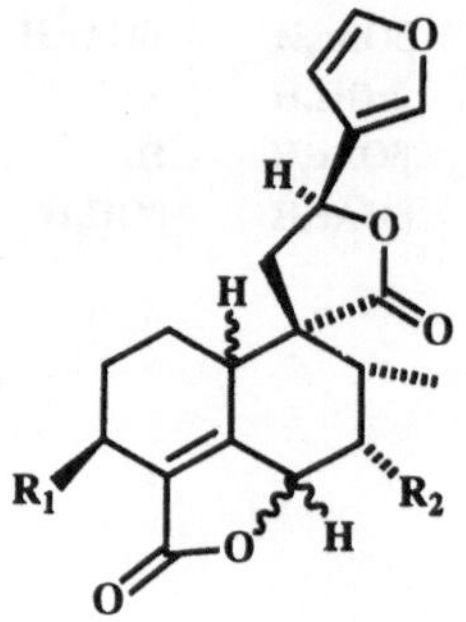

	H-6	H-10	R$_1$	R$_2$
71:	α	α	OH	H
72:	β	β	OH	H
73:	β	β	H	OH
74:	α	β	H	OH
75:	β	β	OH	H

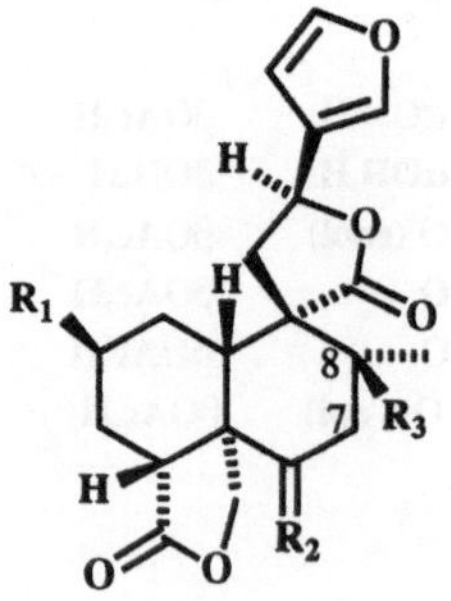

	R$_1$	R$_2$	R$_3$
76:	OH	βOH,H	H
77:	H	βOH,H	H
78:	H	αOH,H	H
79:	H	O	H
80:	H	O	OH
81:	H	O	7,en-
82:	H	βOH,H	OH

References, pp. 184–196

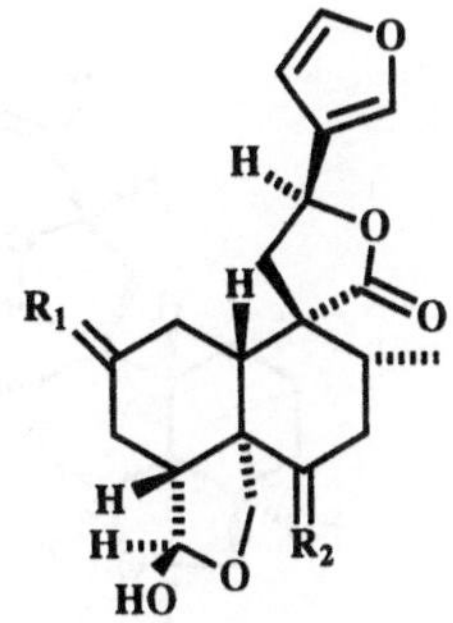

	R_1	R_2
83:	H_2	O
84:	O	O
85:	$\beta OH,H$	O
86:	H_2	$\beta OH,H$

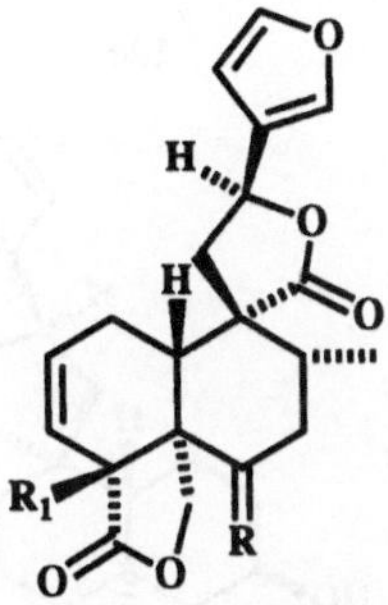

	R	R_1
87:	$\alpha OH,H$	H
88:	$\beta OH,H$	OH
89:	$\beta OH,H$ (2,3 β-epoxi)	OH

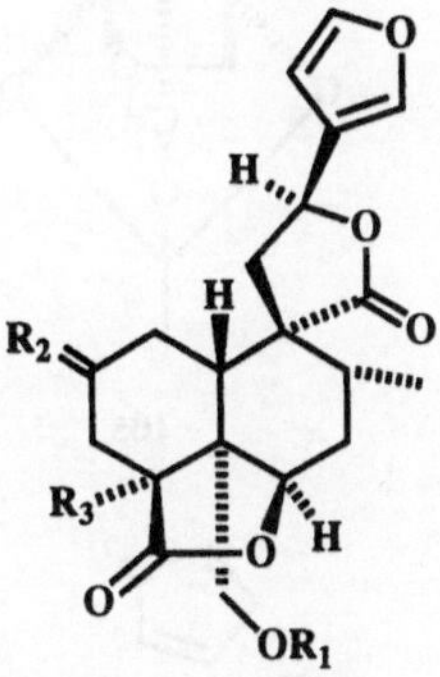

	R_1	R_2	R_3
90:	Ac	H_2	OH
91:	H	H_2	OH
92:	H	O	H

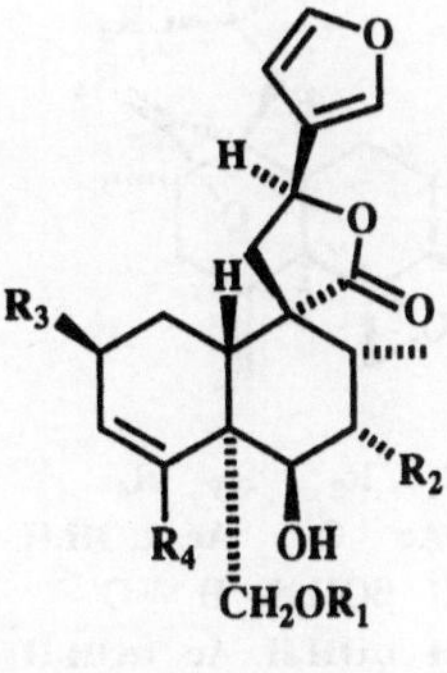

	R_1	R_2	R_3	R_4
93:	Ac	H	H	CH_2OH
94:	H	H	H	CH_2OH
95:	H	H	OH	CH_2OH
96:	Ac	OH	H	CH_2OH
97:	Ac	H	H	CHO

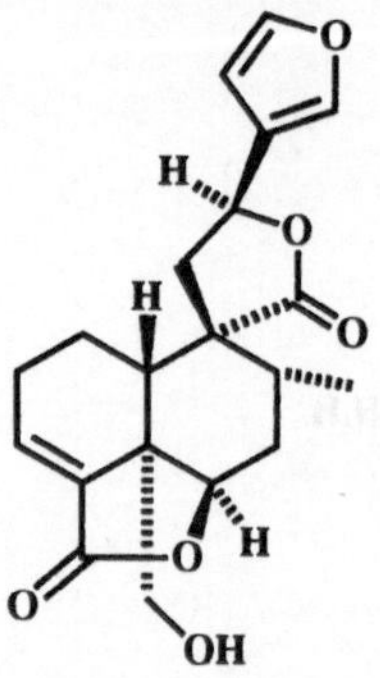

98

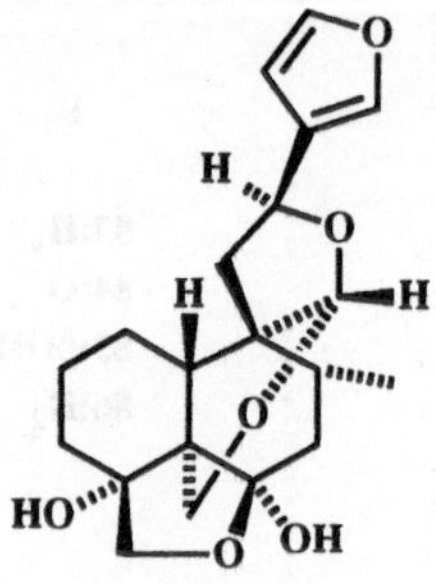

99

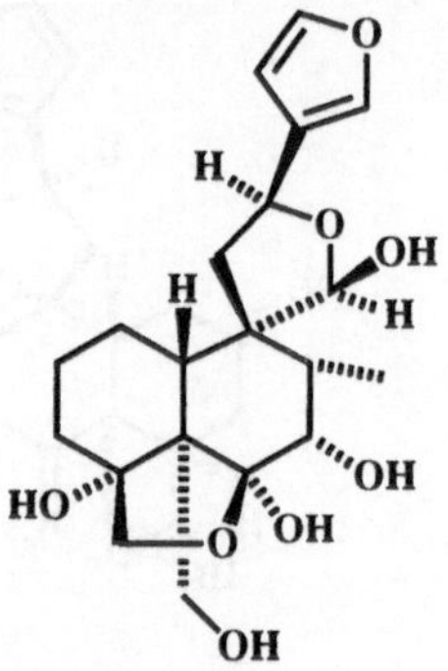

100

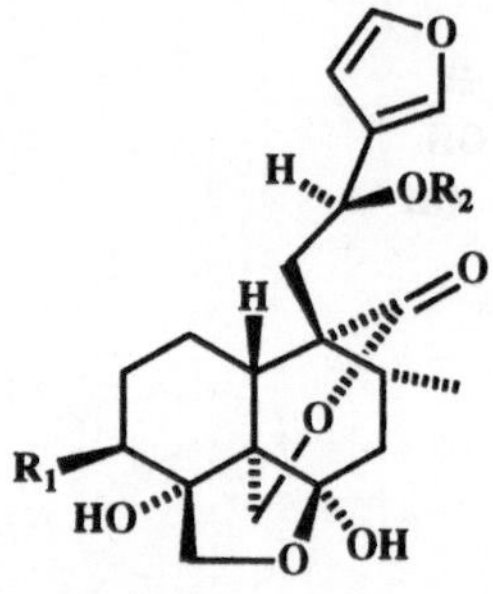

	R_1	R_2
101:	H	H
102:	OH	H
103:	H	Ac

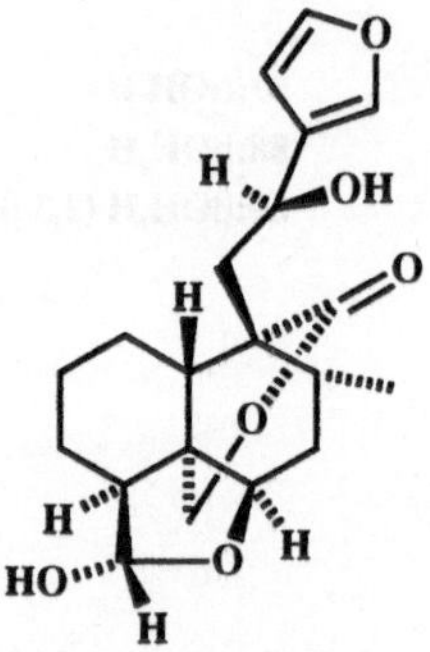

104

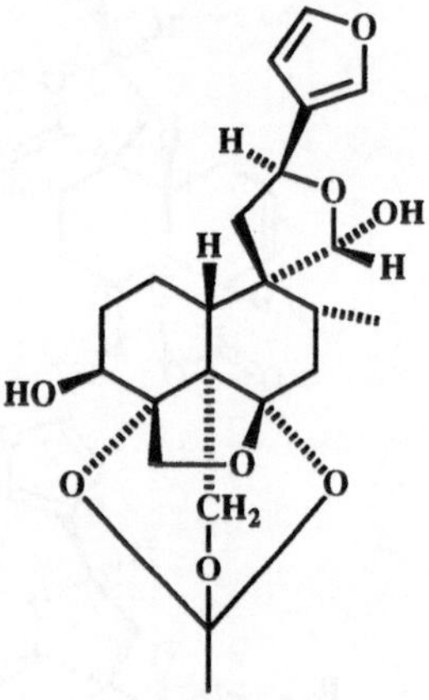

105

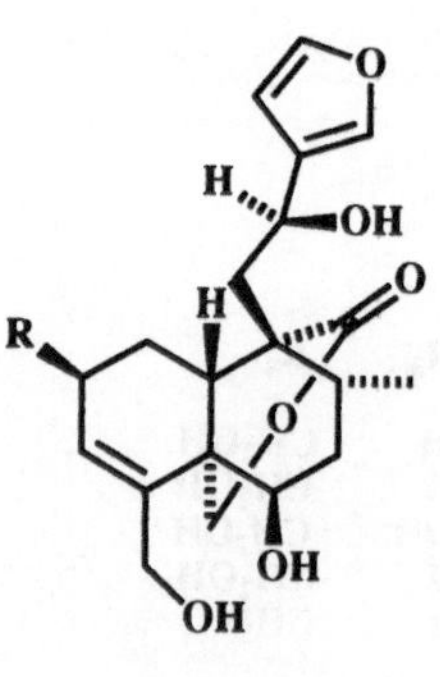

106: R=H
107: R=OH

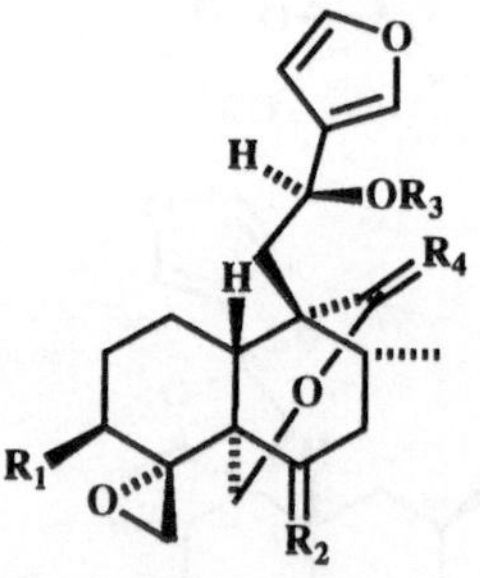

	R_1	R_2	R_3	R_4
108:	OAc	O	Ac	αOH,H
109:	H	βOH,H	H	O
110:	H	αOH,H	Ac	αOH,H

References, pp. 184–196

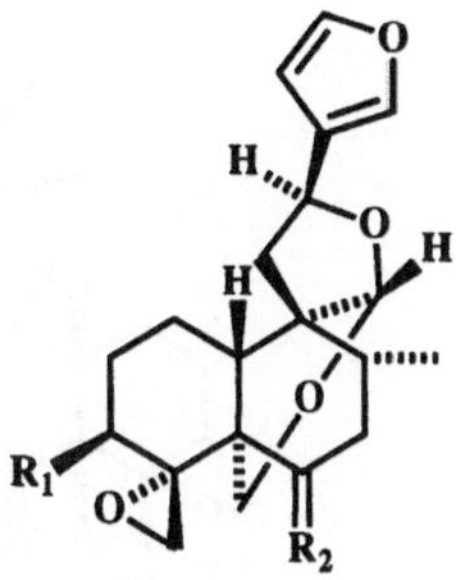

	R_1	R_2
111:	H	O
112:	OAc	O
113:	H	αOAc,H
114:	OAc	βOH,H
115:	OH	βOH,H
116:	OH	O

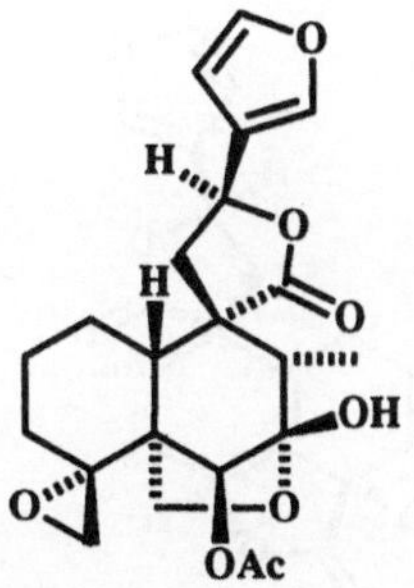

117

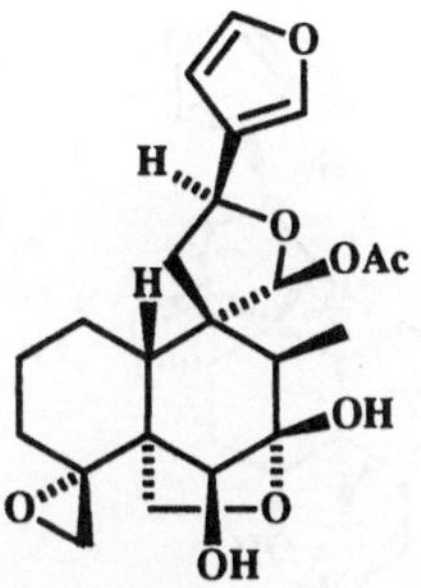

118

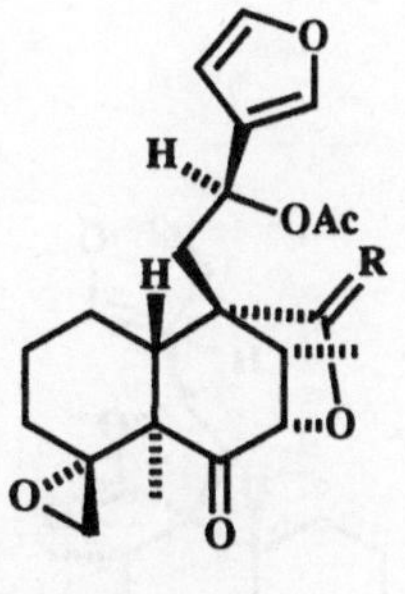

119: R= βOH,H
120: R= αOAc,H

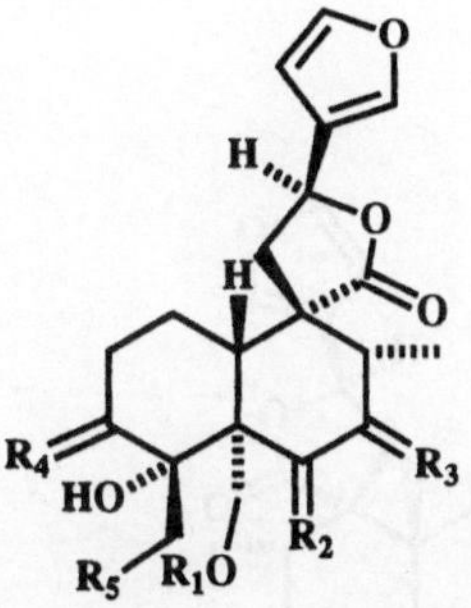

	R_1	R_2	R_3	R_4	R_5
121:	Ac	O	H_2	O	Cl
122:	Ac	αOAc,H	H_2	O	Cl
123:	H	βOH,H	H_2	H_2	OH
124:	Ac	αOH,H	O	H_2	OAc

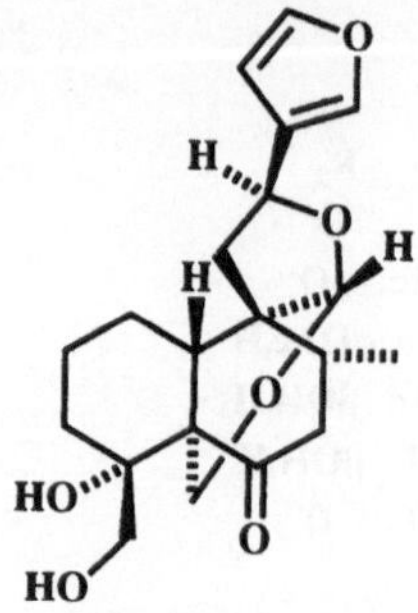

125

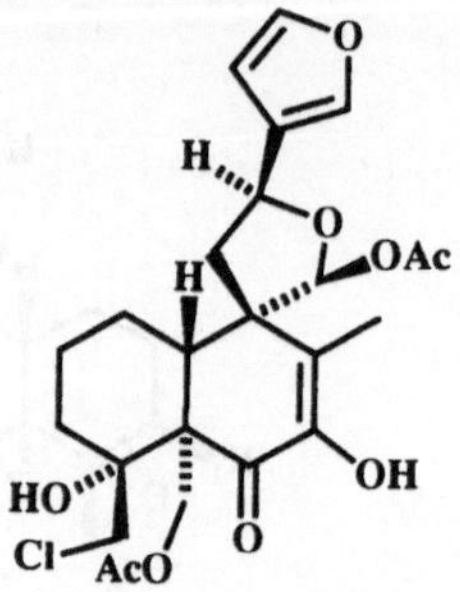

126

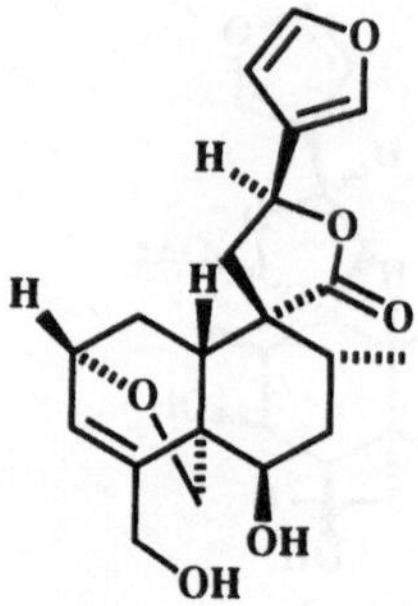

127

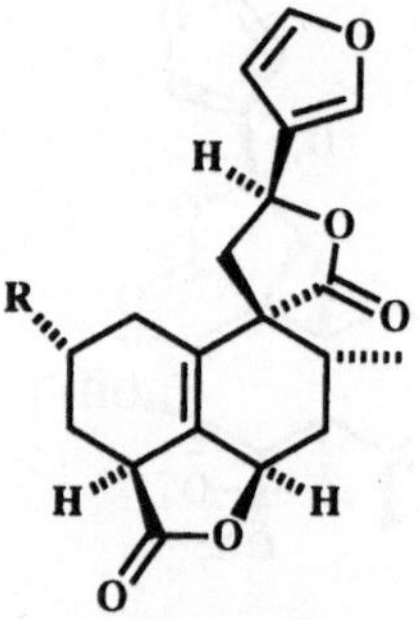

129: R= H
130: R= OAc
131: R= OH

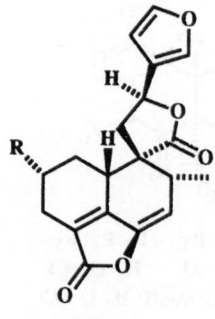

132: R= H
133: R=OH

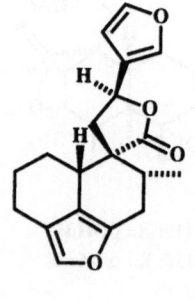

134

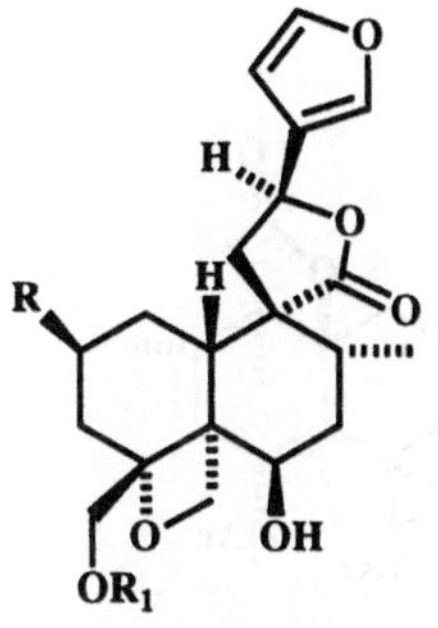

135a: R=R₁=H; b: R=H, R₁=Ac
136: R= OH, R₁=H

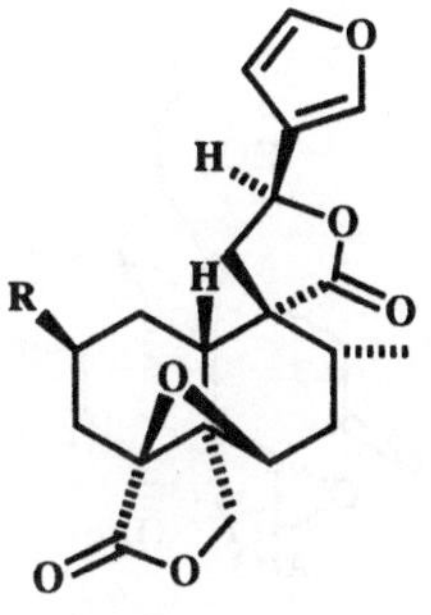

137: R= OH
138: R= H

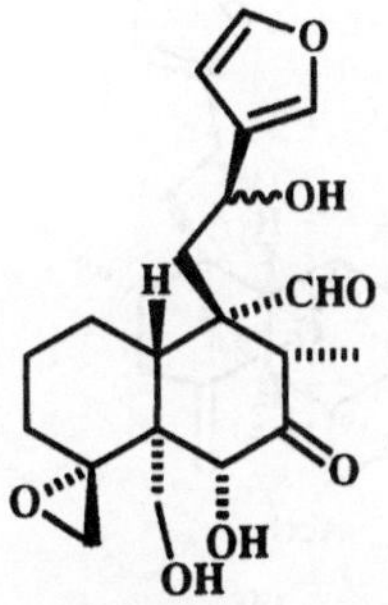

139

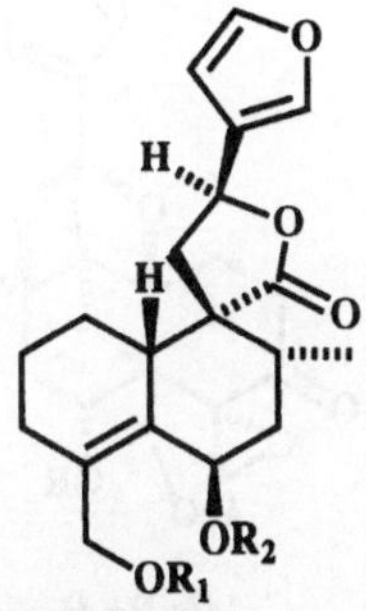

140: R₁=R₂= H
141: R₁=Ac R₂= 2-Acetyl glucose

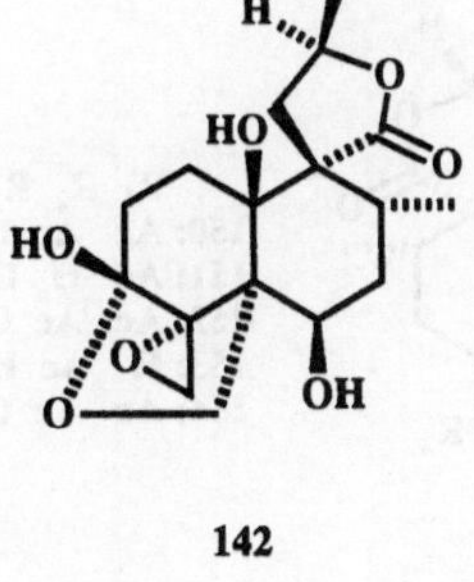

142

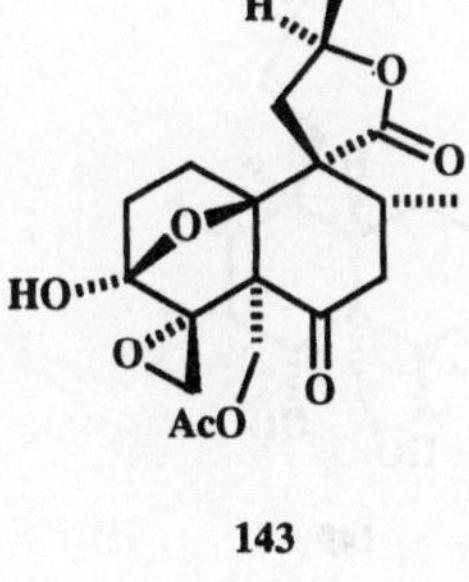

143

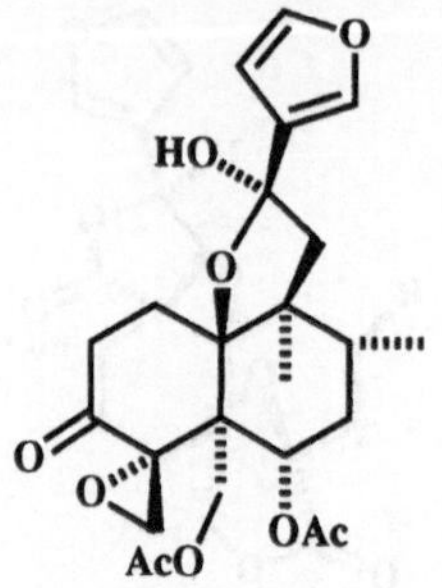

144

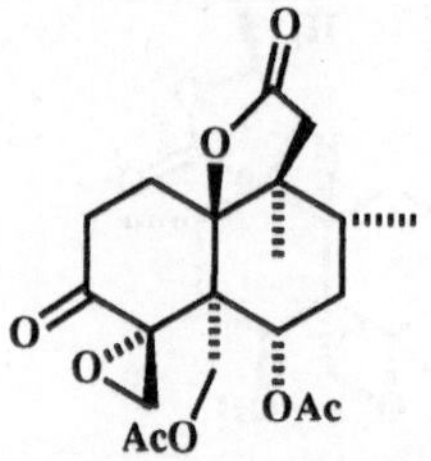

145

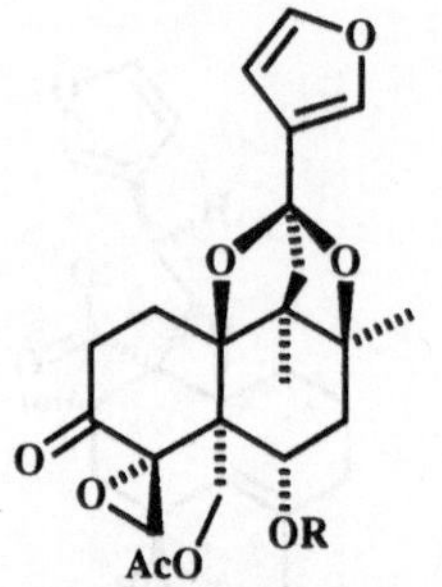

146: R= H
147: R= Ac

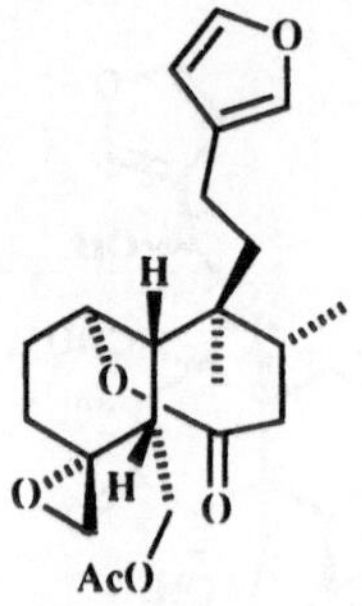

148

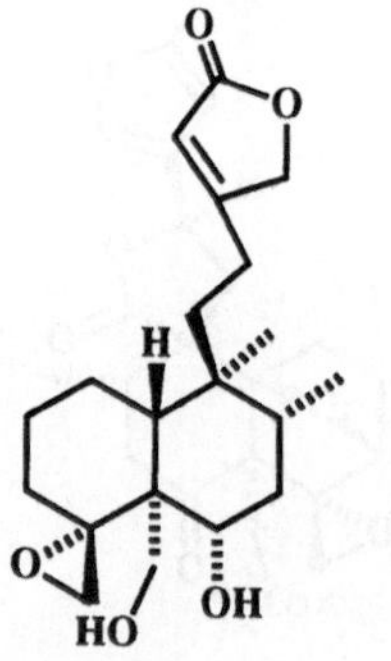

149

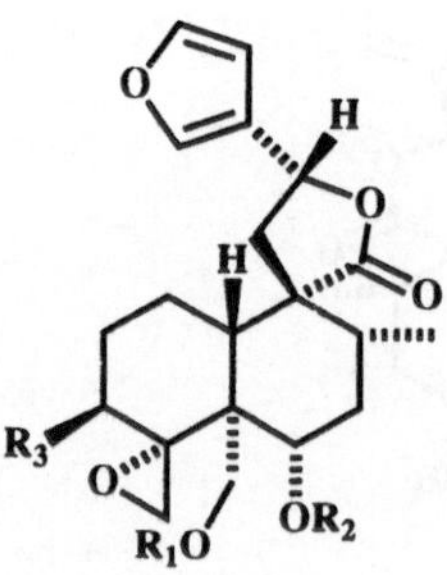

	R₁	R₂	R₃

$$R_1 \quad R_2 \quad R_3$$

150: Ac Ac H
151: Ac H H
152: Ac Ac OAc
153: H Ac H
154: Ac Ac OH

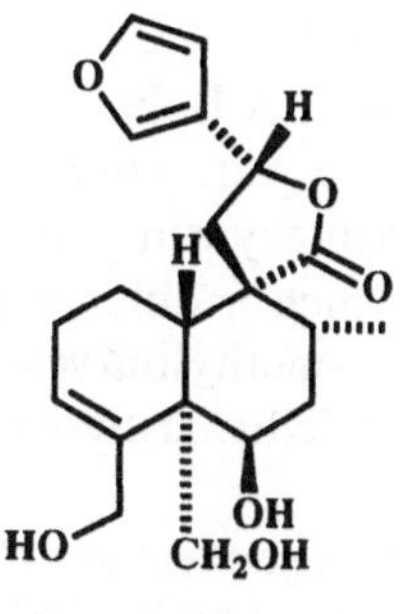

155

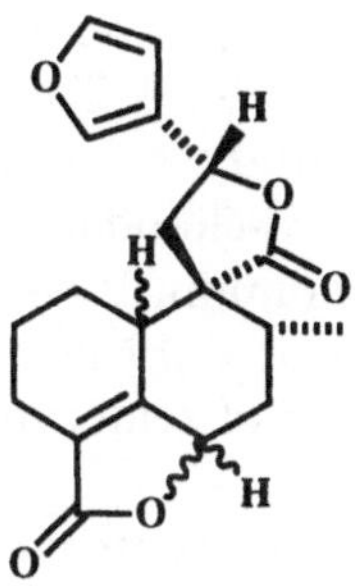

H-6 H-10
156: β   β
157: α   α
158: α   β

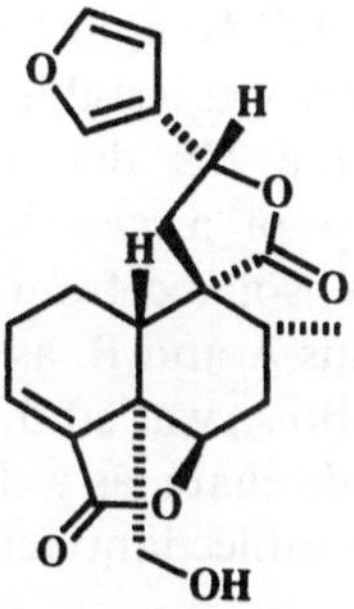

159 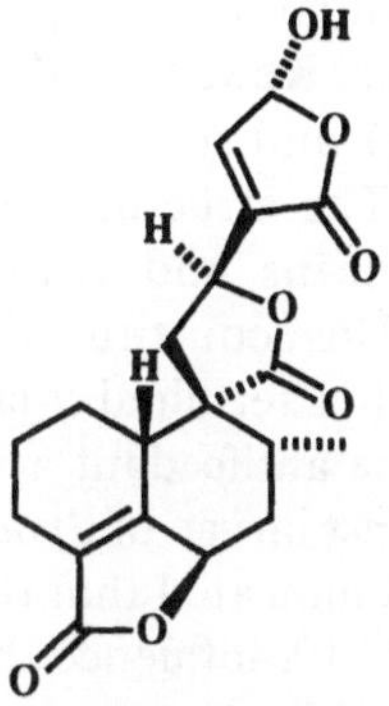   160

Addendum

Interest in the study of the *neo*-clerodane diterpenoid content of plants of the genera *Ajuga, Scutellaria, Teucrium* and *Salvia* (Labiatae) has resulted in isolation of new compounds.

A study of *Ajuga salicifolia* collected in Bulgaria, led to the isolation (*227*) of ajugachin A **11**, a product previously isolated from *A. chamaepitys* var. *chia* (*22*) and its 14,15-dihydro-15-hydroxy derivative (**15**, $R_1 = \beta OH,H; R_2 = COCHMe_2; R_3 = OH$), which the authors considered a possible artefact.

A. macrosperma, a perennial herb from China, has been used in traditional folk medicine to alleviate fever. A phytochemical study of this species (*228, 229*) led to isolation of the *neo*-clerodane diterpenoids ajugamacrins A-E, **220–224** which follow the structural pattern usually found in compounds isolated from Far Eastern *Ajuga* species (*34-39*), i.e. 1,12-dihydroxy-ajugarin I (**22**, 1, 12-diOH), esterified by acetic, isobutyric or 2-methylbutyric acids. *A. pantantha*, also used in Chinese folk medicine, yielded (*229*) ajugapantin A **225** and ajugacumbin B **29**.

The phytochemical study of *Scutellaria cypria* var. *elatior* from Cyprus, yielded (*230*) a *neo*-clerodane diterpene, scutecyprin **226**, structurally related to scutecolumnin B **58** (*42*) from which it differed in the presence of a tiglic ester at C-19 in place of the 2-methylbutyric ester found in **58**. The structure of scutecyprin **226** was deduced by spectroscopic means.

A study of the diterpenoid content of *S. galericulata* collected in Spain led to the isolation (*231*) of jodrellin T **54** and its 14,15-dihydroderivative **55**, previously found in a sample of *S. galericulata* growing in the Royal Botanic Gardens, Kew, UK (*41*). Two new *neo*-clerodane diterpenoids, scutegalin A **227** and scutegalin B **228** (mixture of C-16 epimers), were also obtained. The structure of the new products was established by spectroscopic means and X-ray diffraction of the **229** derivative of scutegalin B. This constitutes the first example of a *neo*-clerodane diterpenoid with a terminal γ-lactol functionality isolated from *Scutellaria* species. The antifeedant activity of scutegalins A and B, as well as that of **229** against larvae of *Spodoptera littoralis* (Bois.) was studied. The results obtained indicated that the nature of the side chain, as well as the ester group at C-19, influence the potency of the antifeedant activity of the compounds (*231*).

From a population of *S. alpina* collected in Bulgaria the *neo*-clerodane diterpenoid scutalpin A was isolated (*232*). Its structure **230** was deduced by spectroscopic means which included heteronuclear multiple quantum coherence (HMQC) and heteronuclear multiple bond connectivity (HMBC). X-ray diffraction analysis established the absolute configuration. A population of *S. alpina* subsp. *javalambrensis* from the Javalambre Mountains (Eastern Spain), yielded (*233*) three new *neo*-clerodane diterpenoids, scutalpins B–D. Their structures and relative stereochemistries as in **231–233**, were established by spectroscopic means which included HMBC spectra and nOe experiments. Scutalpins B and C contain a terminal β-butenolide function and a 8β-axial hydroxy group, whereas scutalpins A and D have a C-8β,C-13 ether bridge and a 13-spiro-15,16-γ lactone. These functionalities have been found in *neo*-clerodane diterpenoids isolated from a Far East species, *S. rivularis* (*43–48*). The structures of scutalpins A–D are of interest from the phytogeographical point of view. Like the *neo*-clerodane diterpenoids isolated from European *Scutellaria* spp (*40–42*) they contain a 4α,18-oxirane group and a functionalized C-19. On the other hand the furofuran or dihydro-furo-furan groups common to the European species, are replaced by terminal groups common to *Scutellaria* spp. from the Far

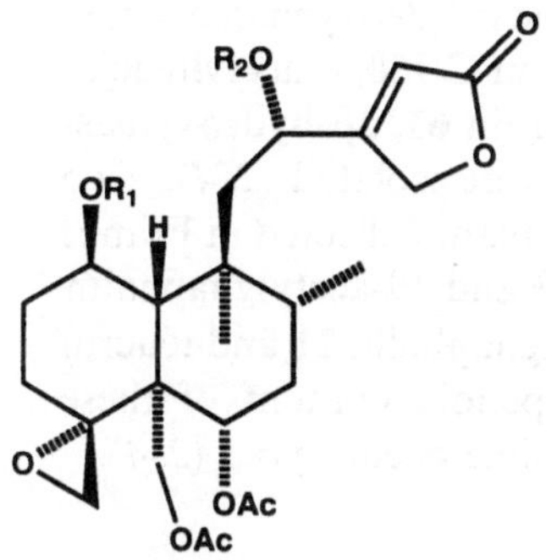

Ajugamacrins	R_1	R_2
220:A:	Ac	COCHMe$_2$
221:B:	Ac	COCHMeEt
222:C:	COCHMe$_2$	COCHMe$_2$
223:D:	COCHMe$_2$	COCHMeEt
224:E:	COCHMe$_2$	COCHMe$_2$
Ajugapantin A		
225:	Ac	Ac

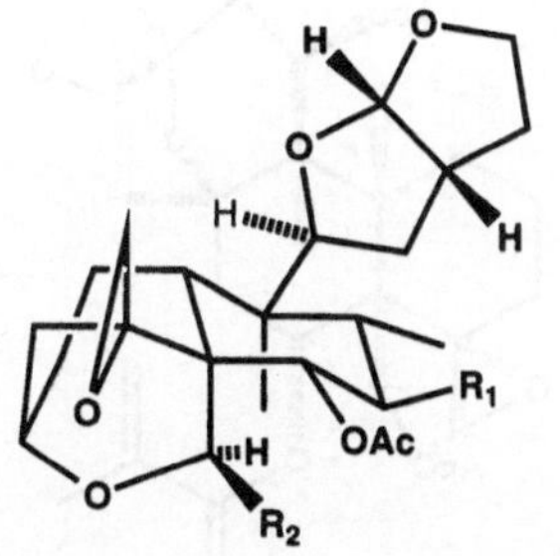

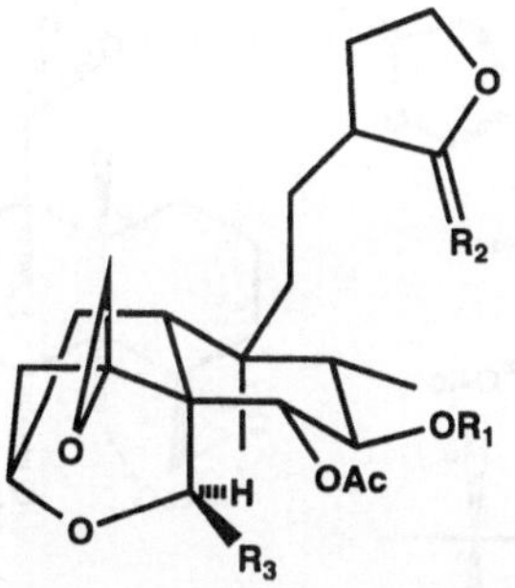

226: R_1 = H R_2 = OCO —

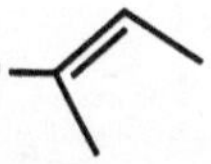

227: R_1 = R_2 = OCO —

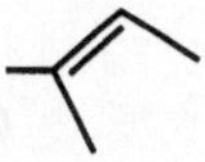

228: R_1 = H R_2 = H, OH (*R* and *S*) R_3 = OCO —

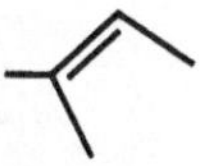

229: R_1 = Ac R_2 = O R_3 = OCO —

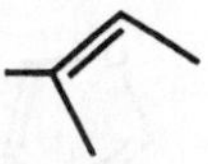

East (*43–48*). Scutalpins B–D have an hydroxy or acetoxy group at C-11. The absolute stereochemistry at this chiral centre was established as *S* in scutalpin C using Horeau's method (*21*). Acetylation of **232** gave scutalpin B which has, therefore, the same chirality at C-11 (*233*).

The *neo*-clerodane diterpenoid content of various *Teucrium* species has been studied. Structures of the known products are those included in the *Teucrium* Table. A reinvestigation of *T. massiliense* led (*234*) to the isolation of the known 19-acetylgnaphalin **21** and teumarin **23**, not found in the first study of this species (*71*). A new product teumassin **234** was also obtained. Its structure was deduced by spectroscopic means and chemical correlation with teumarin **23**. From *T. montbretii* subsp *mont-*

bretii collected in Turkey, the known diterpenoids: 2-deoxychamaedrox-
ide **138**, teuflin **69**, 6-ketoteuscordin **79**, montanin C **150**, 6-acetylteucja-
ponin B **19**, teucrin H$_2$ **77**, 6β-hydroxy-teuscordin **61**, 2β-hydroxyteus-
cordinon **64**, montanin D **135** and teugin **59** were isolated (*234*). Two
populations of *T. belion* were studied (*234*). The plant collected in France
yielded the known diterpenoids eriocephalin **53** and 19-acetylgnaphalin
21. From a population from Majorca, 19-acetylgnaphalin **21** and teucrin
A **73** were obtained. The difference in diterpenoid content of these
populations of *T. belion* was attributed to possible chemotypes (*234*).

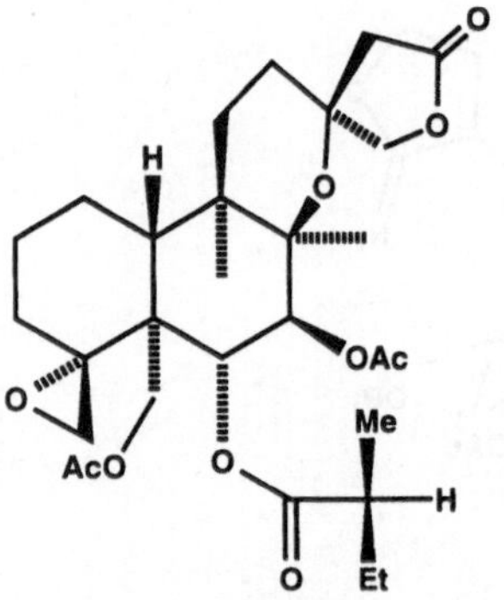

230

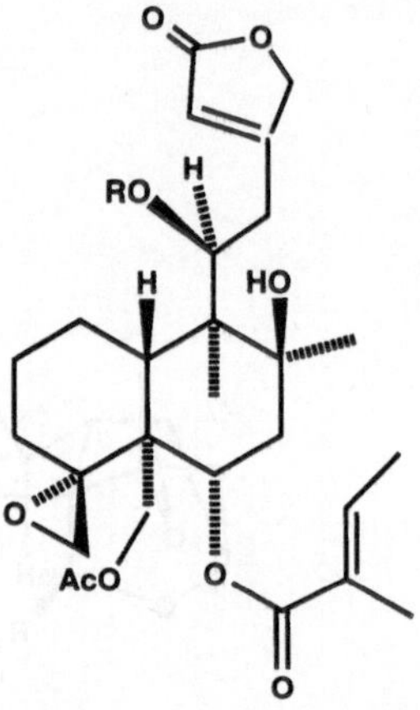

231: R = Ac
232: R = H

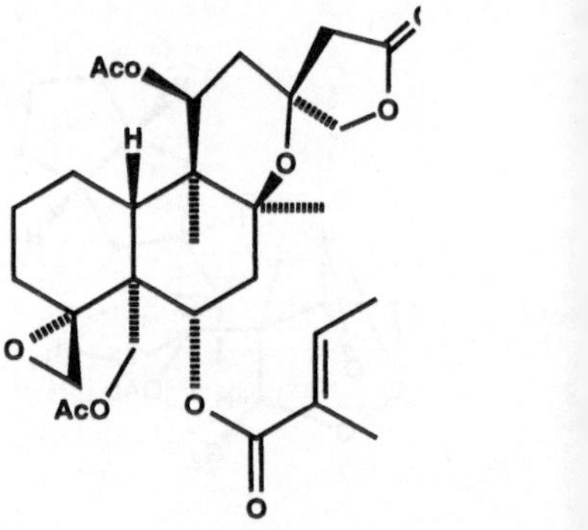

233

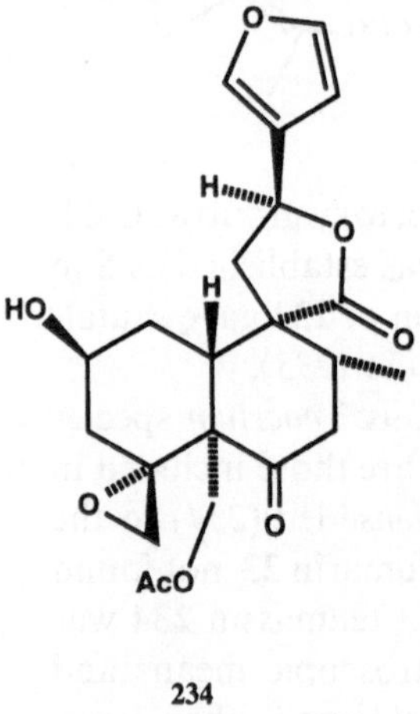

234

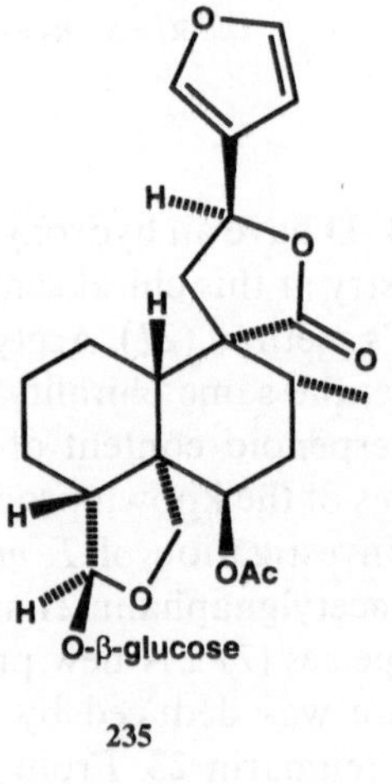

235

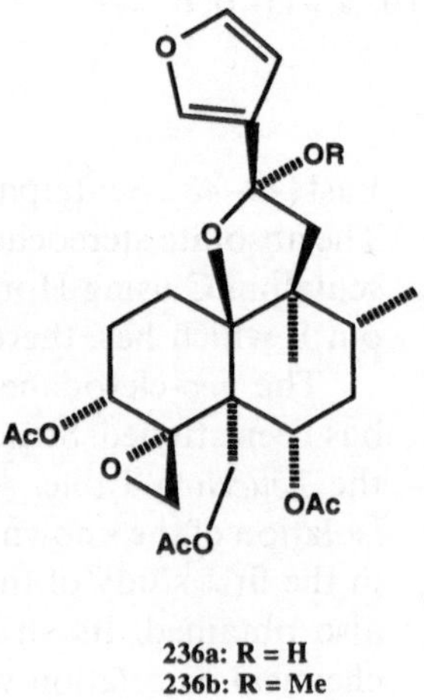

236a: R = H
236b: R = Me

A new study of *T. montanum* (*235*) yielded the *neo*-clerodane diterpenoid montanin H whose structure **110** (7-keto) was established by spectroscopic means. Montanin H has been also isolated from *T. cossonii* (*141*).

From the polar fraction of *T. lamiifolium* the *neo*-clerodane glucoside teulamioside **235** was obtained, together with the known teuspinin **91** and montanin E **123** (*236*). The structure of teulamioside was established by spectroscopic analysis and enzymatic hydrolysis to 6-acetylteukotschyn **86**. Teulamioside **235** and teuflavoside **141** (*123*) are the only *neo*-clerodane glycosides isolated until now from *Teucrium* spp.

Teucrium oliverianum is used in traditional Saudi medicine for the treatment of diabetes. A new study of the aerial parts of this plant (*237*) led to the isolation of three new *neo*-clerodane diterpenoids together with the known teucrolivins A-F (*102–104*). The structures of the new compounds, teucrolins A-C, **236a**, **237** and **238** and of the tetranor-*neo*-clerodane **239**, were established by spectroscopic analysis and X-ray diffraction of the 12-O methyl ether of teucrolin A **236b**.

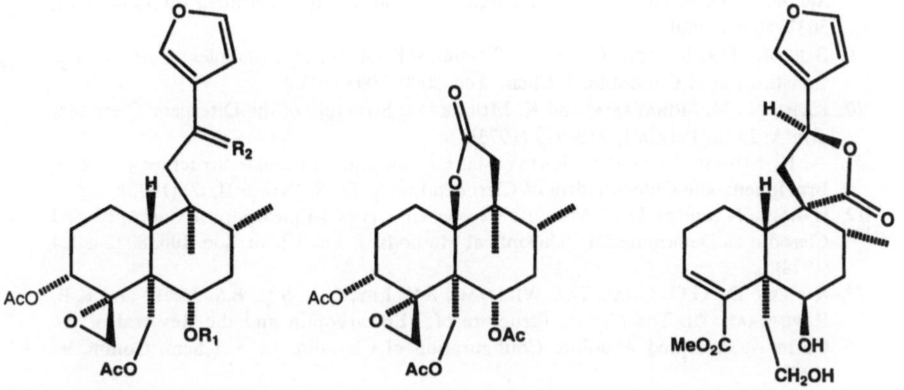

237: R₁ = H   R₂ = OH,H
238: R₁ = Ac   R₂ = O

239

240

A reinvestigation of *T. pernyi* from South East China yielded the known diterpenoids teucvidin **67**, teuflin **69**, montanin D **135** and teuscorodonin **98** not previously found in this plant. The teupernins A **81**, B **80** and C **82** were also isolated, together with a new *neo*-clerodane diterpenoid, teupernin D, whose structure **240** was deduced on spectroscopic evidence (*238*).

References

1. RICHARDSON, P.M.: A Chemical Overview of the Labiatae. Advances in Labiatae Science Paper Abstracts, p. 1. Kew: Royal Botanic Gardens, 1991.
2. MARTINEZ-RIPOLL, M., J. FAYOS, B. RODRIGUEZ, M.C. GARCIA-ALVAREZ, G. SAVONA, F. PIOZZI, M. PATERNOSTRO, and J.R. HANSON: The Absolute Stereochemistry of Some Clerodane Diterpenoids from *Teucrium* Species. J.C.S. Perkin I, 1186–1190 (1981).
3. MERRIT, A.T.L, and S.V. LEY: Clerodane Diterpenoids. Natural Product Reports 243–287 (1992).
4. CANTINO, P.D.: A Phylogenetic Evaluation of Suprageneric Groups in the Labiatae, Among Others. Advances in Labiatae Science. Paper Abstracts, p. 4. Kew: Royal Botanic Gardens, 1991.
5. For revisions see HANSON, J.R.: In: Terpenoids and Steroids Specialist Periodical Reports, Vol. 12 and preceding volumes. London. The Chemical Society; and Nat. Prod. Reports 1–9 (1984–1992).
6. BARTON, D.H.R., H.T. CHEUNG, A.D. CROSS, L.M. JACKMAN, and M. MARTIN-SMITH: Diterpenoid Bitter Principles. Part III. The Constitution of Clerodin. J. Chem. Soc. 5061–5073 (1961).
7. PAUL, I.C., G.A. SIM, T.A. HAMOR, and J.M. ROBERTSON: The Structure of Clerodin: X-Ray Analysis of Clerodin Bromo-Lactone. J. Chem. Soc. 4133–4145 (1962).
8. KOJIMA, Y., and N. KATO: Stereocontrolled Synthesis of Clerodin Homolog – A Synthetic Approach to Structure-Activity Relationships. Tetrahedron Letters 21, 5033–5036 (1980).
9. BARTON, D.H.R., and D. ELAD: Colombo Root Bitter Principles. Part II. The Constitution of Columbin. J. Chem. Soc. 2090–2095 (1956).
10. KATO, N., M. SHIBAYAMA, and K. MUNAKATA: Structure of the Diterpene Clerodendrin A. J.C.S. Perkiń I, 712–719 (1973).
11. —, K. MUNAKATA, and C. KATAYAMA: Crystal and Molecular Structure of the p-Bromobenzoate Chlorohydrin of Clerodendrin A. J.C.S. Perkin II, 69 (1973).
12. HARADA, N., and H. UDA: Absolute Stereochemistry of 3-Epicaryoptin, Caryoptin and Clerodin as Determined by Chiroptical Methods. J. Am. Chem. Soc. 100, 8022–8024 (1978).
13. ROGERS, D., G.G. UNAL, D.J. WILLIAMS, S.V. LEY, G.A. SIM, B.S. JOSHI, and K.R. RAVINDRANATH: The Crystal Structure of 3-Epicaryoptin and the Reversal of the Currently Accepted Absolute Configuration of Clerodin. J.C.S. Chem. Comm. 97 (1979).
14. a: KUBO, I., Y.W. LEE, V. BATOGH-NAIR, K. NAKANISHI, and A. CHAPYA: Structure of Ajugarins. J.C.S. Chem. Comm. 949 (1976). b: TRIVERDI, G., H. KOMURA, I. KUBO, K. NAKANISHI, and B.S. JOSHI: The *ent-neo* Clerodane Absolute Configuration of Ajugarins. J.C.S. Chem. Comm. 885–886 (1979).
15. KUBO, I., M. KIDO, and Y. FUKUYAMA: X-Ray Crystal Structure of 12-BromoAjugarin I and Conclusion on the Absolute Configuration of Ajugarins. J.C.S. Chem. Comm. 897–898 (1980).
16. PIOZZI, F.: The Diterpenoids of Teucrium Species. Heterocycles 15, 1489–1503 (1981).
17. McCRINDLE, R., and K.H. OVERTON: The Chemistry of Cyclic Diterpenoids In: Advances in Organic Chemistry. Methods and Results, Vol. 5, p. 47. New York: Interscience (1965).
18. WILLIS, J.C.: A Dictionary of the Flowering Plants and Ferns, 8th Ed. Cambridge: Cambridge University Press, 1973.

19. CAMPS, F., J. COLL, and A. CORTEL: Allelochemicals on Insect Isolated from Ajuga Plants (Labiatae). Rev. Latinoamer. Quim. **12**, 81–87 (1981).

20. HERNANDEZ, A., C. PASCUAL, J. SANZ, and B. RODRIGUEZ: Diterpenoids from *Ajuga chamaepitys*: two neo-clerodane derivatives. Phytochemistry **21**, 2909–2911 (1982).

21. HOREAU, A., and A. NOUAILLE: Extension et Simplication de la Methode de Determination de la Configuration des Alcools Secondaires par "Dedoublement Partiel" VIII. Aplication a la Genipine. Tetrahedron Letters, 1939–1942 (1971).

22. BONEVA, I.M., B.P. MIKHOVA, P.Y. MALAKOV, G.Y. PAPANOV, H. DUDDECK, and S.L. SPASSOV: Neo-Clerodane Diterpenoids from *Ajuga chamaepitys*. Phytochemistry **29**, 2931–2933 (1990).

23. CAMPS, F., J. COLL, and O. DARGALLO: Neo-Clerodane Diterpenoids from *Ajuga chamaepitys*. Phytochemistry **23**, 2577–2579 (1984).

24. — — — J. RIUS, and C. MIRAVITLLES: Clerodane Diterpenoids from *Teucrium* and *Ajuga* Plants. Phytochemistry **26**, 1475–1479 (1987).

25. . . . . . . , A. CORTEL: New Clerodane Diterpenoids from *Ajuga iva* (Labiatae). Chemistry Letters, 1053–1056 (1982).

26. . . . . . . , O. DARGALLO: Neo-Clerodane Diterpenoids from *Ajuga pseudoiva*. Phytochemistry **23**, 387–389 (1984).

27. MALAKOV, P.Y., G.Y. PAPANOV, M.C. DE LA TORRE, and B. RODRIGUEZ: Neoclerodane Diterpenoids from *Ajuga genevensis*. Phytochemistry **30**, 4083–4085 (1991).

28. — —, PERALES, A., M.C. DE LA TORRE, and B. RODRIGUEZ: The Absolute Stereochemistry of Ajugavensins, Neo-Clerodane Diterpenoids from *Ajuga genevensis*. A Revision of the C-1 Configuration of Ajugavensins A and B. Phytochemistry **31**, 3151–3153 (1992).

29. CAMPS, F., J. COLL, A. CORTEL, and A. MESSEGUER: Ajungareptansin, a New Diterpenoid from *Ajuga reptans* (L.). Tetrahedron Letters, 1709–1712 (1979).

30. KUBO, I., J.A. KLOCKE, J. MINRA, and Y. FUKUYAMA: Structure of Ajugarin IV. J. Chem. Soc., Chem. Comm. 618 (1982).

31. — — —, FUKUYAMA, Y., and A. CHAPYA: Structure of Ajugarin V. Chemistry Letters 223–224 (1983).

32. JONES, P.S., S.V. LEY, N.S. SIMPKINS, and A.J. WHITTLE: Total Synthesis of the Insect Antifeedant Ajugarin I and Degradation Studies of Related Clerodane Diterpenes. Tetrahedron **42**, 6519–6534 (1986).

33. DE GROOT, A.E., and T.A. VAN BEEK: Terpenoid Antifeedants (Part II). The Synthesis of Drimane and Clerodane Insect Antifeedants. Recl. Trav. Chim. Pays-Bas **106**, 1–18 (1987), and references cited therein.

34. MIN, ZH., SH. WANG, Q. ZHENG, B. WU, M. MIZUNO, T. TANAKA, and M. IINUMA: Four New Insect Antifeedant neo-Clerodane Diterpenoids, Ajugacumbins A, B, C and D from *Ajuga decumbens*. Chem. Pharm. Bull. **37**, 2505–2508 (1989).

35. MIN, ZH., M. MIZUNO, Q. WANG, M. IINUMA, and T. TANAKA: Two New neo-Clerodane Diterpenes in *Ajuga decumbens*. Chem. Pharm. Bull. **38**, 3167–3168 (1990).

36. SHINOMURA, H., Y. SASHIDA, and K. OGAWA: Neo-Clerodane Diterpenes from *Ajuga decumbens*. Chem. Pharm. Bull. **37**, 996–998 (1989).

37: — — —, IITAKA, Y.: Ajugamarin, a New Bitter Diterpene from *Ajuga nipponensis* Makino. Tetrahedron Letters **22**, 1367–1368 (1981).

38: — — —, Neo-clerodane Diterpenes from *Ajuga nipponensis*. Chem. Pharm. Bull **37**, 354–357 (1989).

39: — — —, Neo-clerodane Diterpenoids from *Ajuga ciliata* var. villosior. Chem. Pharm. Bull. **37**, 988–992 (1989).

40. ANDERSON, J.C., W.M. BLANEY, M.D. COL E, L.L. FELLOWS, S.V. LEY, R.N. SHEPPARD, and M.S.J. Simmonds: The Structure of Two New Clerodane Diterpenoid Potent Insect Antifeedants from *Scutellaria woronowii* (JUZ); Jodrellin A and B. Tetrahedron Letters **30**, 4737–4740 (1989).

41. Cole, M.D., J.C. Anderson, W.M. Blaney, L.E. Fellows, S.V. Ley, R.N. Sheppard, and M.S.J. Simmonds: Neo-Clerodane Insect Antifeedants from *Scutellaria galericulata*. Phytochemistry **29**, 1793–1796 (1990).

42. De La Torre, M.C., M. Bruno, F. Piozzi, B. Rodriguez, G. Savona, and O. Servettaz: Neo-clerodane Diterpenoids from *Scutellaria columnae*. Phytochemistry **31**, 3639–3641 (1992).

43. Lin, Y.L., Y.H. Kuo, G.H. Lee, and S.M. Peng: Scutellone A, a Novel Diterpene from *Scutellaria rivularis*. J. Chem. Research (S) 320–321 (1987).

44. — —, Cheng, M-Ch., and Y. Wang: Structures of Scutellones D and E Determined by X-Ray Diffraction, Spectral and Chemical Evidence. Neo-clerodane-type Diterpenoids from *Scutellaria rivularis* Wall. Chem. Pharm. Bull. **36**, 2642–2646 (1988).

45. — —: Scutellones C and F; Two New neo-Clerodane Type Diterpenoids from *Scutellaria rivularis*. Heterocycles **27**, 779–783 (1988).

46. — —: Four New neo-Clerodane Type Diterpenoids, Scutellones B, G, H and I, from Aerial Parts of *Scutellaria rivularis*. Chem. Pharm. Bull. **37**, 582–585 (1989).

47. Kizu, H., Y. Imoto, T. Tomimori, K. Tsubono, Sh. Kadota, and T. Kikuchi: Structure of Scuterivulactone D Determined by Two-Dimensional NMR Spectroscopy. A New Diterpenoid from a Chinese Crude Drug 'BAN ZHI LIAN' (*Scutellaria rivularis* Wall). Chem. Pharm. Bull. **35** 1656–1659 (1987).

48. Kikuchi, T., K. Tsubono, Sh. Kadota, H. Kizu, Y. Imoto, and T. Tomimori: Structures of Scuterivulactones C_1 and C_2 by Two-Dimensional NMR Spectroscopy. New Clerodane Type Diterpenoids from *Scutellaria rivularis* Wall. Chem. Letters, 987–990 (1987).

49. Fujita, E., Y. Nagao, and M. Node: Diterpenoids of *Isodon* and *Teucrium* Plants. Heterocycles **5**, 793–838 (1976).

50. Piozzi, F.: The Diterpenoids of *Teucrium* Species. Heterocycles **15**, 1489–1503 (1981).

51. Gacs-Baitz, E., M. Kajtar, G.Y. Papanov, and P.Y. Malakov: Carbon-13 NMR Spectra of Some Furanoid Diterpenes from *Teucrium* species. Heterocycles **19**, 539–550 (1982).

52. Piozzi, F., B. Rodriguez, and G. Savona: Advances in the Chemistry of the Furano Diterpenes from *Teucrium* species. Heterocycles **25**, 807–841 (1987).

53. Bruno, M., F. Piozzi, B. Rodriguez, G. Savona, and O. Servettaz: Ent-Clerodane Diterpenoids from Six Further Species of *Teucrium*. Phytochemistry **24**, 2597–2599 (1985).

54. Brieskorn, C.H., and T. Pleuffer: Labiatenbitterstoffe: Pikropolin und ähnliche Diterpenoide aus Poleigamander. Chem. Ber. **100**, 1998–2010 (1967).

55. Overton, K.H., H.G. Weir, and A. Wylie: The Stereochemistry of the Colombo Root Bitter Principles. J. Chem. Soc (C) 1482–1490 (1966).

56. De La Torre, M.C., B. Rodriguez, G. Savona, and F. Piozzi: Teugnaphalodin, a neo-Clerodane Diterpenoid from *Teucrium gnaphalodes*. Phytochemistry **25**, 171–173 (1986).

57. Savona, G., F. Piozzi, O. Servettaz, B. Rodriguez, J.A. Hueso-Rodriguez, and M.C. De La Torre: Neo-Clerodane Diterpenoids from *Teucrium lepicephalum* and *Teucrium buxifolium*. Phytochemistry **25**, 2569–2572 (1986).

58. De La Torre, M.C., C. Pascual, B. Rodriguez, F. Piozzi, G. Savona, and A. Perales: Neo-Clerodane Diterpenoids from *Teucrium salviastrum*. Phytochemistry **25**, 1397–1403 (1986).

59. DE LA TORRE, M.C., F. FERNANDEZ-GADEA, A. MICHAVILA, B. RODRIGUEZ, F. PIOZZI, and G. SAVONA: Neo-Clerodane Diterpenoids from *Teucrium botrys*. Phytochemistry 25, 2385–2387 (1986).

60. MALAKOV, P.V., I.M. BONEVA, G.Y. PAPANOV, and S.L. SPASSOV: Teulamifin B, a neo-Clerodane Diterpenoid from *Teucrium lamiifolium* and *T. polium*. Phytochemistry 27, 1141–1143 (1988).

61. RODRIGUEZ, B.: Personal communication.

62. SAVONA, G., M. PATERNOSTRO, and F. PIOZZI: New Furanoid Diterpenes from *Teucrium gnaphalodes* L'Her. Tetrahedron Letters, 379–382 (1979).

63. FAYOS, J., M. MARTINEZ-RIPOLL, M. PATERNOSTRO, F. PIOZZI, B. RODRIGUEZ, and G. SAVONA: New Clerodane Diterpenoid from *Teucrium eriocephalum*. J. Org. Chem. 44, 4992–4994 (1979).

64. MARTINEZ-RIPOLL, M., J. FAYOS, B. RODRIGUEZ, M.C. GARCIA-ALVAREZ, G. SAVONA, F. PIOZZI, M. PATERNOSTRO, and J.R. HANSON: The Absolute Stereochemistry of Some Clerodane Diterpenoids from *Teucrium* species. J.C.S. Perkin Trans. I, 1186–1190 (1981).

65. FERNANDEZ, P., B. RODRIGUEZ, J.A. VILLEGAS, A. PERALES, G. SAVONA, F. PIOZZI, and M. BRUNO: Neo-Clerodane Diterpenoids from *Teucrium pyrenaicum* and *T. subspinosum*. Phytochemistry 25, 1405–1409 (1986).

66. MALAKOV, P.Y., and G.Y. PAPANOV: Furanoid Diterpenes from *Teucrium polium*. Phytochemistry 22, 2791–2793 (1983).

67. DE LA TORRE, M.C., B. RODRIGUEZ, M. BRUNO, G. SAVONA, F. PIOZZI, and O. SERVETTAZ: Neo-Clerodane Diterpenoids from *Teucrium micropodiodes*. Phytochemistry 27, 213–216 (1988).

68. GARCIA-ALVAREZ, M.C., J.L. MARCO, B. RODRIGUEZ, G. SAVONA, and F. PIOZZI: Neo-Clerodane Diterpenoids from *Teucrium pyrenaicum*. Phytochemistry 21, 2559–2562 (1982).

69. SAVONA, G., S. PASSANNANTI, M.P. PATERNOSTRO, F. PIOZZI, J.R. HANSON, P.B. HITCHCOCK, and M. SIVERNS: Two New Diterpenoids from *Teucrium fruticans*. J.C.S. Perkin Trans. I, 356–359 (1978).

70: — — —, J.R. HANSON, and M. SIVERNS: 8β-Hydroxy Fruticolone, A Diterpenoid from *Teucrium fruticans*. Phytochemistry 17, 320–322 (1978).

71. —, M. BRUNO, F. PIOZZI, O. SERVETTAZ, and B. RODRIGUEZ: Neo-Clerodane Diterpenoids from *Teucrium massiliense*. Phytochemistry 23, 849–852 (1984).

72. MIYASE, T., H. KAWASAKI, T. NORO, A. EENO, S. FUKUSHIMA, and T. TAKEMOTO: Studies on the Furanoid Diterpenes from *Teucrium japonicum* Houtt. Chem. Pharm. Bull. 29, 3561–3564 (1981).

73. MALAKOV, P.Y., G.Y. PAPANOV, and N.M. MOLLOV: Z. Naturforsch. B34, 1570 (1979) cited by: MALAKOV, P.Y., G.Y. PAPANOV, and J. ZIESCHE, Teupolin III, A Furanoid Diterpene from *Teucrium polium*. Phytochemistry 21, 2597–2598 (1982).

74. MARQUEZ, C., and S. VALVERDE: A New Clerodane Diterpenoid from *Teucrium polium* L. J.C.S. Perkin Trans. I, 2526–2527 (1979).

75. FERNANDEZ, P., B. RODRIGUEZ, G. SAVONA, and F. PIOZZI: Neo-Clerodane Diterpenoids from *Teucrium polium* subsp. *capitatum*. Phytochemistry 25, 181–184 (1986).

76. HANSON, J.R., D.E.A. RIVETT, S.V. LEY, and D.J. WILLIAMS: The X-ray Structure and Absolute Configuration of Insect Antifeedant Clerodane Diterpenoids from *Teucrium africanum*. J.C.S. Perkin Trans. I, 1005–1008 (1982).

77. CARREIRAS, M.C., B. RODRIGUEZ, F. PIOZZI, G. SAVONA, M.R. TORRES, and D. PERALES: A Chlorine-containing and Two 17β-neo-Clerodane Diterpenoids from *Teucrium polium* subsp. *vincentinum*. Phytochemistry 28, 1453–1461 (1989).

78. APSIMON, J.W., and S.F. HALL: Diterpene Chemistry VI. Some Chemistry of 8,14-epoxy Pimaric Acid. Comments on the Biosynthesis of Tetracyclic Diterpenes. Can. J. Chem. **56**, 2156–2162 (1978).

79. PAPANOV, G.Y., and P.Y. MALAKOV: Clerodane Diterpenoids from *Teucrium montanum* subsp. *skorpilii*. Phytochemistry **22**, 2787–2789 (1983).

80. FUJITA, E., I. UCHIDA, and T. FUJITA: Terpenoids. Part XXXII. Structure and Stereochemistry of Teucvin, a Novel Norclerodane-Type Diterpene from *Teucrium viscidum* var. *miquelianum*. J.C.S. Perkin Trans. I, 1547–1555 (1974).

81. UCHIDA, I., T. FUJITA, and E. FUJITA: Terpenoids – XXXIV. Teucvidin, a Minor Norditerpene from *Teucrium viscidum* var. *miquelianum*. Tetrahedron **31**, 841–848 (1975).

82. — —, TAIRA, Z., and K. OSAKI: Crystal Structure Comm. **3**, 569 (1974).

83. MALAKOV, P.Y., G.Y. PAPANOV, and N.M. MOLLOV: Montanin A and B, New Furanoid Diterpenes of Nor-Clerodane Type from *Teucrium montanum* L. Tetrahedron Letters, 2025–2026 (1978).

84. HIKINO, H., Y. HIKINO, and I. YOSHIOKA: Structure and Autoxidation of Atractylon. Chem. Pharm. Bull. **10**, 641–642 (1962).

85. DOMINGUEZ, G., M.C. DE LA TORRE, and B. RODRIGUEZ: Transformation of Neoclerodane Diterpenoids into 19-Norneoclerodane Derivatives. J. Org. Chem. **56**, 6595–6600 (1991).

86. HALTON, R.A., and R.M. KENNEDY: Stereochemical Requirements for Fragmentation of Homoallylic Epoxy Alcohols. Tetrahedron Letters **25**, 4455–4458 (1984).

87. DE LA TORRE, M.C., P. FERNANDEZ, and B. RODRIGUEZ: Thermal Rearrangements of Some Neo-Clerodane Diterpenoids. Tetrahedron **43**, 4679–4684 (1987).

88. PAPANOV, G.Y., P.Y. MALAKOV, and F. BOHLMANN: Tenscordinon, a Furanoid Diterpene from *Teucrium scordium*. Phytochemistry **20**, 170–171 (1981).

89. — —, 2-Keto-19-hydroxyteuscordin, a Neo-Clerodane Diterpene from *Teucrium scordium*. Phytochemistry **24**, 297–299 (1985); and references cited therein.

90. PINHEY, J.T., R.F. SIMPSON, and I.L. BATEY: The Constituents of *Olearia heterocarpa*. The Structure of Olearin, a Diterpene Dilactone of the Cascarillin Group. Aust. J. Chem. **25**, 2621–2637 (1972).

91. BRUNO, M., G. SAVONA, C. PASCUAL, and B. RODRIGUEZ: Teugin, a neo-Clerodane Diterpenoid from *Teucrium fragile*. Phytochemistry **20**, 2259–2261 (1981).

92. MALAKOV, P.Y., M.C. DE LA TORRE, B. RODRIGUEZ, and G.Y. PAPANOV: Some Chemical Transformations of the Neo-Clerodane Diterpene Teubotrin. Tetrahedron **47**, 10129–10136 (1991).

93. SAVONA, G., M.C. GARCIA-ALVAREZ, and B. RODRIGUEZ: Dihydroteugin, a Neo-Clerodane Diterpenoid from *Teucrium chamaedrys*. Phytochemistry **21**, 721–723 (1982).

94. MARCO, J.L., B. RODRIGUEZ, C. PASCUAL, G. SAVONA, and F. PIOZZI: Teuscorodin, Teuscorodonin and 2-Hydroxy-teuscorolide, Neo-Clerodane Diterpenoids from *Teucrium scorodonia*. Phytochemistry, **22**, 727–731 (1983).

95. HUESO-RODRIGUEZ, J.A., F. FERNANDEZ-GADEA, C. PASCUAL, B. RODRIGUEZ, G. SAVONA, and F. PIOZZI: Neo-Clerodane Diterpenoids from *Teucrium lanigerum*. Phytochemistry **25**, 175–180 (1986).

96. EGUREN, L., A. PERALES, J. FAYOS, B. RODRIGUEZ, G. SAVONA, and F. PIOZZI: New Neoclerodane Diterpenoid Containing an Oxetane Ring Isolated from *Teucrium chamaedrys*. X-Ray Structure Determination. J. Org. Chem. **47**, 4157–4160 (1982).

97. GARCIA-ALVAREZ, M.C., G. LUKACS, A. NESZMELYI, F. PIOZZI, B. RODRIGUEZ, and G. SAVONA: Structure of Teucroxide. Application of Natural-Abundance ^{13}C–^{13}C Coup-

ling Constants Observed via Double-Quantum Coherence. J. Org. Chem. **48**, 5123–5126 (1983).

98. MALAKOV, P.Y., G.Y. PAPANOV, N.M. MOLLOR, and S.L. SPASSOV: Montanin D, a New Furanoid Diterpene of Clerodane Type from *Teucrium montanum*. L. Z. Naturforsch. **33b**, 1142–1144 (1978).

99. BRUNO, M., F. PIOZZI, G. SAVONA, B. RODRIGUEZ, M.C. DE LA TORRE, and O. SERVETTAZ: 2-Deoxy-chamaedroxide, a Neo-Clerodane Diterpenoid from *Teucrium divaricatum*. Phytochemistry **26**, 2859–2861 (1987).

100. WANI, M.C., H.L. TAYLOR, M.E. WALL, P. COGGON, and A.T. MCPHAIL: Plant Antitumor Agents. VI. Isolation and Structure of Taxol, a Novel Antileukemic and Antitumor Agent from *Taxus brevifolia*. J. Am. Chem. Soc. **93**, 2325 (1971).

101. DE LA TORRE, M.C., B. RODRIGUEZ, M. BRUNO, G. SAVONA, F. PIOZZI, A. PERALES, M.R. TORRES, and O. SERVETTAZ: Two C-10 Oxygenated Neo-Clerodane Diterpenoids from *Teucrium pestalozzae*. Phytochemistry **29**, 2229–2233 (1990).

102. BRUNO, M., A.A. OMAR, A. PERALES, F. PIOZZI, B. RODRIGUEZ, G. SAVONA, and M.C. DE LA TORRE: Neo-Clerodane Diterpenoids from *Teucrium oliverianum*. Phytochemistry **30**, 275–282 (1991).

103. DE LA TORRE, M.C., M. BRUNO, F. PIOZZI, G. SAVONA, A.A. OMAR, A. PERALES, and B. RODRIGUEZ: Two Neo-Clerodane Diterpenoids Containing an Unusual 2,6-Dioxabicyclo(2.2.1)Heptane Structural Moiety. Tetrahedron **47**, 3463–3470 (1991).

104. DE LA TORRE, M.C., M. BRUNO, F. PIOZZI, G. SAVONA, B. RODRIGUEZ, and A.A. OMAR: Teucrolivins D–F, Neo-Clerodane Derivatives from *Teucrium oliverianum*. Phytochemistry **30**, 1603–1606 (1991).

105. BRUNO, M., R. ALCAZAR, M.C. DE LA TORRE, F. PIOZZI, B. RODRIGUEZ, G. SAVONA, A. PERALES, and N.A. ARNOLD: Diterpenoids from *Teucrium gracile* and *Teucrium fruticans*, *Neo*-Clerodane and Seco-neo-Clerodane Derivatives. Phytochemistry **31**, 3531–3534 (1992).

106. FAYOS, J., F. FERNANDEZ-GADEA, C. PASCUAL, A. PERALES, F. PIOZZI, M. RICO, B. RODRIGUEZ, and G. SAVONA: Correct Structures of Montanin C. Teupolin I, and 12-epi-Teucvin, Three (12R)-Neoclerodane-20,12-olides Isolated from *Teucrium* species. J. Org. Chem. **49**, 1789–1793 (1984).

107. PASCUAL, C., P. FERNANDEZ, M.C. GARCIA-ALVAREZ, J.L. MARCO, F. FERNANDEZ-GADEA, M.C. DE LA TORRE, J.A. HUESO-RODRIGUEZ, B. RODRIGUEZ, M. BRUNO, M. PATERNOSTRO, F. PIOZZI, and G. SAVONA: The C-12 and C-20 Configurations of Some Neo-Clerodane Diterpenoids Isolated from *Teucrium* species. Phytochemistry **25**, 715–718 (1986).

108. LOURENCO, A., M.C. DE LA TORRE, B. RODRIGUEZ, N. HARADA, H. ONO, H. UDA, M. BRUNO, F. PIOZZI, and G. SAVONA: The Absolute Stereochemistry at C-12 in 12-Hydroxylated Neo-Clerodane Diterpenoids. Tetrahedron **48**, 3925–3934 (1992); and references cited therein.

109. DE LA TORRE, M.C., B. RODRIGUEZ, F. PIOZZI, G. SAVONA, M. BRUNO, and M.C. CARREIRAS: Neo-Clerodane Diterpenoids from *Teucrium abutiloides*. Phytochemistry **29**, 579–584 (1990).

110. LABBE, C., M.I. POLANCO, and M. CASTILLO: 12-Epi-Teuscordonin and Other Neoclerodanes from *Teucrium bicolor*. Journal of Natural Products **52**, 871–874 (1989).

111. HANDONG, S., CH. XINGLIANG, W. TIANEN, P. LUTAI, L. ZHONGWEN, and CH. DEYUAN: A neo-Clerodane Diterpenoid from *Teucrium bidentatum*. Phytochemistry **30**, 1721–1723 (1991).

112. MARQUEZ, C., R.M. RABANAL, S. VALVERDE, L. EGUREN, A. PERALES, and J. FAYOS:

Diterpenes from *Teucrium capitatum* L., X-ray Crystal and Molecular Structure of Capitatin. Tetrahedron Letters **21**, 5039–5042 (1980).

113. ——: Diterpenes from *Teucrium capitatum* L. X-ray Crystal and Molecular Structure of Lolin. Tetrahedron Letters **22**, 2823–2826 (1981).

114. SAVONA, G., F. PIOZZI, M. BRUNO, G. DOMINGUEZ, B. RODRIGUEZ, and O. SERVETTAZ: Teucretol, A neo-Clerodane Diterpenoid from *Teucrium creticum*. Phytochemistry **26**, 3285–3288 (1987).

115. BRUNO, M., F. PIOZZI, G. SAVONA, M.C. DE LA TORRE, and B. RODRIGUEZ: Neo-Clerodane Diterpenoids from *Teucrium canadense*. Phytochemistry **28**, 3539–3541 (1989).

116. SAVONA, G., F. PIOZZI, M.C. DE LA TORRE, O. SERVETTAZ, and B. RODRIGUEZ: A Valencane Sesquiterpenoid from *Teucrium carolipani*. Phytochemistry **26**, 571–572 (1987).

117. DOMINGUEZ, X.A., A. MERIJANIAN, B.I. GONZALEZ, A. ZAMUDIO, and A.L. SALAZAR: Estructura de la Eugarzasadina, una nueva *nor*-Diterpenlactona aislada del *Teucrium-cubense*, Jacq. (Labiacea). Rev. Latinoamer. Quim. **5**, 225–229 (1974).

118. PAPANOV, G.Y., and P.Y. MALAKOV: Furanoid Diterpenes in the Bitter Fraction of *Teucrium chamaedrys* L. Z. Naturforsch. **35b**, 764–766 (1980).

119. FERNANDEZ-GADEA, F., C. PASCUAL, B. RODRIGUEZ, and G. SAVONA: 6-Epiteucrin A, A Neo-Clerodane Diterpenoid from *Teucrium chamaedrys*. Phytochemistry **22**, 723–725 (1983).

120. RODRIGUEZ, M.C., J. BARLUENGA, C. PASCUAL, B. RODRIGUEZ, G. SAVONA, and F. PIOZZI: Neo-Clerodane Diterpenoids from *Teucrium chamaedrys*: The Identity of Teucrin B with Dihydroteugin. Phytochemistry **23**, 2960–2961 (1984).

121. MALAKOV, P.Y., and G.Y. PAPANOV: Teuchamaedrin C, A Neo-Clerodane Diterpenoid from *Teucrium chamaedrys*. Phytochemistry **24**, 301–303 (1985).

122. RODRIGUEZ, M.C., J. BARLUENGA, G. SAVONA, F. PIOZZI, O. SERVETTAZ, and B. RODRIGUEZ: Isoteuflidin, a Neo-Clerodane Diterpenoid from *Teucrium chamaedrys*, and Revised Structures of Teucrins F and G. Phytochemistry **23**, 1465–1469 (1984).

123. SAVONA, G., F. PIOZZI, O. SERVETTAZ, B. RODRIGUEZ, F. FERNANDEZ-GADEA, and M. MARTIN-LOMAS: A neo-Clerodane Glucoside and neo-Clerodane Diterpenoids from *Teucrium flavum* subsp. *glaucum*. Phytochemistry **23**, 843–848 (1984).

124. SAVONA, G., M.P. PATERNOSTRO, F. PIOZZI, J.R. HANSON, P.B. HITCHCOCK, and S.A. THOMAS: Teuflidin, a Norclerodane Diterpenoid from *Teucrium flavum*. J.C.S. Perkin I, 1080–1083 (1978).

125. . . . . . : The Structure of Teuflin, a Diterpenoid from *Teucrium flavum*; X-ray Crystallographic and Spectroscopic Determination. J.C.S. Perkin Trans. I, 1915–1917 (1979).

126. TONN, C.E., J.A. CALTABIANO, J.C. GIANELLO, and O.S.L. GIORDANO: Two neo-Clerodane Diterpenoids from *Teucrium grisebachii*. Rev. Latinoamer. Quim **21**, 52–54 (1990).

127. GACZ-BAITZ, E., L. RADICS, G.B. OGANESSIAN, and V.A. MONTSAKANIAN: Teucrins H1–H4, Novel Clerodane-Type Diterpenes from *Teucrium hyrcanicum*. Phytochemistry **17**, 1967–1973 (1978).

128. ZHI-DA, M., X. NING, ZH. PEI, ZH. SHON-XUN, W. CHONG-SHU, and ZH. QI-TAI: A neo-Clerodane Diterpene from *Teucrium japonicum*. Phytochemistry **30**, 4175–4177 (1991).

129. SIMOES, F., B. RODRIGUEZ, M. BRUNO, F. PIOZZI, G. SAVONA, and N.A. ARNOLD: Neo-Clerodane Diterpenoids from *Teucrium kotschyanum*. Phytochemistry **28**, 2763–2768 (1989).

130. SIMOES, F., B. RODRIGUEZ, F. PIOZZI, G. SAVONA, M. BRUNO, and N.A. ARNOLD: Isoteucrin H4, A 19-nor-Clerodane Diterpenoid of Biogenetic Interest from *Teucrium kotschyanum*. Heterocycles **28**, 111–115 (1989).

131. BONEVA, I.M., P.Y. MALAKOV, and G.Y. PAPANOV: 12-Epiteupolin II, a Neo-Clerodane Diterpenoid from *Teucrium lamiifolium*. Phytochemistry **27**, 295–297 (1988).

132. FERNANDEZ-GADEA, F., B. RODRIGUEZ, G. SAVONA, and F. PIOZZI: Isoeriocphalin and 20-Deacetyleriocephalin, neo-Clerodane Diterpenoids from *Teucrium lanigerum*. Phytochemistry **23**, 1113–1118 (1984).

133. SAVONA, G., F. PIOZZI, O. SERVETTAZ, F. FERNANDEZ-GADEA, and B. RODRIGUEZ: Teumarin, a neo-Clerodane Diterpenoid from *Teucrium marum*. Phytochemistry **23**, 611–613 (1984).

134. DE LA TORRE, M.C., B. RODRIGUEZ, M. BRUNO, G. SAVONA, F. PIOZZI, and O. SERVETTAZ: Neo-Clerodane Diterpenoids from *Teucrium micropodioides*. Phytochemistry **27**, 213–216 (1988).

135. DE LA TORRE, M.C., M. BRUNO, G. SAVONA, F. PIOZZI, B. RODRIGUEZ, and O. SERVETTAZ: Neo-Clerodane Diterpenoids from *Teucrium pestalozzae*, *T. odontites* and *T. microphyllum*. Phytochemistry **29**, 988–989 (1990).

136. MALAKOV, P.Y., G.Y. PAPANOV, N.M. MOLLOV, and S.L. SPASSOV: Montanin C, a New Furanoid Diterpene from *Teucrium montanum* L. Z. Naturforsch. **33b**, 789–791 (1978).

137. MALAKOV, P.Y., and G.Y. PAPANOV: Montanin G, a Neo-Clerodane Diterpenoid from *Teucrium montanum*. Z. Naturforsch. **42b**, 1000–1002 (1987).

138. BRUNO, M., G. DOMINGUEZ, A. LOURENCO, F. PIOZZI, B. RODRIGUEZ, G. SAVONA, M.C. DE LA TORRE, and N.A. ARNOLD: Neo-Clerodane Diterpenoids from *Teucrium gracile*. Phytochemistry **30**, 3693–3697 (1991).

139. SEXMERO, M.J., M.C. DE LA TORRE, B. RODRIGUEZ, M. BRUNO, F. PIOZZI, and G. SAVONA: Neo-Clerodane Diterpenoids from *Teucrium oxylepis* subsp. *marianum*. Phytochemistry **30**, 4079–4082 (1991).

140. NING, X., M. ZHI-DA, ZH. SHON-XUN, W. BING, ZH. QI-TAI, and ZH. PEI: Neo-Clerodane Diterpenoids from *Teucrium pernyi*. Phytochemistry **30**, 1963–1966 (1991).

141. ALCAZAR, R., M.C. DE LA TORRE, B. RODRIGUEZ, M. BRUNO, F. PIOZZI, G. SAVONA, and N.A. ARNOLD: Neo-Clerodane Diterpenoids from Three Further Species of *Teucrium*. Phytochemistry **31**, 3957–3960 (1992).

142. DE LA TORRE, M.C., B. RODRIGUEZ, F.M. RIZK, M. BRUNO, and F. PIOZZI: Montanin C from *Teucrium polium* var. *album*. Fitoterapia **LIX**, 129–130 (1988).

143. MALAKOV, P.Y., G.Y. PAPANOV, and N.M. MOLLOV: Furanoid Diterpenes in the Bitter Fraction of *Teucrium polium* L. Z. Naturforsch. **34b**, 1570–1572 (1979).

144. EGUREN, L., A. PERALES, J. FAYOS, G. SAVONA, M. PATERNOSTRO, F. PIOZZI, and B. RODRIGUEZ: New Clerodane Diterpenoid from *Teucrium polium* subsp. *aureum*. X-ray Structure Determination. J. Org. Chem. **46**, 3364–3367 (1981).

145. DE LA TORRE, M.C., F. PIOZZI, A.F. RIZK, B. RODRIGUEZ, and G. SAVONA: 19-Acetylteupolin IV, a Neo-Clerodane Diterpenoid from *Teucrium polium* subsp. *pilosum*. Phytochemistry **25**, 2239–2240 (1986).

146. MARCO, J.L., B. RODRIGUEZ, G. SAVONA, and F. PIOZZI: Diterpenoids from *Teucrium scorodonia*, Three neo-Clerodane Derivatives. Phytochemistry **21**, 2567–2569 (1982).

147. PAPANOV, G.Y., and P.Y. MALAKOV: New Furanoid Diterpenes from *Teucrium scordium* L. Z. Naturforsch. **36b**, 112–113 (1981).

148. . . . . . : 6β-Hydroxyteuscordin and 2β, 6β-Dihydroxyteuscordin, Two New Diterpenoids from *Teucrium scordium* L. Z. Naturforsch. **37b**, 519–520 (1982).

192   L. RODRÍGUEZ-HAHN, B. ESQUIVEL, and J. CÁRDENAS

149. JAKUPOVIC, J., R.N. BARUAH, F. BOHLMANN, and W. QUACK: New Clerodane Derivatives from *Teucrium scordium*. Planta Medica, 341–342 (1985).
150. SAVONA, G., M. PATERNOSTRO, F. PIOZZI, and B. RODRIGUEZ: New Clerodane Diterpenoids from *Teucrium spinosum* L. Heterocycles **14**, 193–195 (1980).
151. NODE, M., M. SAI, and E. FUJITA: Isolation of the Diterpenoid Teuflin (6-Epiteucvin) from *Teucrium viscidum* var. *miquelianum*. Phytochemistry **20**, 757–760 (1981).
152. SAVONA, G., F. PIOZZI, B. RODRIGUEZ, C. PASCUAL, and O. SERVETTAZ: 2β-Hydro-xyteucvidin from *Teucrium webbianum*. Phytochemistry **25**, 2857–2859 (1986).
153. DE LA TORRE, M.C., N. EZER, B. RODRIGUEZ, G. SAVONA, F. PIOZZI, and O. SERVETTAZ: Neo-Clerodane Diterpenoids, Flavonoids and Other Constituents from Some *Teucrium* Species. Fitoterapia **LIX**, 70–72 (1988).
154. BENTHAM, G.: Labiatae, In: BENTHAM, G., and J.D. HOOKER (eds.): *Genera Plantarum*, Vol. 2, pp. 1160–1223. London: Reeve & Co. 1876.
155. RODRIGUEZ-HAHN, L., B. ESQUIVEL, J. CARDENAS, and T.P. RAMAMOORTHY: Distribution of Diterpenes in Salvia. Advances in Labiate Science, 335–347 (1992).
156. EPLING, C.: A revision of *Salvia* subgenus *Calosphace*. Repert. Spec. Nov. Regn. Beih. **110**, 1–383 (1939).
157. RODRIGUEZ-HAHN, L., B. ESQUIVEL, A.A. SANCHEZ, C. SANCHEZ, J. CARDENAS, and T.P. RAMAMOORTHY: Diterpenos Abietánicos de Salvias Mexicanas. Rev. Latinoamer. Quim. **20**, 105–110 (1989).
158. . . . . . .: Nuevos Diterpenos de Salvias Mexicanas. Rev. Latinoamer. Quim. **18**, 104–109 (1987).
159. FERNANDEZ, M.C., B. ESQUIVEL, J. CARDENAS, A.A. SANCHEZ, R.A. TOSCANO, and L. RODRIGUEZ-HAHN: Clerodane and Aromatic Seco-Clerodane Diterpenoids from *Salvia rhyacophila*. Tetrahedron **47**, 7199–7208 (1991).
160. RODRIGUEZ, B., C. PASCUAL, and G. SAVONA: The Correct Structure of Salvifaricin, a *cis-neo*-Clerodane Diterpenoid from *Salvia farinacea*. Phytochemistry **23**, 1193–1194 (1984).
161. SANCHEZ, A.A., B. ESQUIVEL, A. PERA, J. CARDENAS, M. SORIANO-GARCIA, A. TOSCANO, and L. RODRIGUEZ-HAHN: Lasianthin, a Neo-Clerodane Diterpenoid from *Salvia lasiantha*. Phytochemistry **26**, 479–482 (1987).
162. JONES, R.H., C.L. ANGELL, T. ITO, and R.J.D. SMITH: The Carbonyl Stretching Bands in the Infrared Spectra of Unsaturated Lactones. Can. J. Chem. **37**, 2007–2022 (1959).
163. JOLAD, S.D., J.J. HOFFMANN, K.H. SCHRAM, J.R. COLE, M.S. TEMPESTA, and R.B. BATES: Constituents of *Eremocarpus setigerus* (Euphorbiaceae), A New Diterpene, Eremone, and Hautriwaic Acid. J. Org. Chem. **47**, 1353–1358 (1982).
164. RODRIGUEZ-HAHN, L., G. MARTINEZ, and J. ROMO: Estructura del Acido Melisodórico, un Diterpeno Aislado de *Salvia melissodora* Lag. Rev. Latinoamer. Quim. **4**, 93–100 (1973).
165. RODRIGUEZ-HAHN, L., A. GARCIA, B. ESQUIVEL, and J. CARDENAS: Structure of Kerlinic Acid from *Salvia keerlii*. Chemical Correlation with Melisodoric Acid. Can. J. Chem. **65**, 2687–2690 (1987).
166. TSAI, T.Y., A. MINTA, and K. WIESNER: A Simple Synthesis of Carodenolides and Their Less Toxic Isomers via Furyl Intermediates. Heterocycles **12**, 1397–1402 (1979).
167. HERZ, W., A.M. PILOTTI, A.C. SODERHOLM, I.K. SHUHAMA, and W. VICHNEWSKI: New *ent*-Clerodane-type Diterpenoids from *Baccharis trimera*. J. Org. Chem. **42**, 3913–3917 (1977).
168. ESQUIVEL, B., A. VALLEJO, R. GAVIÑO, J. CARDENAS, A.A. SANCHEZ, T.P. RAMAMOOR-THY, and L. RODRIGUEZ-HAHN: Clerodane Diterpenoids from *Salvia melissodora*. Phytochemistry **27**, 2903–2905 (1988).

169. ESQUIVEL, B., J. CARDENAS, L. RODRIGUEZ-HAHN, and T.P. RAMAMOORTHY: The Diterpenoid Constituents of *Salvia fulgens* and *Salvia microphylla*. J. of Natural Products **50**, 738–740 (1987).

170. ESQUIVEL, B., M. HERNANDEZ, T.P. RAMAMOORTHY, J. CARDENAS, and L. RODRIGUEZ-HAHN: Semiatrin, a Neo-Clerodane Diterpenoid from *Salvia semiatrata*. Phytochemistry **25**, 1484–1486 (1986).

171. ESQUIVEL, B., L.M. HERNANDEZ, J. CARDENAS, T.P. RAMAMOORTHY, and L. RODRIGUEZ-HAHN: Further *Ent*-Clerodane Diterpenoids from *Salvia melissodora*. Phytochemistry **28**, 561–566 (1989).

172. ESQUIVEL, B., A. MENDEZ, A. ORTEGA, M. SORIANO-GARCIA, A. TOSCANO, and L. RODRIGUEZ-HAHN: Neo-Clerodane-type Diterpenoids from *Salvia keerlii*. Phytochemistry **24**, 1769–1772 (1985).

173. CUEVAS, G., O. COLLERA, F. GARCIA, J. CARDENAS, E. MALDONADO, and A. ORTEGA: Diterpenes from *Salvia breviflora*. Phytochemistry **26**, 2019–2021 (1987).

174. SORIANO-GARCIA, M., R.A. TOSCANO, B. ESQUIVEL, M. HERNANDEZ, and L. RODRIGUEZ-HAHN: Structure and Stereochemistry of 2-Hydroxy-12(S)-hydroxyneoclerodane-3,13(14)-diene-15,16; 19,20-diolide (Semiatrin), a Diterpene. Acta Cryst. **C43**, 272–274 (1987).

175. ORTEGA, A., J.F. BLOUNT, and P.S. MANCHAND: Salvinorin, a New *trans*-Neoclerodane Diterpene from *Salvia divinorum* (Labiatae). J. Chem. Soc. Perkin Trans. I, 2505–2508 (1982).

176. VALDES III, L.J., W.M. BUTLER, G.M. HATFIELD, A.G. PAUL, and M. KOREEDA: Divinorin A, a Psycotropic Terpenoid, and Divinorin B from Hallucinogenic Mexican Mint *Salvia divinorum*. J. Org. Chem. **49**, 4716–4720 (1984).

177. KOREEDA, M., L. BROWN, and L.J. VALDES III: The Absolute Stereochemistry of Salvinorins. Chemistry Letters, 2015–2018 (1990).

178. SAVONA, G., M.P. PATERNOSTRO, F. PIOZZI, J.R. HANSON, P.B. HITCHCOCK, and S.A. THOMAS: Salviarin, a New Diterpenoid from *Salvia splendens*. J.C.S. Perkin I, 643–646 (1978).

179. SAVONA, G., M.P. PATERNOSTRO, F. PIOZZI, and J.R. HANSON: Splendidin, a New *Trans*-Clerodane from *Salvia splendens*. J.C.S. Perkin I, 533–534 (1979).

180. SAVONA, G., M. BRUNO, M. PATERNOSTRO, J.L. MARCO, and B. RODRIGUEZ: Salviacoccin, a Neo-Clerodane Diterpenoid from *Salvia coccinea*. Phytochemistry **21**, 2563–2566 (1982).

181. BRUNO, M., G. SAVONA, F. FERNANDEZ-GADEA, and B. RODRIGUEZ: Diterpenoids from *Salvia greggii*. Phytochemistry **25**, 475–477 (1986).

182. GARCIA-ALVAREZ, M.C., M. HASAN, A. MICHAVILA, F. FERNANDEZ-GADEA, and B. RODRIGUEZ: Epoxysalviacoccin, a Neo-Clerodane Diterpenoid from *Salvia plebeia*. Phytochemistry **25**, 272–274 (1986).

183. ESQUIVEL, B., J. CARDENAS, T.P. RAMAMOORTHY, and L. RODRIGUEZ-HAHN: Clerodane Diterpenoids of *Salvia lineata*. Phytochemistry **25**, 2381–2384 (1986).

184. BRIESKORN, C.H., and T. STEHLE: Labiaten-Bitterstoffe; eine neue Verbindung des Clerodantyps. Chem. Ber. **106**, 922–928 (1973).

185. SORIANO-GARCIA, M., B. ESQUIVEL, R.A. TOSCANO, and L. RODRIGUEZ-HAHN: Structure and Stereochemistry of (8S,12R)-*cis*-Clerodane-1,3,13(16),14-tetraene-15,16-epoxy-12(17); 18,19-diolide (Linearolactone), a Diterpene. Acta Cryst. **C43**, 1565–1567 (1987).

186. SAVONA, G., D. RAFFA, M. BRUNO, and B. RODRIGUEZ: Salvifarin and Salvifaricin, Neo-Clerodane Diterpenoids from *Salvia farinacea*. Phytochemistry **22**, 784–786 (1983).

187. Eguren, L., J. Fayos, A. Perales, G. Savona, and B. Rodriguez: Salvifarin, X-Ray Structure Determination of a cis-Neo-Clerodane Diterpenoid from *Salvia farinacea.* Phytochemistry **23**, 466–467 (1984).

188. Esquivel, B., J. Ochoa, J. Cardenas, T.P. Ramamoorthy, and L. Rodriguez-Hahn: Clerodane-type Diterpenoids from *Salvia sousae.* Phytochemistry **27**, 483–486 (1988).

189. Esquivel, B., O. Esquivel, J. Cardenas, A.A. Sanchez, T.P. Ramamoorthy, R.A. Toscano, and L. Rodriguez-Hahn: Clerodane and Seco-Clerodane Diterpenoids from *Salvia reptans.* Phytochemistry **30**, 2335–2338 (1991).

190. Jimenez, M., E.D. Moreno, and E. Diaz: Diterpenos de la *Salvia gensneraefolia* I. Estructuras de las Gensnerofolinas A y B. *Rev. Latinoamer. Quim.* **10**, 166–171 (1979).

191. Rodriguez-Hahn, L., B. Esquivel, and J. Cardenas: New Diterpenoid Skeletons of Clerodanic Origin from Mexican *Salvia* Species, in "Trends in Organic Chemistry", 99–111 (1992).

192. Maldonado, E., J. Cardenas, B. Salazar, R.A. Toscano, A. Ortega, C.K. Jankowski, A. Aumelas, and M.R. Van Calsteren: Salvianduline C, a 5,6-Seco-Clerodane Diterpenoid from *Salvia lavanduloides.* Phytochemistry **31**, 217–220 (1992).

193. Rodriguez-Hahn, L., R. O'Reilly, B. Esquivel, E. Maldonado, A. Ortega, J. Cardenas, and R.A. Toscano: Tilifodiolide, Tetralin-type Diterpenoid of Clerodanic Origin from *Salvia tiliaefolia.* J. Org. Chem. **55**, 3522–3525 (1990).

194. Ortega, A., J. Cardenas, A. Toscano, E. Maldonado, A. Aumelas, M.R. van Calsteren, and C. Jankowski: Salviandulines A and B. Two Secoclerodane Diterpenoids from *Salvia lavanduloides.* Phytochemistry **30**, 3357–3360 (1991).

195. Zdero, C., A.A. Ahmed, F. Bohlmann, and G.M. Mungai: Diterpenes and Sesquiterpene Xylosides from East African *Conyza* Species. Phytochemistry **29**, 3167–3172 (1990).

196. Gonzalez, A.G., J.R. Herrera, J.G. Luis, A.G. Ravelo, M.L. Rodriguez, and E. Ferro: Cardiophillidin, a Seco-*ent*-Clerodane Diterpenoid from *Salvia cardiophylla.* Tetrahedron Letters **29**, 363–366 (1988).

197. Pandey, U.C., A.K. Singhal, N.C. Barua, R.P. Sharma, J.N. Baruah, K. Watanabe, P. Kulanthaivel, and W. Herz: Stereochemistry of Strictic Acid and Related Furano Diterpenes from *Conyza japonica* and *Grangea maderaspatana.* Phytochemistry **23**, 391–397 (1984).

198. Misra, R., R.C. Pandey, and S. Dev: The Absolute Stereochemistry of Hardwickiic Acid and Its Congeners. Tetrahedron Letters, 2681–2684 (1968).

199. Flores, E.: Estudio Fitoquimico de *Salvia thymoides* Benth. B.Sc. Thesis.

200. Hernandez, S.: Estudio Fitoquimico de *Salvia purpurea* Cav. B.Sc. Thesis.

201. Singh, P., M.C. Sharma, K.C. Joshi, and F. Bohlmann: Diterpenes Derived from Clerodanes from *Pulicaria angustifolia.* Phytochemistry **24**, 190–192 (1985).

202. Esquivel, B., J. Cardenas, A. Toscano, M. Soriano-Garcia, and L. Rodriguez-Hahn: Structure of Salvigenolide, a Novel Diterpenoid with a Rearranged Neo-Clerodane Skeleton from *Salvia fulgens.* Tetrahedron **41**, 3213–3217 (1985).

203. Rodriguez-Hahn, L., B. Esquivel, A.A. Sanchez, J. Cardenas, O.G. Tovar, M. Soriano-Garcia, and A. Toscano: Puberulin and Isopuberulin, Benzonorcaradiene and Benzocycloheptatriene Diterpenoids of Clerodanic Origin from *Salvia puberula.* J. Org. Chem. **53**, 3933–3936 (1988).

204. Cardenas, J., B. Esquivel, R.A. Toscano, and L. Rodriguez-Hahn: Languiduline, a Diterpenoid with an Unusual Structure from *Salvia languidula.* Heterocycles **27**, 1809–1812 (1988).

205. Sanchez, A.A., B. Esquivel, T.P. Ramamoorthy, J. Cardenas, and L. Rodriguez-

HAHN: *Neo*-clerodane and Languidulane Diterpenoids from *Salvia zimapanensis*, unpublished results.

206. CARDENAS, J., T. PAVON, B. ESQUIVEL, A. TOSCANO, and L. RODRIGUEZ-HAHN: Salvilanguidulines, Four New Diterpenoids Isolated from *Salvia languidula* with an Unusual Epoxy Spiro γ-Lactone. Tetrahedron Letters **33**, 581–584 (1992).

207. STANDLEY, P.C., and L.O. WILLIAMS: Labiatae in Flora of Guatemala. Part IX. Fieldiana. Botany **24**, 237 (1973).

208. TSCHESCHE, R., and H.U. PLENIO: Über Marrubiasid und Marrubialacton, zwei Diterpenderivative mit ungelagertem Labdangerüst aus *Leonurus marrubiastrum* L. Chem. Ber. **106**, 2929–2942 (1973).

209. TSCHESCHE, R., and B. STREUFF: Über drei neue Diterpen-Derivate aus *Leonurus marribiastrum* L., 2. Chem. Ber. **111**, 2130–2142 (1978).

210. MCCRINDLE, R., E. NAKAMURA, and A.B. ANDERSON: Constituents of Solidago Species. Part VII. Constitution and Stereochemistry of the *cis*-Clerodanes from *Solidago arguta* Ait. and of Related Diterpenoids. J.C.S. Perkin I 1590–1597 (1976), and references cited therein.

211. NISHINO, CH., S. MANABE, M. KAZUI, and T. MATSUZAKI: Piscicidal *cis*-Clerodane Diterpenes from *Solidago altissima* L". Absolute Configurations of 5α,10α-*cis*-Clerodanes. Tetrahedron Letters **25**, 2809–2812 (1984).

212. MANABE, S., and CH. NISHINO: Stereochemistry of *cis*-Clerodane Diterpenes. Tetrahedron **42**, 3461–3470 (1986).

213. BRIESKORN, C.H., and R. HOFMANN: Labiatenbitterstoffe: Ein Clerodanderivat aus *Leonurus cardiaca* L. Tetrahedron Letters, 2511–2512 (1979).

214. NIWA, M., and S. YAMAMURA: Stereostructures of Several Solidagolactones (Elongatolides). Tetrahedron Letters **22**, 2789–2792 (1981).

215. POPA, D.P., T.M. ORGIYAN, Z. SAMEK, and L. DOLEJS: Structure of Stachysolone. Khim. Prir. Soedin, 295–299 (1972); Chem. Abstr. **77**, 152384f (1972).

216. POPA, D.P., and T.M. ORGIYAN: Stereochemistry of Stachysolone. Khim. Prir. Soedin. 735–738 (1972); Chem. Abstr. **78**, 136447u (1973).

217. POPA, D.P., T.M. ORGIYAN, and K.S. KHARITOV: Structure of Annuanone. Khim. Prir. Soedin. 324–330 (1974); Chem. Abstr. **81**, 120805k (1974).

218. POPA, D.P., and T.M. ORGIYAN: Minor Diterpenoids from *Stachys annua*. Khrim. Prir. Soedin. 406 (1974); Chem. Abstr. **82**, 28538y (1975).

219. ADINOLFI, M., G. BARONE, R. LANZETTA, G. LAONIGRO, L. MANGONI, and M. PARRILLI: Diterpenes from *Stachys recta*. Journal of Natural Products **47**, 541–543 (1984).

220. FAZIO, C., S. PASSANNANTI, M.P. PATERNOSTRO, and F. PIOZZI: Neo-Clerodane Diterpenoids from *Stachys rosea*. Phytochemistry **31**, 3147–3149 (1992).

221. MISRA, R., and S. DEV: Partial Synthesis of Kolavelool and Hardwickiic Acid. Tetrahedron Letters, 2685–2686 (1968).

222. SIMMONDS, M.S.J., W.M. BLANEY, S.V. LEY, G. SAVONA, M. BRUNO, and B. RODRIGUEZ: The Antifeedant Activity of Clerodane Diterpenoids from *Teucrium* Phytochemistry **28**, 1069–1071 (1989).

223. BELLES, X., F. CAMPS, J. COLL, and M.D. PIULACHS: Insect Antifeedant Activity of Clerodane Diterpenoids Against Larvae of *Spodoptera littoralis* (Boisd.) (Lepidoptera). Journal of Chemical Ecology **11**, 1439–1445 (1985).

224. KATON, M. TAKAHASHI, M. SHIBAYAMA, and K. MUNAKATA: Antifeeding Active Substances for Insects in *Clerodendron tricotomum* Thunb. Agric. Biol. Chem. **36**, 2579–2582 (1972).

225. SIMMONDS, M.S.J., W.M. BLANEY, B. ESQUIVEL, and L. RODRIGUEZ-HAHN: Unpublished results.

226. COLE, M.D., P.D. BRIDGE, J.E. DELLAR, L.E. FELLOWS, M.C. CORNISH, and J.C. ANDERSON: Antifungal Activity of neo-Clerodane Diterpenoids from Scutellaria. Phytochemistry 30, 1125–1127 (1991).
227. BOZOV, P.I., G.Y. PAPANOV, P.Y. MALAKOV, M.C DE LA TORRE, and B. RODRIGUEZ: A Clerodane Diterpene from Ajuga salicifolia. Phytochemistry 34, 1173–1175 (1993).
228. SHEN, X., A. ISOGAI, K. FURIHATA, H. SUN, and A. SUZUKI: Two neo-Clerodane Diterpenoids from Ajuga macrosperma. Phytochemistry 33, 887–889 (1993).
229. SHEN, X., A. ISOGAI, K. FURIHATA, H. SUN, and A. SUZUKI: Neo-Clerodane Diterpenoids from Ajuga macrosperma and Ajuga pantantha. Phytochemistry 34, 1091–1094 (1993).
230. BRUNO, M., M.C. DE LA TORRE, F. PIOZZI, B. RODRIGUEZ, G. SAVONA, and N.A. ARNOLD: A neo-Clerodane Diterpenoid from Scutellaria cypria var. elatior. Phytochemistry 33, 931–932 (1993).
231. RODRIGUEZ B., M.C. DE LA TORRE, M. BRUNO, F. PIOZZI, G. SAVONA, M.S. J. SIMMONDS, W. M. BLANEY, and A. PERALES: Neo-Clerodane Insect Antifeedants from Scutellaria galericulata. Phytochemistry 33, 309–315 (1993).
232. BOZOV, P.I., P.Y. MALAKOV, G.Y. PAPANOV, M.C. DE LA TORRE, B. RODRIGUEZ, and A. PERALES: Scutalpin A, a neo-Clerodane Diterpene from Scutellaria alpina. Phytochemistry 34, 453–456 (1993).
233. DE LA TORRE, M.C., B. RODRIGUEZ, M. BRUNO, P.Y. MALAKOV, G.Y. PAPANOV, F. PIOZZI, and G. SAVONA: Neo-Clerodane Diterpenoids from Scutellaria alpina subsp. javalambrensis. Phytochemistry 34, 1589–1594 (1993).
234. BRUNO, M., F. PIOZZI, B. RODRIGUEZ, G. SAVONA, M.C. DE LA TORRE, and O. SERVETTAZ: Neo-Clerodane Diterpenes from Teucrium species. Phytochemistry 31, 4366–4367 (1992).
235. MALAKOV, P.Y., G.Y. PAPANOV, and I.M. BONEVA: Neo-Clerodane Diterpenoids from Teucrium montanum. Phytochemistry 31, 4029–4030 (1992).
236. MALAKOV, P.Y., G.Y. PAPANOV, I.M. BONEVA, M.C. DE LA TORRE, and B. RODRIGUEZ: Teulamioside, a neo-Clerodane Glucoside from Teucrium lamiifolium. Phytochemistry 34, 1095–1098 (1993).
237. AL-YAHYA, M.A., I. MUHAMMAD, H.H. MIRZA, F.S. EL-FERALY, and A.T. MCPHAIL: Neoclerodane Diterpenoids and Their Artifacts from Teucrium oliverianum. J. of Natural Products 56, 830–842 (1993).
238. DI-AN, S., and L. GUANG-YI: Teupernin D, a neo-Clerodane Diterpenoid from Teucrium pernyi. Phytochemistry 33, 716–717 (1993).

(Received November 12, 1992)

Author Index

Page numbers printed in *italics* refer to References

Subject Index

Fortschritte der Chemie organischer Naturstoffe
Progress in the Chemistry of Organic Natural Products

Founded by L. Zechmeister
Edited by W. Herz, G.W. Kirby, R.E. Moore, W. Steglich, and Ch. Tamm

Volume 62

1993. 52 figures. VIII, 330 pages.
Cloth DM 280,–, öS 1960,–*). ISBN 3-211-82402-2

Contents: S.V. Bhat: Forskolin and Congeners. • L. Minale, R. Riccio, and F. Zollo: Steroidal Oligoglycosides and Polyhydroxysteroids from Echinoderms. • Author Index. • Subject Index.

Volume 61

1993. 4 figures. X, 206 pages.
Cloth DM 220,–, öS 1540,–*). ISBN 3-211-82388-3

Contents: D.G.I. Kingston, A.A. Molinero, and J.M. Rimoldi: The Taxane Diterpenoids. • Author Index. • Subject Index.

Volume 60

1992. 59 figures. VIII, 243 pages.
Cloth DM 184,–, öS 1290,–*). ISBN 3-211-82374-3

Contents: I. Wahlberg and A.-M. Eklund: Cyclized Cembranoids of Natural Occurrence. • M. Petitou and C.A.A. van Boeckel: Chemical Synthesis of Heparin Fragments and Analogues. • Author Index. • Subject Index. • General Index Vols. 21–60.

*) Price reduction for subscribers: 10%

Special Offer: Special reduced price (20% reduction) for the complete Series Vols. 1–60 incl. Cumulative Index to Vols. 1–20.

Prices are subject to change without notice

Springer-Verlag Wien New York

Sachsenplatz 4–6, P.O.Box 89, A-1201 Wien · 175 Fifth Avenue, New York, NY 10010, USA
Heidelberger Platz 3, D-14197 Berlin · 3-13, Hongo 3-chome, Bunkyo-ku, Tokyo 113, Japan

Monatshefte für Chemie / Chemical Monthly

The Monatshefte für Chemie / Chemical Monthly, founded 1880, are one of the worlds oldest scientific journals. They are edited by a recently reorganized Editorial Board, in close cooperation with the Austrian Academy of Sciences. Following the traditional policy, the Monatshefte für Chemie / Chemical Monthly continue to publish contributions from all areas of chemistry in the forms of full papers as well as short communications. Besides reporting on the progress of chemical research in Austria, the journal publishes contributions from all countries of the world.

Subscription Information:
1995. Volume 126 (12 issues)
DM 1272,– , öS 8904,–, plus carriage charges.
ISSN 0026-9247, Title No. 706

Prices are subject to change without notice

Springer-Verlag Wien New York

Sachsenplatz 4–6, P.O.Box 89, A-1201 Wien · 175 Fifth Avenue, New York, NY 10010, USA
Heidelberger Platz 3, D-14197 Berlin · 3-13, Hongo 3-chome, Bunkyo-ku, Tokyo 113, Japan